Political/Military Applications of Bayesian Analysis

Westview Replica Editions

The concept of Westview Replica Editions is a response to the continuing crisis in academic and informational publishing. Library budgets for books have been severely curtailed. Ever larger portions of general library budgets are being diverted from the purchase of books and used for data banks, computers, micromedia, and other methods of information retrieval. Interlibrary loan structures further reduce the edition sizes required to satisfy the needs of the scholarly community. Economic pressures on the university presses and the few private scholarly publishing companies have severely limited the capacity of the industry to properly serve the academic and research communities. As a result, many manuscripts dealing with important subjects, often representing the highest level of scholarship, are no longer economically viable publishing projects--or, if accepted for publication, are typically subject to lead times ranging from one to three years.

Westview Replica Editions are our practical solution to the problem. We accept a manuscript in camera-ready form, typed according to our specifications, and move it immediately into the production process. As always, the selection criteria include the importance of the subject, the work's contribution to scholarship, and its insight, originality of thought, and excellence of exposition. The responsibility for editing and proofreading lies with the author or sponsoring institution. We prepare chapter headings and display pages, file for copyright, and obtain Library of Congress Cataloging in Publication Data. A detailed manual contains simple instructions for preparing the final typescript, and our editorial staff is always available to answer questions.

The end result is a book printed on acid-free paper and bound in sturdy library-quality soft covers. We manufacture these books ourselves using equipment that does not require a lengthy make-ready process and that allows us to publish first editions of 300 to 600 copies and to reprint even smaller quantities as needed. Thus, we can produce Replica Editions quickly and can keep even very specialized books in print as long as there is a demand for them.

About the Book and Author

Political/Military Applications of Bayesian Analysis: Methodological Issues

Douglas E. Hunter

This book demonstrates how Bayesian analysis, a mathematical technique for revising the probabilities of hypotheses as new evidence becomes available, can be used by military and political analysts to answer questions both about decisions and about possible future events. Such questions include the following: Will the head of a certain government decide to sign a treaty with its former opponent? What will one state eventually decide to do about attacking its neighbor? Will a given state attack another within six months? Which political party will win the upcoming elections in a particular country?

An analyst using Bayes first specifies a set of mutually exclusive and collectively exhaustive possible answers which constitute the hypotheses for the analysis. The probability of each hypothesis is then assessed, based on existing knowledge of the situation. As new evidence is received, the use of Baye's formula leads to the revision of the original probabilities.

Dr. Hunter, professor of intelligence at the Defense Intelligence College, provides a detailed explanation of the application of Bayesian analysis and a very careful examination of numerous methodological problems. He also provides a basic introduction to probability theory so that those unfamiliar with it can fully understand the text.

Political/Military Applications of Bayesian Analysis
Methodological Issues

Douglas E. Hunter

Westview Press / Boulder, Colorado

A Westview Replica Edition

Published in 1984 in the United States of America by
Westview Press, Inc.
5500 Central Avenue
Boulder, Colorado 80301
Frederick A. Praeger, President and Publisher

Library of Congress Catalog Card Number: 83-10202
ISBN 0-86531-954-5

Printed and bound in the United States of America

10 9 8 7 6 5 4 3 2 1

Dedicated to Captain Richard W. Bates, USN (Ret)
and
Jack Zlotnick

Contents

Preface

Bayesian analysis, as treated in this book, will be viewed as a technique of statistical inference used to revise the probabilities of political/military hypotheses concerning a situation after evidence relating to the hypotheses is received. For example, an analyst might be interested in the possibility that state A might attack state B within the next three weeks. The analyst could set up a Bayesian analysis with these two hypotheses:

H_1 = State A is planning to attack State B within the next three weeks;

H_2 = State A is not planning to attack State B within the next three weeks.[1]

Based on all the evidence at hand when the hypotheses were formulated, the analyst might assess the initial probabilities of the hypotheses as

$P(H_1) = 0.8,$

$P(H_2) = 0.2.$

The analyst then would wait for evidence relating to state A's intentions to be collected. Suppose that the first item of evidence was that state A had brought up river bridging equipment to a river between the states. The analyst would make two probability judgments, asking first how likely it would be for state A to bring up the equipment if it were planning to attack within three weeks, and second how likely it would be for state A to bring up the equipment if it were not planning to attack within three weeks. These judgments would be subjective in nature, but still would have to be justified in light of the political/military situation existing at the time. After these judgments were made, Bayes's formula would provide the mathematically correct revised probabilities for H_1 and H_2. Subsequent items of evidence would result in further revisions of the probabilities of H_1 and H_2.

A logical question to ask at this point is this: "Why bother making the judgments described and why bother using Bayes's formula? Why not just intuitively revise the probabilities of H_1 and H_2 after each item of evidence is received?" The third chapter on the major advantages of Bayesian analysis will discuss this question in greater detail. The short answer appropriate here is that there is a fair amount of psychological research indicating that an intuitive revision of probabilities tends not to be as great as the evidence would warrant; people tend to be "conservative" when it comes to revising their initial probabilities in the light of evidence. If Bayes's formula is used, the argument goes, this conservatism is overcome. Moreover, use of Bayes's theorem provides a formal procedure for revising the probability assessments of hypotheses about a situation in such a way that the evidence upon which the revisions are based (1) is more systematically considered than it would be intuitively and (2) is given more weight than it would be if it were intuitively aggregated and judged. In a sense, more "information" is extracted from the evidence if a formal Bayesian procedure is used than if an intuitive process is used.

The basic purpose of this book is to explore, with care and in some detail, many of the methodological problems that are involved in the everyday, practical application of Bayesian analysis to political/military studies. Many of these problems are subtle and can lead to an incorrect analysis if not recognized and avoided. Whenever possible, solutions are offered to the methodological problems raised. Some highly technical and complex solutions have not been explored because it is felt that these solutions simply are too complicated for everyday usage. These solutions are, of course, footnoted for those who wish to explore the issues further.

To prepare a reader for the technical discussion of problems, an introduction to probability theory is presented. This introduction is as simple and straightforward as possible, and is based on the author's six years of experience in teaching a basic probability course. Following this introduction in the first chapter, Bayes's formula is developed in the second chapter and a tabular solution format presented, which permits a quick hand-calculator approach to solving typical Bayesian problems.

After discussing the major advantages of Bayesian analysis in the third chapter, the remaining chapters, the heart of the book, present many of the methodological problems with Bayesian analysis and their possible solutions, if any. For each problem, a format is used which permits both a succinct overview of what sometimes are complex discussions and a rapid review, at a later date, of the major issues. This format consists of four parts: a short descriptive phrase to identify the problem, a more exact statement of the problem, a summary of the solutions (if any) to the problem, and the (occasionally lengthy) discussion of the problem.

This book concerns itself almost entirely with normal Bayesian analysis, and does not explore the more recently developed form known as hierarchical Bayesian analysis; its methodological problems are only alluded to. The best guide to the latter is Decision and Design's Handbook for Decision Analysis.[2] The book titled Expert-Generated Data[3] also treats this more advanced form of Bayes.

The latter book also provides an overview of some of the other analytical techniques employing subjectively assessed data which have been used in recent years.

NOTES

1. As we will show later (in Problem 3 of Chapter Four), although these two hypotheses are typical of what we find in the Bayesian literature, they are defectively worded. More exactly, the second hypothesis contains implicit, "buried" subhypotheses, and the existence of these "buried" subhypotheses will cause the results of the Bayesian analysis to be somewhat misleading.

2. BARCLAY, SCOTT et. al. (1977) Handbook for Decision Analysis. Decisions and Designs, Inc. Technical Report TR-77-6-30, September, McLean, Virginia. ADA049221.

3. HOPPLE, GERALD W., and KUHLMAN, JAMES A. (eds.) (1981) Expert-Generated Data: Applications in International Affairs. Boulder, Colorado: Westview Press.

Acknowledgements

The contributions of numerous people are reflected in this book. Dr. Gerald W. Hopple read the manuscript several times and made many very helpful comments. Dorothy Williams read the manuscript after I had revised it further and made a number of useful suggestions. The book mainly grew out of a graduate seminar titled "Advanced Topics in Bayesian Analysis" which I taught once a year for some five years at the Defense Intelligence College. I encouraged my seminar students to document their contributions in class to the year-to-year cumulative process of analyzing many of the methodological problems involved with Bayesian analysis. Those students whose contributions are cited are the following: Norman Allred, Earl Blake, Jr., Kenneth Boll, Michael Burke, Lloyd Feldman, Martin Joyce III, Timothy Laur, Beverly Lohre, James McGowan, Thomas McIntyre, Duane Miller, Jerome Rapin, Gerald Shea, Sylvia Smith, and Robert ("Bob") Wright. I regret that there are a few places where I cannot identify the student responsible for the particular point being made. My thanks to all of my seminar students through the years, whether they have been mentioned or not. My particular appreciation goes to seminar student Bob Wright who made a very careful analysis of the last draft I prepared for the seminar on the problems of Bayes and provided many pages of very valuable comments.

Several auditors in the course also contributed through their oral or written comments. They include Richards J. Heuer, Jr., Dr. John Kringen, Jeffrey Wise, and Gus Schoone.

The first two chapters are based on an introductory course in probability theory which I taught many times through the years. Student Michael Brock made an insightful comment noted in chapter one, and students Joseph ("Joe") May and Keith Oatman each made a contribution to the later chapters.

Whenever I cite someone's contribution in the text, I am agreeing with or supporting the idea expressed, unless otherwise noted. Of course, I accept the responsibility for any errors of fact or judgment.

I have been encouraged in the writing of this book and aided in a number of very helpful ways by my past immediate supervisors, including LTC Oliver O. Wainwright, USA (Ret), Lt Col James R. Carruth, USAF, COL Dennis K. Hodsdon, USA, and Lt Col James T. Jones, USAF. Other persons who have been very helpful include the college's provost, Dr. Robert L. DeGross, the former director of research, CDR Bruce Watson, USN, past academic deans Col Raymond Goelz, USAF (Ret), Col Billie L. Hensley, USAF (Ret) and the present academic dean, Col Allen E. Wolf, USAF.

The typing of this sort of text is extremely onerous, but three fine typists have worked on it. YN2 Susan Monahan Cavins, USN, typed the first draft and SP4 Pamela Eye, USA, the second. The third, and final draft (with numerous changes, so that in some ways it could be called the fifth or sixth draft) which was delivered to the publisher was extremely well typed by MSgt O.C. Moreland, Jr., USAF. My very great thanks go to him for a task far above and beyond the normal duty requirements. William Diggs and Forest R. ("Ron") Noel were responsible for all the drawings of the figures in the book.

My wife, Elaine MacEacheren Hunter, supported me in many ways in preparing the manuscript, and S. Hunter has provided constant companionship throughout the long years of preparation.

I also am grateful to all of the people at Westview Press who helped in the publication of this book, particularly Dean Birkenkamp and Victoria Yogman.

One great regret I have is that I could not specifically reference the contributions of Jack Zlotnick. His name is mentioned at a number of points, and I think that I have been fair in acknowledging his ideas, but without specific citations because his writings were in the unclassified portions of still classified documents. My seminar students and I were impressed by the subtlety of his thought and the importance of his work. As a mark of my respect for his seminal writings, this book is dedicated in part to him.

The other person to whom this book is dedicated is Captain Richard W. Bates, USN (Ret), a former Commandant of the Defense Intelligence College. No Commandant has contributed more to the development of the college. This book is dedicated to him as a sign of the great respect I and so many others have for him, and also in gratitude for the many ways in which he has aided me in my career.

Douglas E. Hunter

1
Some Basic Elements of Probability Theory

Although many readers of this book will have had at least an introductory course in probability or statistics, I have assumed no prior exposure to probability. Therefore, this first chapter provides an informal coverage of the rather basic elements of probability that are needed to understand the remainder of the book. The reader who already has covered these elementary topics still should read through this chapter, although quickly, because my use of probability tree diagrams is more central to the development of the theory than is true in many treatments, and is critical not only for the next chapter but the remainder of the manuscript as well.

SOME CONCEPTS OF SET THEORY

A set is a well defined collection of objects; the objects are called the elements of the set. We can talk about the set of students sitting in a small classroom, and easily identify the elements about which we are talking. The collection of air molecules in a classroom, however, is not "well defined" at all because we can't identify exactly the elements to which we are referring; for most purposes, it would not be useful to try to speak of this collection as a set.

A subset is a portion of a defined set.

A set can be specified by listing all of its elements within parentheses. For example, the set of numbers one through six can be called set S_1 and written as $S_1 = (1, 2, 3, 4, 5, 6)$.

Because we will want to perform certain operations on sets, it is usual when dealing with a given problem or issue to define a specific collection of elements as the so-called universal set, which contains all the elements about which we need be concerned. The universal set will be denoted here by T (where T stands for the totality of elements). Once the universal set T is defined for a given problem, it is understood that all sets subsequently discussed are subsets of T. (Different universal sets will be defined for different problems, of course.)

Some other helpful notions about sets include the following. A *Venn* diagram is a pictorial representation of sets and subsets.

For example, consider Figure 1.1 where the universal set T consists of the real numbers one through six, i.e. T = (1, 2, 3, 4, 5, 6) and sets A and B (both subsets of T) are A = (2, 4, 6) and B = (3, 4, 5, 6).

FIGURE 1.1
Example of a Venn Diagram

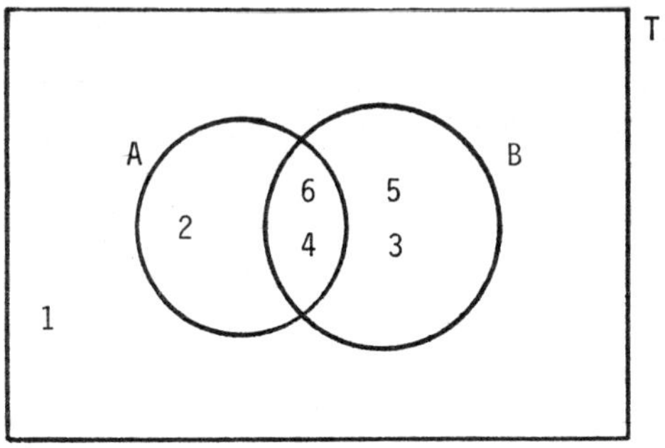

The <u>intersection</u> of two sets A and B is defined as the set of elements in common to both A and B; it consists of those elements in the "overlap" of A and B. It will be written here as A & B, where "&" is the "and" sign. (The standard notation is $A \cap B$.)[1] "A & B" can be said aloud as "A and B" or "A intersection B." In Figure 1.1, A & B = (4, 6).

Suppose that we are dealing with a different situation as portrayed in Figure 1.2. In this case, the set of elements in common to both A and B contains no elements; it is the so-called "empty" set or "<u>null</u>" set. Here, it is denoted by the letter Ø. For Figure 1.2, we write A & B = Ø.

FIGURE 1.2
A Venn Diagram for Non-overlapping Sets

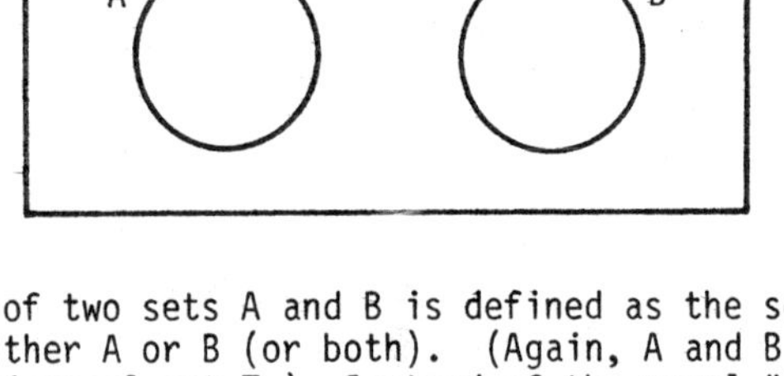

The <u>union</u> of two sets A and B is defined as the set of elements in either A or B (or both). (Again, A and B are subsets of the same universal set T.) Instead of the usual "U" sign for the union, we will use the word "or." So, "A or B" can be said aloud as "A union B" or as "A or B." Note that the "or" here is <u>not</u> an exclusive type of "or," but instead means "and/or." That is, the common notion of, say, "tea or coffee," where one or the other but not both is offered, is _not_ the way "or" is used by mathematicians and logicians. Here, then, "tea or coffee" means tea or coffee or both, i.e., "or" means "and/or." "A or B" means the set of elements in A or B or both, where each element is counted only once. A or B is the "combination" of the elements in sets A and B. In Figure 1.1 above, A or B = (2, 3, 4, 5, 6). Since the element "1" is outside of both A and B, it is not in their union. It is, of course, part of the universal set T.

Another necessary notion is that of the complement of a set A, which is defined as the set of elements in T which are not in A. Here, it is written as "not A." In Figure 1.1, not A = (1, 3, 5).

All of the concepts discussed above about sets require no mention of the concept of probability. Since it is so central to our discussion, we now will turn to it.

SOME CONCEPTS OF PROBABILITY THEORY

Basic Notions

In the study of probability theory, we find that we usually are interested in sets whose elements are the possible outcomes of experiments or are the possible outcomes when natural phenomena are observed. A set of _all_ possible outcomes of an experiment or observation is termed a _sample space_. In probability theory, then, the universal set T will consist of the totality of possible outcomes of a given experiment or observation. For example, if a die is tossed once, a sample space is the set of numbers one through six, i.e., T = (1, 2, 3, 4, 5, 6).[2] If a coin is tossed once, the sample space is T = (h, t), where h = heads, t = tails. If there are a finite number of elements in T, it is said that T is a _finite sample space_. Only finite sample spaces are dealt with in this book. Each individual outcome in a particular sample space T is termed a "_sample point_."

Also necessary is the notion of an "event." An _event_ is a set defined on a given sample space T, i.e., an event is a subset of a given sample space T. Often, a verbal description is used to characterize the event. For example, using Figure 1.1, if we now interpret (i.e., specify) that the numbers one through six represent the sample space T of the experiment of throwing a fair[3] die one time, then we can speak of set A = (2, 4, 6), or the event A of throwing an even outcome. An event which contains only one sample point is called a "simple event," i.e., a simple event consists of just one outcome (= sample point) of an experiment or observation. (We therefore can use the terms "simple event" and "sample point" interchangeably -- each represents exactly _one_ outcome.)

Further, we say that event A "happens" or "occurs" when the outcome of the experiment or observation is one of the simple events in A. In terms of the A = (2, 4, 6) of the previous paragraph, we say that event A occurs if one of the outcomes (i.e., simple events) "2," "4," or "6" occurs. This is reasonable language. That is, if you are told "if you throw a fair die one time, then the event of an 'even outcome' will occur if you get a '2' or a '4' or a '6' as the result of your toss," you hardly would disagree. So, again, an event is said to occur when one of the simple events (= sample points) in it occurs.

Assigning Probabilities

Using the languge just discussed, we want to be able to assign probabilities to events.[4] One common way of approaching this topic is to use the notions of relative frequency and symmetry to assign the probabilities of the occurrence of simple events; the probability of the occurrence of a given event will be shown to involve adding up the probabilities of the simple events of which the event is composed (i.e., the simple events contained in the given event).

Turning first, then, to assigning the probabilities of the occurrence of simple events by using the notions of relative frequency and symmetry, let us return to the example where a fair die is tossed one time. The applicable Venn diagram is shown as Figure 1.1. If we want to assign the probability of the simple event of throwing a "3," a reasonably intuitive approach is to say, where P("3") means the probability of throwing a "3," that

$$P(\text{"3"}) = \frac{\text{number of times "3" happens (i.e., occurs) in "n" throws}}{\text{"n" throws}}$$

as "n" becomes large.

If we actually were to take a high quality die and throw it 600 times, we would expect that for such a large "n" that P("3") would be close to 1/6. The general principle that we would be using would be what we will call

$$\text{RULE 1: } P(\text{simple event}) = \frac{\text{number of times the simple event happens in "n" trials}}{\text{number of trials, "n"}}$$

as "n" becomes large.

(By trials, we mean experiments or observations.) This principle can be used on "unfair" dies, of course, or in other situations. For example, if a die were "loaded," we might find that in 600 tosses, "3" turned up only 10 times. Then, we would say that $P(\text{"3"}) = \frac{10}{600} = \frac{1}{60}$, i.e., P("3") is approximately $\frac{10}{600}$. As "n" becomes larger and larger, we would have more and more confidence in the value we would assign to P("3") by using Rule 1. That is, we expect to get closer and closer to the "true" value as the number of tosses becomes larger and larger. (Statistics courses provide guidance as to how large "n" must become in order to have a specified degree of certainty in the accuracy of the calculated value.)

This approach also could be used to assign the probability of something like a given thumbtack pointing down when thrown once, where the sample space is T=(up, down). For a given thumbtack, we once found

$$P(\text{down}) = \frac{\text{number of times thumbtack ends up pointing down}}{\text{number of times thumbtack is thrown}}$$

$$= \frac{114}{200} = 0.57.$$

That is, in 200 tosses the thumbtack ended up pointing down 114 times. We then assigned a value of 0.57 for the probability the thumbtack would end up pointing down.

Very often, instead of having to carry out an experiment like throwing a die or a thumbtack in order to come up with values that can be assigned to the various simple events, we simply can appeal to the symmetry of the situation and easily decide on the value that we "should" assign (because if Rule 1 were used, that is the value we would expect to get).[5]

For example, if we were dealing with the experiment of a fair die being tossed repeatedly, we would expect that Rule 1 would yield probabilities close to "1/6" for the probability of the simple event of die face "1" showing, and the same probability for each of the other die faces showing. We would reason this way because there are six outcomes, of which "1" is one outcome, and, since "fairness" implies that each outcome is as likely to occur as another, "1" "should" occur one-sixth of the time. The same reasoning would hold for outcomes 2-6. We therefore would <u>assign</u> probabilities of:

P(1) = 1/6, P(2) = 1/6, P(3) = 1/6, P(4) = 1/6, P(5) = 1/6, P(6) = 1/6.

Note that, as a result:

P(1) + P(2) + P(3) + P(4) + P(5) + P(6) = 1.

This always must be the case; that is, when we assign probabilities to the elements (= outcomes = sample points = simple events) of a sample space, we must do so in a way that the sum of the probabilities of all the simple events will equal 1.

One consequence of this requirement is that if we are dealing with a situation where we have, say, "n" simple events, and for reasons of fairness/symmetry we expect that all the simple events are equally likely to occur, then the probability of occurrence of each <u>must</u> be $\frac{1}{n}$. When we add up the probabilities of the "n" simple events where each has a probability of $\frac{1}{n}$ we get $\frac{1}{n} + \frac{1}{n} + \frac{1}{n} + \cdots = \frac{n}{n} = 1$, which is what we <u>must</u> have (i.e., the sum must be "1").

Thus, we have

<u>RULE 2</u>: $P(\text{simple event}) = \dfrac{1}{\text{total number of possible outcomes (i.e., simple events), "n"}}$

when all "n" outcomes (i.e., simple events) are equally likely to occur.

Note very carefully that this second rule for assigning probabilities will work only when, because of symmetry, it is clear that all the outcomes are equally likely. It could not be used for a thumbtack because this object is not symmetrical with respect to the outcomes of "up" or "down." Nor could it be used for a loaded die. If we are dealing with fair dice or coins, however, we can use Rule 2. For example, in the case of a fair coin, Rule 2 says that for one throw

$$P(\text{heads}) = \frac{1}{\text{total number of possible outcomes (simple events)}}$$

where all the outcomes are equally likely, so

$$P(\text{heads}) = \frac{1}{2} = 0.5,$$

a value about which we can be comfortable.

Very often in elementary probability courses, the examples are chosen so that in the experiments the assumptions of fairness and symmetry can be made and so that all of the outcomes are equally likely. Rule 2 then can be used to assign the probabilities for each of the simple events.

We are, of course, interested in assigning probabilities to events as well as to simple events. For example, returning to the Venn diagram presented earlier (Figure 1.1), we might ask what the probability is that event A will happen. Several things should be noted. First, as we saw earlier, A, the event of an even outcome, is said to happen (occur) if the outcome of the experiment is one of the outcomes i.e., simple events, in A, i.e., "2" or "4" or "6." That is, A happens (occurs) if the outcome (simple event) "2" or "4" or "6" happens (occurs). Now if Rule 2 is used, we can assign the values P("2") = 1/6, P("4") = 1/6, P("6") = 1/6 because we are assuming that the die is "fair" and therefore that all the outcomes are equally likely. The question is, since event A is not a simple event, but is made up of the three simple events, "2," "4," and "6," what is a reasonable definition for the probability of event A? The answer is

<u>RULE 3</u>: The probability that an event A will occur is <u>defined</u> as the sum of the probabilities of the simple events of which it is composed.

In this case, then, where P(A) = probability of A,

$$P(A) = P(\text{"2"}) + P(\text{"4"}) + P(\text{"6"}) = 1/6 + 1/6 + 1/6 = 1/2 = 0.5.$$

This answer is intuitively satisfying; if someone asked you what the probability of throwing an even outcome (i.e., event A) would be, assuming a fair die, you would say "1/2." This Rule 3 definition that the probability that an event A will occur is the sum of the probabilities of the simple events of which it is composed is the key to assessing probabilities. When we use Rule 3, we can assign probabilities to the simple events in A by using either Rule 1 or Rule 2.

In addition, there is a Rule 4 that is helpful for finding the probability of an event A when the sample space T has "n" outcomes and each is as likely to occur as any other. According to Rule 2, for such a sample space T,

$$P(\text{particular simple event}) = \frac{1}{n}.$$

Then if we assume that event A has "n_1" simple events in it (meaning that A can happen in "n_1" ways), since the probability of each simple event is $\frac{1}{n}$, we have "$\frac{1}{n}$" for the probability of the first simple event in A, "$\frac{1}{n}$" for the probability of the second simple event in A etc., for a total of "n_1" times. Hence,

$$P(A) = \underbrace{\frac{1}{n} + \frac{1}{n} + \frac{1}{n} + \cdots + \frac{1}{n}}_{\text{"}n_1\text{" terms}} = \frac{n_1}{n}.$$

We thus can say for an event A, we have

RULE 4:

$$P(A) = \frac{\text{number of ways, "}n_1\text{", the event A can happen (i.e., the number "}n_1\text{" of simple events contained in A) (i.e., the number "}n_1\text{" of outcomes in A)}}{\text{total number "n" of possible outcomes (i.e., simple events) in T}}$$

if all "n" outcomes (i.e., simple events) in T are equally likely to occur.

Actually, given the number of "i.e.'s" in the above equation, there are at least three different but equivalent ways we can express Rule 4; all are found in elementary probability texts, and the version used in any particular situation is a matter of taste and convenience. The first way of saying Rule 4 is to say:

RULE 4A:

$$P(A) = \frac{\text{number of ways, "}n_1\text{", A can happen}}{\text{total number "n" of possible outcomes in T}}.$$

if all "n" outcomes in T are equally likely to occur.

RULE 4B says:

$$P(A) = \frac{\text{the number "}n_1\text{" of simple events in A}}{\text{total number "n" of simple events in T}}$$

if all "n" simple events in T are equally likely to occur.

RULE 4C says:

$$P(A) = \frac{\text{the number "}n_1\text{" of outcomes in A}}{\text{total number "n" of possible outcomes in T}}$$

if all "n" outcomes in T are equally likely to occur.

Now Rule 4 (and its versions 4A, 4B, and 4C) is a very handy rule to use. A few examples are in order. Using Rule 4A, we can say that if we have an urn with 1 green, 2 black, and 3 red balls in it and we draw out one ball, so that all 6 outcomes of a ball being drawn are equally likely to occur, then if R is a red ball,

$$P(R) = \frac{\text{number of ways the event of drawing a red ball can happen}}{\text{total number of possible outcomes in T}}$$

$$= \frac{3}{6} = \frac{1}{2} = 0.5.$$

Suppose we have a well shuffled deck of 52 cards, where we can assume any one card is as likely to be drawn as any other. Then, if T is defined as the sample space of all 52 possible outcomes of drawing one card, we can say that the probability of drawing a king, P(K), is by Rule 4A

$$P(K) = \frac{\text{number of ways the event "K" can happen}}{\text{total number of possible outcomes in T}} = \frac{4}{52} = \frac{1}{13} = 0.077$$

(because there are 52 equally likely possible outcomes and there are 4 ways of getting a king).

Similarly, if B is the drawing of a black card,

$$P(B) = \frac{\text{number of ways the event "B" can happen}}{\text{total number of possible outcomes in T}} = \frac{26}{52} = \frac{1}{2} = 0.5.$$

A more interesting example is to find, if a fair die is tossed twice, the probability of the sum of the die faces being "7." Table 1.1 shows a sample space of 36 possible results for this experiment.

Since the assumption that the die is "fair" has been made, it seems reasonable to argue that all possible outcomes on the first toss are equally likely to occur, and all possible outcomes on the

TABLE 1.1
A Sample Space when a Fair Die is Tossed Twice

		Results of Second Toss					
		1	2	3	4	5	6
	1	1,1	1,2	1,3	1,4	1,5	1,6
Results	2	2,1	2,2	2,3	2,4	2,5	2,6
of	3	3,1	3,2	3,3	3,4	3,5	3,6
First	4	4,1	4,2	4,3	4,4	4,5	4,6
Toss	5	5,1	5,2	5,3	5,4	5,5	5,6
	6	6,1	6,2	6,3	6,4	6,5	6,6

second toss are equally likely to occur, so that each of the 36 combined possible outcomes is as likely to occur as another. Then Rule 4 can be used and we can calculate

$$P(\text{"7"}) = \frac{\text{number of ways the event "7" can happen}}{\text{total number of possible outcomes in T}} = \frac{6}{36} = \frac{1}{6} = 0.167$$

where we calculate the numerator by noting that there are 6 ways we can get a "7," namely (1,6), (2,5), (3,4), (4,3), (5,2), and (6,1).

Suppose in assessing the probability of an event that we drop the assumption that all the simple events are equally likely. This means that we cannot use Rule 4 to assign probabilities to events. How, then, can we find the probability of an event A? Rule 3 still applies, of course. That is, the probability of an event A equals the sum of the probabilities of the simple events of which it is composed. Rule 1 can be used to assign the probabilities for the simple events in A. Consider, for example, the non-symmetrical object shown in Figure 1.3.

FIGURE 1.3
A Three Dimensional Non-symmetrical Object

Then, use of Rule 1 might lead to the assignment of these probabilities to faces "1," "2," "3," and "4":

P("1") = 1/10, P("2") = 3/10, P("3") = 5/10, P("4") = 1/10.

Then, if we asked P(Even (number) face shows), we would calculate that

$$P(\text{Even face shows}) = P(\text{"2"}) + P(\text{"4"})$$
$$= 3/10 + 1/10 = 4/10 = 0.4.$$

Some Probability Theorems

Now let us look at a number of relatively simple probability theorems that it will be necessary to use.

The first is the previously presented result of

RULE 5: $P(T) = 1$.

This result really was presented above and probably is intuitively reasonable to the reader. We again consider the previously presented Venn diagram (Figure 1.1). Applying Rule 3,

$$P(T) = P(\text{"1"}) + P(\text{"2"}) + P(\text{"3"}) + P(\text{"4"}) + P(\text{"5"}) + P(\text{"6"})$$
$$= 1/6 + 1/6 + 1/6 + 1/6 + 1/6 + 1/6 = 1.$$

Note that the probabilities that were assigned to simple events "1" through "6" were based on use of Rule 2. The use of Rule 2, in essence, insured that the probabilities assigned would be such that P(T) = 1.[6] If, however, Rule 2 could not be used to assign probabilities to simple events, so that Rule 1 had to be used, we would have to be careful to make sure that when we assigned the probabilities to the simple events it was the case that P(T) = 1. That is, all treatments of probability require that the sum of the probabilities assigned to the simple events be "1," i.e., that P(T) = 1.

Rule 6 follows quickly. It is that

RULE 6: $0 \leq P(\text{event } A) \leq 1$.

This is true because the largest A could be is "T," so P(A) at most could equal P(T), i.e., 1. On the other hand, the smallest A could be would be the null set, Ø, and P(Ø) is defined as the number "0".[7] So, P(A) could be as small as 0. Hence

$$0 \leq P(A) \leq 1.$$

Turning now to more general concerns, you probably know, or will find it reasonable, that the probability that an event occurs plus the probability that it does not occur equals the number "1." If an event E does not occur, then its complement, not E, must occur, as the Figure 1.4 Venn diagram helps illustrate.

FIGURE 1.4
The Venn Diagram for Events "E" and "not E"

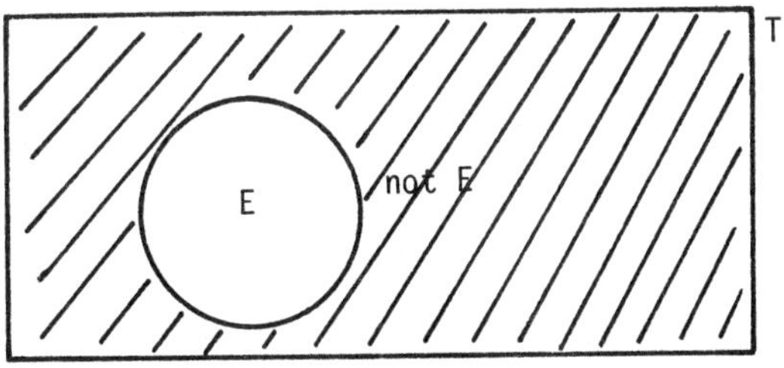

Thus, the probability that an event E does not happen is simply P(not E). Then, the claim is that

<u>RULE 7</u>: P(E) + P(not E) = 1.

For example, if we return to our Venn diagram for the fair die toss, suppose B is the set of outcomes where the die shows "3" or more. This is shown as Figure 1.5. Then B = (3, 4, 5, 6), while

FIGURE 1.5
A Venn Diagram for the Outcomes of Throwing a Fair Die Once

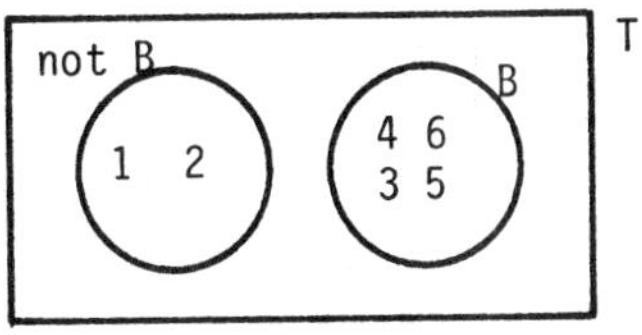

not B = (1, 2). The claim is that P(B) + P(not B) = 1. This is clearly the case since the application of Rule 3 and Rule 2 yields

P(B) = P("3") + P("4") + P("5") + P("6")
     = 1/6 + 1/6 + 1/6 + 1/6 = 4/6

and P(not B) = P("1") + P("2") = 1/6 + 1/6 = 2/6,
so that P(B) + P(not B) = 4/6 + 2/6 = 6/6 = 1.

A slightly rigorous treatment of this is given in footnote 8 below.

One reason for developing this formula

P(E) + P(not E) = 1

is that very often it is extremely difficult to calculate P(E) or P(not E) directly, but easy to calculate the other. If it is easy to calculate P(E), we can calculate P(not E) by using the above formula to get

P(not E) = 1 - P(E).

If it's easy to calculate P(not E), on the other hand, we can use the formula P(E) = 1 - P(not E) to calculate P(E).

A Formula for P(A or B)

It turns out that it also will be useful to get formulas for P(A or B) and P(A & B) in terms of their constituent elements. We deal in this section with P(A or B).

Turning once more to the die example of Figure 1.1, suppose we guess that

P(A or B) = P(A) + P(B).

This might look like a reasonable guess. It is easy to check to see if our guess is correct. Using Rule 3

P(A or B) = The sum of the probabilities of the simple events of which the event "A or B" is composed
= P("2") + P("4") + P("6") + P("3") + P("5")

while P(A) = The sum of the probabilities of the simple events of which the event A is composed
= P("2") + P("4") + P("6")

and P(B) = The sum of the probabilities of the simple events of which the event B is composed,
= P("3") + P("5") + P("4") + P("6").

The question is, then, does P(A or B) = P(A) + P(B), or, substituting, does,

P("2") + P("4") + P("6") + P("3") + P("5") = P("2") + P("4") + P("6") + P("3") + P("5") + P("4") + P("6")?

A little inspection shows that we have two extra terms on the right, namely P("4") and P("6"). Why didn't we get equality; that is, why didn't P(A or B) = P(A) + P(B)? The answer hopefully is clear: When we added P(A) to P(B), we counted the probabilities of the simple events "4" and "6" once when we calculated P(A) = P("2") + P("4") + P("6"), but we also counted the probabilities of the same simple events "4" and "6" a second time when we calculated P(B) = P("3") + P("5") + P("4") + P("6"). But, as we saw when we looked at A or B = (2, 3, 4, 5, 6), we should count the probabilities of "4" and "6" only once. Simply put, we counted the overlap, A & B, of A and B twice. Now, the solution may be clear; fix up the formula by saying

P(A or B) = P(A) + P(B) - P(A & B).

By putting in the term " - P(A & B)" we will subtract off the double-counted probabilities of the simple events in the overlap of A and B, i.e., the overlap A & B. To see that this works, we can use this formula to check on P(A or B) in the example. The claim is that

$$P(A \text{ or } B) = P(A) + P(B) - P(A \,\&\, B).$$

Substituting on the left and right,

$$\begin{aligned} P(\text{"2"}) + P(\text{"4"}) + P(\text{"6"}) + P(\text{"3"}) + P(\text{"5"}) &= P(\text{"2"}) + P(\text{"4"}) \\ &\quad + P(\text{"6"}) + P(\text{"3"}) \\ &\quad + P(\text{"5"}) + P(\text{"4"}) \\ &\quad + P(\text{"6"}) - P(\text{"4"}) \\ &\quad - P(\text{"6"}) \\ &= P(\text{"2"}) + P(\text{"4"}) \\ &\quad + P(\text{"6"}) + P(\text{"3"}) \\ &\quad + P(\text{"5"}). \end{aligned}$$

Now, we really do have equality. Although this is not a rigorous proof, it should be intuitively obvious that the formula

$$P(A \text{ or } B) = P(A) + P(B) - P(A \,\&\, B)$$

is valid, and a rigorous probability course will provide a proof. The following is an example of the use of what we will call

<u>RULE 8</u>: P(A or B) = P(A) + P(B) - P(A & B).

Suppose we draw one card from a deck and let Q = the event of drawing a Queen and R = the event of drawing a red card.

Then, P(Q or R) = P(Q) + P(R) - P(Q & R).

The Venn diagram is shown in Figure 1.6, where

QC = Queen of Clubs
QS = Queen of Spades
QH = Queen of Hearts
QD = Queen of Diamonds.

FIGURE 1.6
A Venn Diagram for the Outcomes of Drawing One Card from a Deck

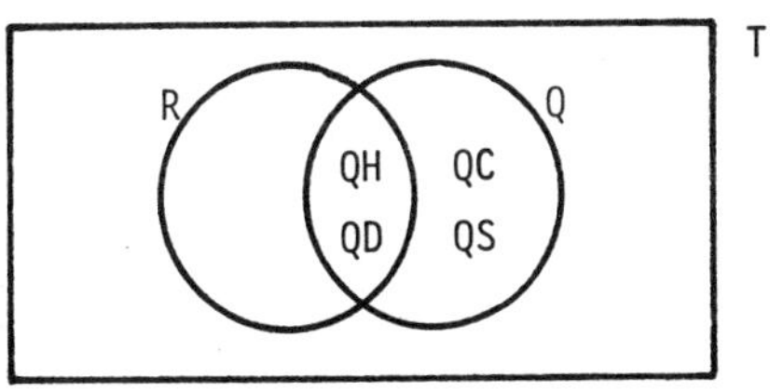

So, P(Q or R) = P(Q) + P(R) - P(Q & R)
= P(Q) + P(R) - P(QH, QD) (listing the elements in Q&R)
= P(Q) + P(R) - (P(QH) + P(QD)) (using Rule 3)

= 4/52 + 26/52 - 2/52 = 28/52 = 0.54.

We might ask what P(A or B) is if the picture is like Figure 1.7, i.e., if A & B = Ø, where A and B are non-empty. (We say

FIGURE 1.7
A Venn Diagram with Mutually Exclusive Sets

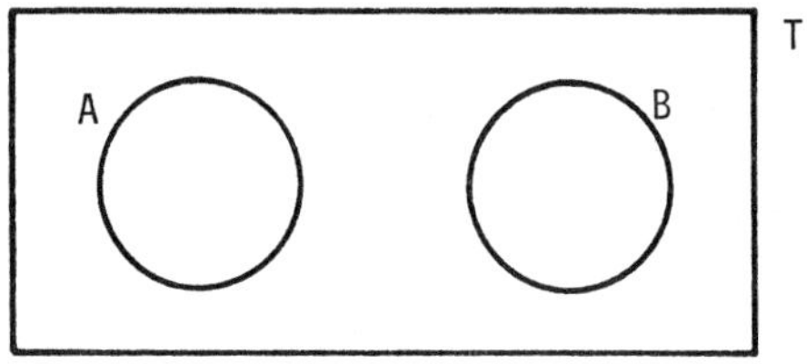

that A and B are mutually exclusive in such a situation; this is because only one can happen at a time -- see below.) If we use the formula P(A or B) = P(A) + P(B) - P(A&B) in such a situation, we find that the last term, P(A & B) = P(Ø). That is, as noted, A & B = the empty (or null) set Ø in this situation. Since there are no simple events in Ø, P(Ø) is defined as 0, as we noted above. Hence, the formula becomes

P(A or B) = P(A) + P(B) - 0 = P(A) + P(B).

So, we have

<u>RULE 9</u>: P(A or B) = P(A) + P(B) when A and B are mutually exclusive.

As just noted, the standard terminology for this pictured situation (Figure 1.7), is to say that A and B are mutually exclusive. In this situation, only one of the events A and B can happen at a time. This is because, as the picture shows, since A and B are defined on the same sample space T, if A happens (i.e., if one of the simple events in A occurs), B certainly can't occur at the same time because that same simple event can't occur simultaneously in B - there is no overlap between A and B. Similarly, if B happens, A can't happen. Thus, if A and B are mutually exclusive, only one can happen at a time! Furthermore, as Figure 1.7 implies, if A and B are mutually exclusive, then A & B = Ø.

A little reflection should convince the reader that all of the following are equivalent (i.e., any one implies all the others):

(1) Figure 1.7,

(2) A & B = Ø,
(3) A and B are mutually exclusive,
(4) With respect to sets (events) A and B, it is the case that only one can happen at a time.

As an example of the use of Rule 9 when dealing with mutually exclusive events, let QH = the event of drawing the Queen of hearts and B = the event of drawing a black card. Then, in drawing one card from a deck of cards, QH and B are mutually exclusive since only one can happen at a time. Figure 1.8 shows the picture.

FIGURE 1.8
A Venn Diagram for Drawing One Card from a Deck when Some Events are Mutually Exclusive

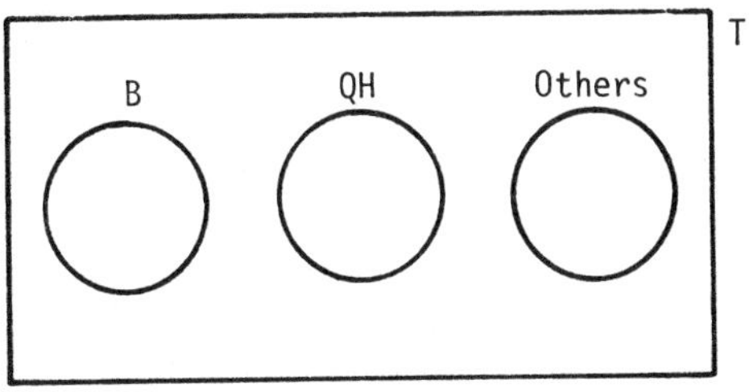

Then
P(QH or B) = P(QH) + P(B) = 1/52 + 26/52 = 27/52 = 0.52.

Quite frankly, my advice is to forget Rule 9 and always use Rule 8. If it turns out that you are dealing with mutually exclusive events, the last term will drop out. In this particular example, use of the general formula yields

$$\begin{aligned} P(QH \text{ or } B) &= P(QH) + P(B) - P(QH \,\&\, B) = 1/52 + 26/52 - P(\emptyset) \\ &= 1/52 + 26/52 - 0 = 27/52 = 0.52. \end{aligned}$$

Nothing is lost by always using Rule 8. But using Rule 9 may lead you to forget that it is only true when A and B are mutually exclusive, and perhaps to use it erroneously when A and B are not mutually exclusive.[8]

<u>A Formula for P(A or B or C or D or ···)</u>

We can generalize the formula for P(A or B) when A and B are mutually exclusive. We say that a series of events A, B, C, D, ··· are mutually exclusive if only one can happen at a time. The picture for this situation is displayed as Figure 1.9.

FIGURE 1.9
A Venn Diagram for a Series of Mutually Exclusive Events

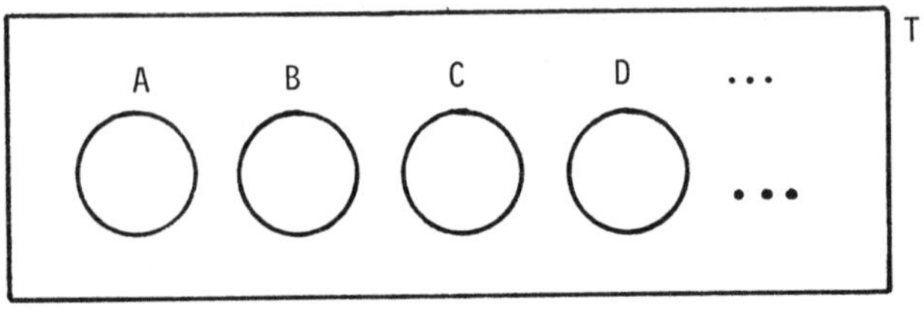

We generalize the definition of A or B by defining A or B or C or D or ··· as the set of elements belonging to one or more of A, B, C, D, ···. Thus, A or B or C or D or ··· is the "union" of all the simple events in A, B, C, D, ···. We then have

RULE 10: If A, B, C, D, ··· are mutually exclusive events, then P(A or B or C or D or ···) = P(A) + P(B) + P(C) + P(D) + ···.

This situation is illustrated in Figure 1.9. The formula in Rule 10 is so simple because there are no overlapping sets and hence no probabilities of intersections to subtract.

Several other notions are helpful, too. Suppose we have a series of mutually exclusive events A, B, C, D, ···. Then, the Venn diagram is as shown in Figure 1.9. The events are said to be exhaustive if one of them has to happen. This means, of course, that they contain all the simple events in T. If there were even one simple event outside of A, B, C, D, ··· then it conceivably could happen instead of one of A, B, C, D ···.

If A, B, C, D, ··· are mutually exclusive and exhaustive, then it is reasonable to argue that A or B or C or D or ··· = T, since every simple event is in one of A, B, C, D ···. But since (A or B or C or D or ···) = T, P(A or B or C or D or ···) = P(T) = 1. Since we asserted just above that if A, B, C, D, ··· were mutually exclusive, it followed that P(A or B or C or D or ···) = P(A) + P(B) + P(C) + P(D) + ··· it follows that if A, B, C, D, ··· also are exhaustive, we must have P(A) + P(B) + P(C) + P(D) + ··· = 1.

Summarizing, we have this definition:

Events A, B, C, D, ··· are mutually exclusive and exhaustive if
a. Only one can happen at a time and
b. One of them must happen.

Furthermore, we have shown

RULE 11: If events A, B, C, D, ··· are mutually exclusive and exhaustive then
P(A or B or C or D or ···) = P(A) + P(B) + P(C) + P(D) + ··· = 1.

(Note that all the simple events together are mutually exclusive and exhaustive.)

Some examples may clarify these notions. Suppose that we draw one card from a deck of cards. Let B = the event of drawing a black card and R = the event of drawing a red card where T = the 52 possible outcomes of drawing one card from a deck of cards. Then, the Venn diagram is the one shown as Figure 1.10.

FIGURE 1.10
A Venn Diagram for Drawing One Card from a Deck with Mutually Exclusive and Exhaustive Events

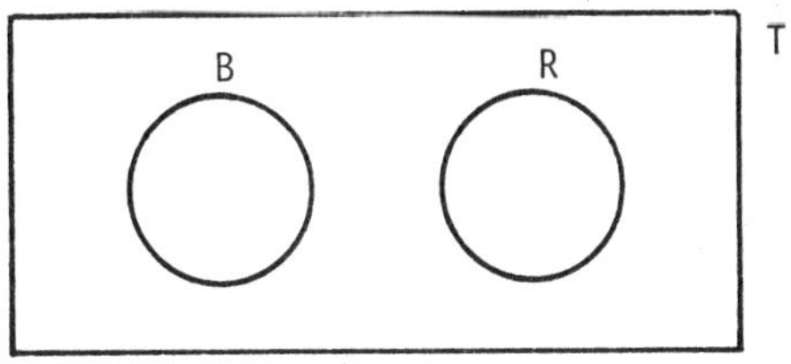

Events B and R are mutually exclusive and exhaustive, and P(B or R) = 1.

Consider Figure 1.11. Events A and B are neither mutually exclusive nor exhaustive; hence,

P(A or B) ≠ P(A) + P(B) where ≠ means "is not equal to."

FIGURE 1.11
A Venn Diagram for Drawing One Card from a Deck with Events Neither Mutually Exclusive nor Exhaustive

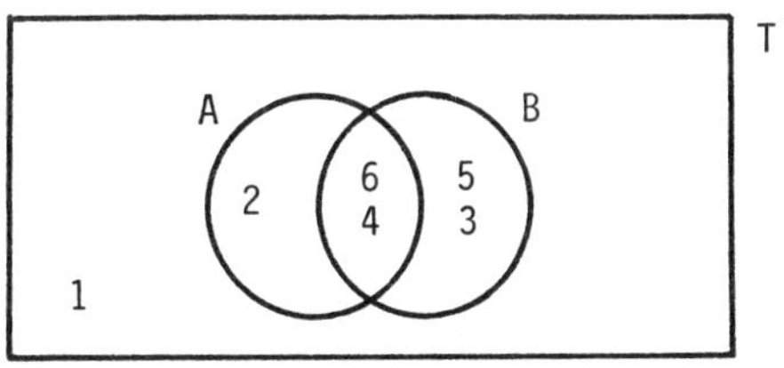

In contrast, if the Venn diagram is the one depicted in Figure 1.12 then events A and B are mutually exclusive, but not exhaustive ("1" is outside both). Further,

P(A or B) = P(A) + P(B), but it is <u>not true</u> that P(A or B) = 1.

FIGURE 1.12
A Venn Diagram for Drawing One Card from a Deck with Events Mutually Exclusive, but not Exhaustive

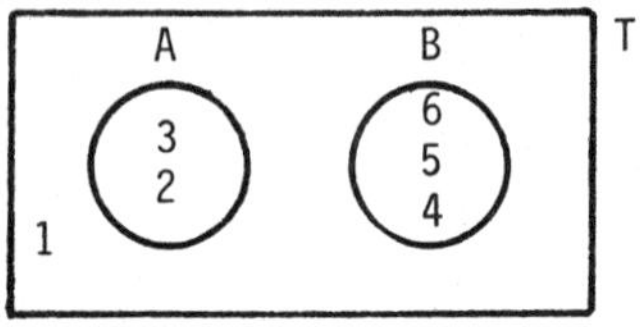

Next, consider the card drawing situation where one card is drawn from a standard deck and events H = heart, C = club, S = spade, D = diamond. Then we have Figure 1.13. We can say that events H, S, and D are mutually exclusive, but not exhaustive. So P(H or S or D) = P(H) + P(S) + P(D) but P(H or S or D) ≠ 1. On the other hand, events H, C, S, and D are mutually exclusive and exhaustive, so P(H or C or S or D) = P(H) + P(C) + P(S) + P(D) = 1.

FIGURE 1.13
A Venn Diagram for Drawing One Card from a Deck

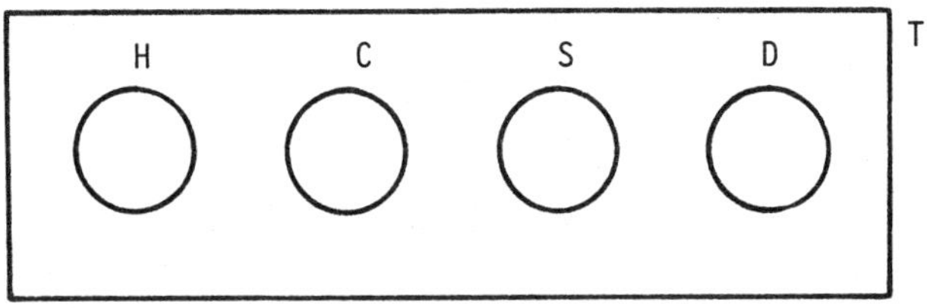

Summary of the Discussion to this Point

RULE 1:

$$P(\text{simple event}) = \frac{\text{number of times the simple event happens in "n" trials}}{\text{number of trials, "n"}}$$

as "n" becomes large.

RULE 2:

$$P(\text{simple event}) = \frac{1}{\text{total number of possible outcomes (i.e., simple events), "n"}}$$

when all "n" outcomes (i.e., simple events) are equally likely to occur.

RULE 3:

P(event A) = The sum of the probabilities of the simple events of which it is composed

RULE 4:

$$P(A) = \frac{\text{number of ways, "}n_1\text{", event A can happen (i.e., the number "}n_1\text{" of simple events contained in A) (i.e., the number "}n_1\text{" of outcomes in A)}}{\text{total number "n" of possible outcomes (i.e., simple events) in T}}$$

if all "n" outcomes (simple events) in T are equally likely to occur.

RULE 4A:

$$P(A) = \frac{\text{number of ways, "}n_1\text{", A can happen}}{\text{total number "n" of possible outcomes in T}}$$

if all "n" outcomes in T are equally likely to occur.

RULE 4B:

$$P(A) = \frac{\text{the number "}n_1\text{" of simple events in A}}{\text{total number "n" of simple events in T}}$$

if all "n" simple events in T are equally likely to occur.

RULE 4C:

$$P(A) = \frac{\text{the number "}n_1\text{" of outcomes in A}}{\text{total number "n" of possible outcomes in T}}$$

if all "n" outcomes in T are equally likely to occur.

RULE 5: $P(T) = 1$.

RULE 6: $0 \leq P(\text{event A}) \leq 1$.

RULE 7: $P(E) + P(\text{not } E) = 1$.

RULE 8: $P(A \text{ or } B) = P(A) + P(B) - P(A \,\&\, B)$.

RULE 9: $P(A \text{ or } B) = P(A) + P(B)$ when A and B are mutually exclusive.

Equivalent ways of conceptualizing the term "mutually exclusive are these:

a) A and B are mutually exclusive if

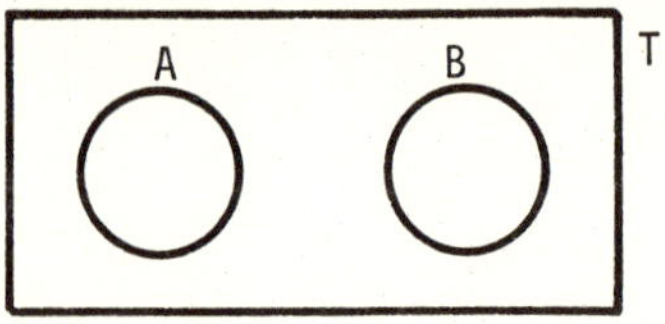

b) A and B are mutually exclusive if A & B = Ø.

c) A and B are mutually exclusive if only one can happen at a time.[9]

Definition: Events A, B, C, D, ··· (defined on the same sample space, T, of course) are mutually exclusive if only one can happen at a time.

Definition: A or B or C or D or ··· = the set of points belonging to one or more of events A, B, C, D, ···.

RULE 10: If events A, B, C, D, ··· are mutually exclusive, then P(A or B or C or D or ···) = P(A) + P(B) + P(C) + P(D) + ···.

Definition: Events A, B, C, D, ··· are mutually exclusive and exhaustive if

a. Only one can happen at a time and
b. One of them must happen.

RULE 11: If events A, B, C, D, ··· are mutually exclusive and exhaustive, then

$$P(A \text{ or } B \text{ or } C \text{ or } D \text{ or } \cdots) = P(A) + P(B) + P(C) + P(D) + \cdots = 1.$$

Development of the Formula for P(A & B)

As we noted earlier, we also want to develop a formula for the probability of A & B in terms of the constituent elements. The way this sometimes is done is to introduce the notion of conditional probability.

Suppose that we have the situation in the Figure 1.1 Venn diagram we have used before. Then, A = (2, 4, 6) = the event of throwing an even number on a die toss, and B = (3, 4, 5, 6) = the event of throwing 3 or more on a die toss.

Now, suppose someone tosses a (fair) die, looks at the result, and says "Event A has occurred." As we noted earlier, to say that event A has occurred is to say that one of the outcomes, i.e., simple events, in A has occurred; hence the person really is saying that exactly one of the outcomes of "2," "4," or "6" has occurred. Suppose we ask what the probability is that B also has occurred, given this information that A has occurred. Notationally, we write this as P(B/A) and say it aloud as "the probability of B given that A has occurred." Suppose we try to reason out what P(B/A) should equal. The first thing to note is that, since we have been told that A has occurred, it means that the sample space we want to consider in calculating the probability that B also has occurred has "shrunk" to a new space T'(see Figure 1.14) because exactly one of the simple events of

FIGURE 1.14
A Venn Diagram where the Universal set T' is an Even Outcome (of the Experiment of Throwing a Fair Die Once)

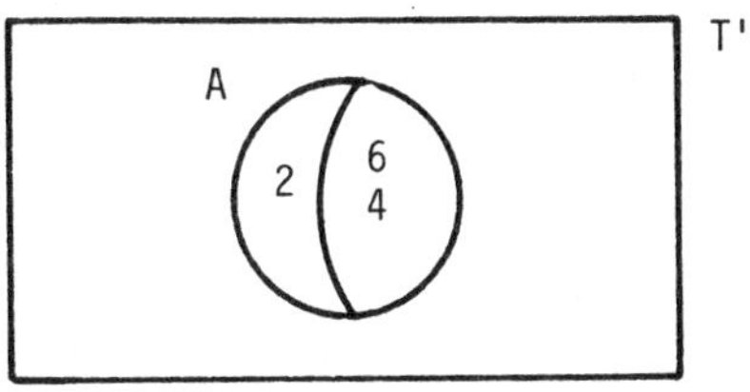

"2," "4," or "6" has occurred. In this sample space, it is reasonable to assume that all the outcomes are equally likely (because they were equally likely in the original sample space). Hence, we can use Rule 4B and write

$$P(B/A) = \frac{\text{the number "}n_1\text{" of simple events in the event "B also has occurred given that A has occurred"}}{\text{total number "n" of simple events in T'}}.$$

This phasing is rather awkward, and can be replaced by this equivalent phrasing (since the event "B also has occurred given that A has occurred" can be rewritten as "B also can happen if A has happened"):

$$P(B/A) = \frac{\text{the number "}n_1\text{" of simple events in the event "B also can happen if A has happened"}}{\text{total number "n" of simple events in T'}}.$$

If we look at the numerator, which still is phrased awkwardly, a little thought may convince the reader that the number "n_1" of simple events in the event "B also can happen if A has happened" is "2," because the event "B also can happen if A has happened" occurs if either the simple event "4" occurs or the simple event "6" occurs. (These are the only elements of B found in the new sample space T'.) But, these simple events are what comprise "A & B"; thus the event "B also can happen if A has happened" is the same as event "A & B". Since these are identical events, we can write

$$P(B/A) = \frac{\text{the number "}n_1\text{" of simple events in A \& B}}{\text{the total number "n" of simple events in T'}}.$$

Looking at the above denominator for a moment, the total number "n" of simple events in T' is simply the number of simple events in A, since A = T'. So, we can write

$$P(B/A) = \frac{\text{the number "}n_1\text{" of simple events in A \& B}}{\text{the number "n" of simple events in A}}.$$

Now for a little trick. Any fraction can have its numerator and denominator be divided by the same finite, non-zero number and still equal its original value. If we divide the numerator and denominator of the above fraction by "the total number of simple events in T," a finite, non-zero number, we get

$$P(B/A) = \frac{\dfrac{\text{the number "}n_1\text{" of simple events in A \& B}}{\text{the total number of simple events in T}}}{\dfrac{\text{the number "n" of simple events in A}}{\text{the total number of simple events in T}}}.$$

This is a very complicated looking fraction, but if the reader again looks back at Rule 4B, it hopefully will be clear that the numerator of this big fraction is just P(A & B). That is,

$$P(A \,\&\, B) = \frac{\text{the number "}n_1\text{" of simple events in A \& B}}{\text{the total number of simple events in T}}.$$

Also, the denominator of this big fraction is just P(A). That is, again using Rule 4B,

$$P(A) = \frac{\text{the number "n" of simple events in A}}{\text{the total number of simple events in T}}.$$

Thus, we have shown that

$$P(B/A) = \frac{P(A \,\&\, B)}{P(A)}.$$

By the way, usually in probability courses all this intuitive discussion is eliminated and the reader is simply given the <u>definition</u> that

$$P(B/A) = \frac{P(A \,\&\, B)}{P(A)}.$$

An easier way to remember this formula is to note that since A & B = B & A, it must be the case that P(A & B) = P(B & A). Hence

$$P(B/A) = \frac{P(A \,\&\, B)}{P(A)} = \frac{P(B \,\&\, A)}{P(A)}, \text{ i.e., } P(B/A) = \frac{P(B \,\&\, A)}{P(A)}.$$

If we use this latter version of the formula for the probability of B given that A has occurred, a mnemonic device to use to remember the formula is to say that "the probability that the first event will occur given that the second event has occured is equal to the probability of the intersection of the first and second events divided by the probability of the second event." Use of this "first, second" concept gets the user away from being hung-up on specific letters of the alphabet.

As noted originally, the purpose of developing the notion of conditional probability was to get a formula for P(A & B). If we look at the formula we just developed,

$\frac{P(A \,\&\, B)}{P(A)} = P(B/A)$, then if both sides of the formula are multiplied by P(A) (where we assume $P(A) > 0$), we get

$P(A) \cdot \frac{P(A \,\&\, B)}{P(A)} = P(A) \cdot P(B/A)$, i.e. (when we cancel the P(A)'s on the left side of the equation),

$$P(A \,\&\, B) = P(A) \cdot P(B/A).$$

Using the notion of first and second events, this formula says that the probability of the occurrence of a first and second event equals the probability of the first event times the (conditional) probability that the second event will occur given that the first event has occurred. Paul G. Hoel notes that "either one of the two events may be called the first event, since this is merely convenient language for discussing them, and there is no time order implied in the way they occur. Even though there is no time order implied for the two events ... it is customary to call this conditional probability 'the probability that [B] will occur when it is known [given] that [A] has occurred.'"[10]

Note that we can calculate P(A & B) as

$P(A \,\&\, B) = P(A) \cdot P(B/A)$. But since $P(A \,\&\, B) = P(B \,\&\, A) = P(B) \cdot P(A/B)$ we have

$$P(A \,\&\, B) = P(A) \cdot P(B/A) = P(B) \cdot P(A/B).$$

In calculating P(A & B) in a particular problem, then, we can try to use

$$P(A \,\&\, B) = P(A) \cdot P(B/A)$$

and if we don't have the data for this version of the formula we can turn to the second version of

$$P(A \,\&\, B) = P(B) \cdot P(A/B)$$

and see if we have the data. I prefer to use the version

$$P(A\&B) = P(A) \cdot P(B/A)$$

if at all possible because it lends itself to the "first, second" mnemonic language noted above.

The Notion of a Probability Tree

We turn now to a discussion of a very powerful tool for setting up and solving elementary probability problems. This is the concept of a probability tree. Later, we also will find it to be a powerful tool for explaining Bayesian analysis.

Suppose we have a situation where we have an urn containing three red balls and two black balls, and we draw sequentially two balls, not replacing the first ball drawn. Suppose we want to calculate the probability of drawing two red balls. We can use a probability tree to set up and solve this problem. We draw a tree from left to right, starting at a point (= node), and setting forth all the alternatives. For the first drawing we proceed as in Figure 1.15 to diagram the possible outcomes where

R_1 = a red ball on the first draw

and B_1 = a black ball on the first draw.

FIGURE 1.15
A Tree Diagram of the Results of the First Urn Draw

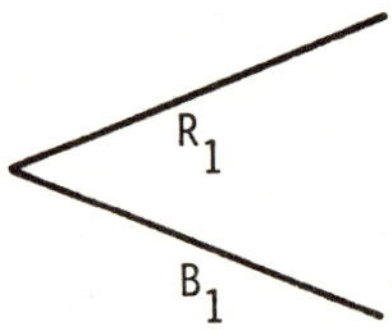

Thus, we use a subscript of "1" to denote the first drawing. We also speak of the "branch" on the tree where R_1 happens, and the "branch" where B_1 happens.

Note that these possible outcomes of R_1 and B_1 are mutually exclusive and exhaustive - only one can happen at a time, and one must happen. In a properly constructed probability tree, from any node the alternatives listed must be mutually exclusive and exhaustive.

For the second drawing, we proceed to diagram the possible outcomes (Figure 1.16) where

R_2 = a red ball on the second draw

and B_2 = a black ball on the second draw.

FIGURE 1.16
A Tree Diagram of the Results of the Second Urn Draw

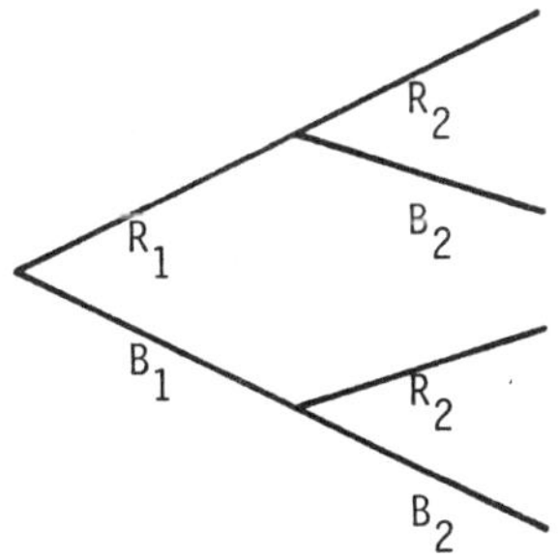

Thus, we use a subscript of "2" to denote the second drawing. Note that the possible outcomes of R_2 and B_2 are mutually exclusive and exhaustive, as the alternatives from any node must be. In a tree, a sequence of branches is called a path. So, we can speak of the topmost path, where R_1 and R_2 have happened, the next path where R_1 and B_2 have happened, etc.

The tree as drawn models all the possible color combination results of the experiment of drawing two balls, one after the other, from an urn. As the tree in Figure 1.17 shows, there are four paths, each corresponding to a possible color combination:

R_1 and R_2 (i.e., red on the first draw and red on the second draw)

R_1 and B_2

B_1 and R_2

B_1 and B_2.

FIGURE 1.17
A Tree Diagram of the Results of the First and Second Urn Draws Showing Color Combinations

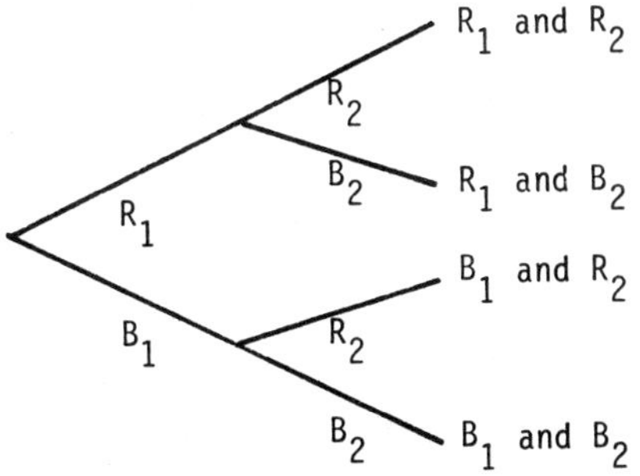

These are, as noted, the four separate color combination outcomes. But, they really are a shorthand for a larger list of outcomes, because we have not distinguished among the colored balls. If we did, there would be 20 outcomes, which we can list in matrix fashion in Table 1.2 (remember, if a ball is drawn first, it can't be drawn a second time because it is left out).

TABLE 1.2
A Sample Space for Drawing Two Balls from the Urn

Result of second draw

Result of First Draw	R	R	R	B	B
R		R,R	R,R	R,B	R,B
R	R,R		R,R	R,B	R,B
R	R,R	R,R		R,B	R,B
B	B,R	B,R	B,R		B,B
B	B,R	B,R	B,R	B,B	

So, if we distinguish among balls we have $5 \cdot 4 = 20$ possible outcomes. Moreover, it seems that each outcome is as likely as another. That is, any one of the 5 balls is as likely to be drawn on the first draw as any other. Furthermore, if we look at the first row, after the first one of the 3 red balls has been drawn, each of the remaining 4 balls is as likely to be drawn as any other. This gives 4 equiprobable combinations for the first row of the matrix. If we look at the second row, after the second one of the 3 red balls has been drawn, each of the remaining 4 balls is as likely to be drawn as any other. This gives 4 more equiprobable combinations for the second row , and it seems reasonable to see these as equiprobable with the first row. If we continue with this line of reasoning for all 5 rows we end up with $5 \cdot 4 = 20$ equiprobable events, and thus reasonably will assign a probability of 1/20 to each.

One question is, what is the relation between this matrix and our tree (Figure 1.18), which just shows color combinations? To

FIGURE 1.18
A Tree Diagram of the Results of the First and Second Urn Drawings Labeling Color Combination Results

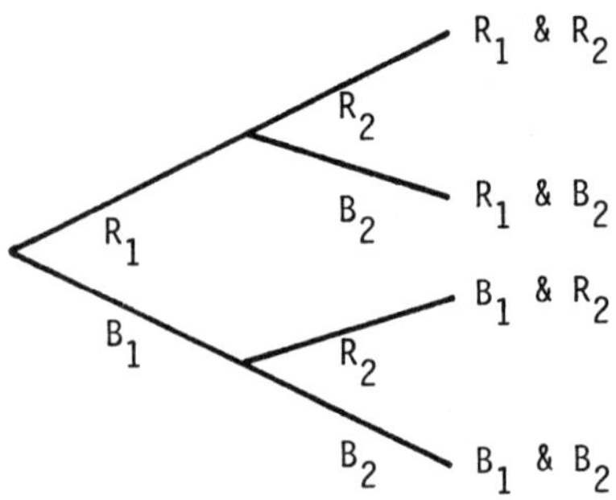

help to answer this question, note that on this tree we now have labeled the top path as R_1 & R_2 instead of R_1 and R_2 as we did in Figure 1.17. R_1 & R_2 must have a definite mathematical meaning, of course. It really is proper notation because, if R_1 is defined as the set of all outcomes where a red ball is drawn first, and R_2 is defined as the set of all outcomes where a red ball is drawn second, then R_1 & R_2 will be the set of all outcomes where red is drawn first and red (also) is drawn second - exactly what the top path is supposed to show.

R_1, R_2, and R_1 & R_2 are marked on the matrix shown here as Table 1.3 (a repeat of Table 1.2 and a Venn diagram if we think of our matrix entries as elements in a Venn diagram).

TABLE 1.3
Results of the First and Second Urn Drawings

Results of First Draw	Results of Second Draw: R	R	R	B	B
R		R,R	R,R	R,B	R,B
R	R,R		R,R	R,B	R,B
R	R,R	R,R		R,B	R,B
B	B,R	B,R	B,R		B,B
B	B,R	B,R	B,R	B,B	

(Outlined regions in the matrix: R_1 = the first three rows; R_2 = the first three columns; $R_1\&R_2$ = the R,R entries.)

As can be seen, the composite event R_1 & R_2 is composed of six simple events; since we decided above that each entry (= simple event) in the matrix (= Venn diagram) would be assigned a probability of 1/20, we must have, by Rule 3,

$$P(R_1\&R_2) = 1/20 + 1/20 + 1/20 + 1/20 + 1/20 + 1/20 = 6/20.$$

In the same fashion, we can use the matrix (= Venn diagram) to assign all four "path probabilities." The results are shown in Figure 1.19.

The next question is this: Is there some way the tree can be used to avoid the tedious calculation that must be carried out to find the four path probabilities? Yes, there is. Look at the top path for a moment. We have assigned a value of $P(R_1 \& R_2) = 6/20$ based on the analysis of the outcomes of which R_1 & R_2 is composed. If we look further at this probability $P(R_1 \& R_2)$, we may remember that we developed a formula for the probability of the intersection of two events, i.e., $P(A\&B) = P(A)\cdot P(B/A)$. Thus, it should be the case, where R_1 and R_2 are defined as they are on the same sample space, that

FIGURE 1.19
A Probability Tree Diagram Showing the Path Probabilities

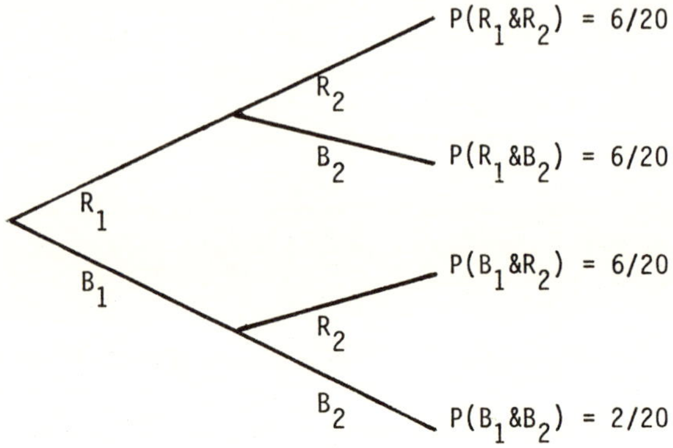

$$P(R_1 \& R_2) = P(R_1) \cdot P(R_2/R_1).$$

Now we found, by Rule 3, that $P(R_1 \& R_2) = 6/20$. Looking again at the figure, R_1 is composed of 12 outcomes, each with a probability of 1/20 so $P(R_1) = 12/20$. On the other hand, $P(R_2/R_1)$ is harder to locate. Going back to our motivational discussion of the formula for $P(B/A)$, we can say that if R_1 is given, then there are 12 possible outcomes; of these, there are 6 where R_2 also occurs. Since all the outcomes are equally likely, $P(R_2/R_1) = 6/12$. To check on the consistency of this analysis, we should calculate

$$P(R_1) \cdot P(R_2/R_1)$$

and compare the result to our previously calculated $P(R_1 \& R_2)$. Using the just derived probabilities, $P(R_1) \cdot P(R_2/R_1) = 12/20 \cdot 6/12 = 6/20$, which is what we calculated earlier for $P(R_1 \& R_2)$.

But, we still are using the same large matrix (= Venn diagram), and the question remains whether we can use the tree to calculate more easily $P(R_1 \& R_2)$. Hoel shows us the way: the point is to view $P(R_1 \& R_2) = P(R_1) \cdot P(R_2/R_1)$ as saying that the result of a two-stage (or "N"-stage) experiment equals the product of the probabilities of two (or "N") <u>one</u>-stage experiments.[11] In this case, if we go back to our experiment of drawing two balls, and look at the first drawing (the first one-stage experiment to consider), then the tree is shown in Figure 1.20 and we say that, since there are 5 balls in the urn, each as likely to be drawn as another, and there are 3 red balls, by Rule 4c

FIGURE 1.20
A Tree Diagram of the First-Stage Drawing

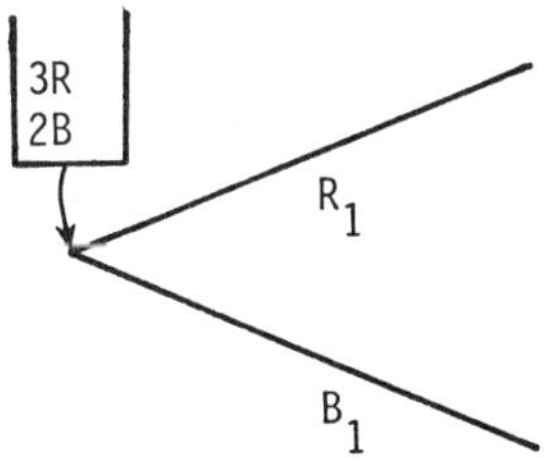

$$P(R_1) = \frac{\text{number of ways of getting a red ball on the first draw}}{\text{total number of possible outcomes in T}}$$

= 3/5, where the numerator "3" comes from saying that any one of the 3 red balls could be drawn. Similarly,

$P(B_1) = 2/5$.

We can put these probabilities on the tree (see Figure 1.21).

FIGURE 1.21
A Probability Tree Diagram of the Results of the First-Stage Experiment

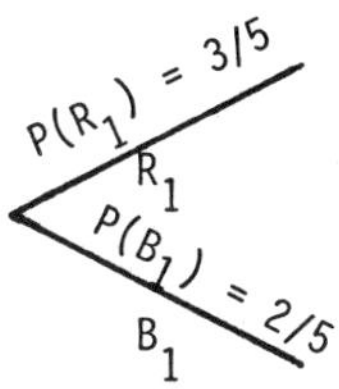

Note that figuring out $P(R_1)$ this way gives us the result of 3/5, while figuring out $P(R_1)$ using the 20 outcome matrix yields $P(R_1) = 12/20$. Fortunately, the results are identical, i.e. 3/5 = 12/20. This result always will hold. That is, instead of using a large multistage outcome matrix, we simply can calculate the probabilities of the alternatives from the first node in a tree by looking at the outcomes at that first stage of the multistage process. The results will be the same as if we had done all the work involved with using the multistage Venn diagram or matrix. The reader can check easily the single-stage result of $P(B_1) = 2/5$ by looking at the 20 entry large matrix.

As noted, we can put these values on the tree (Figure 1.21). Since R_1 and B_1 are mutually exclusive and exhaustive, Rule 11 says it must be true that $P(R_1 \text{ or } B_1) = P(R_1) + P(B_1) = 1$.

Since $P(R_1) = 3/5$

and $P(B_1) = 2/5$, we have

$P(R_1) + P(B_1) = 3/5 + 2/5 = 5/5 = 1$, as we are supposed to have. This reflects the previously noted important principle: <u>In a properly constructed probability tree, from any node the alternatives must be mutually exclusive and exhaustive and hence the sum of the probabilities of the branches from that node must be "1".</u>

The tree thus far models all 4 possible color combinations, the (path) probabilities of all 4 combinations, and the probabilities of the alternatives from the first node, where those initial probabilities were calculated just on the basis of the first drawing (see Figure 1.22 based on Figures 1.19 and 1.21).

FIGURE 1.22
A Probability Tree Diagram of the Results of the First-Stage Experiment and of the Path Probabilities

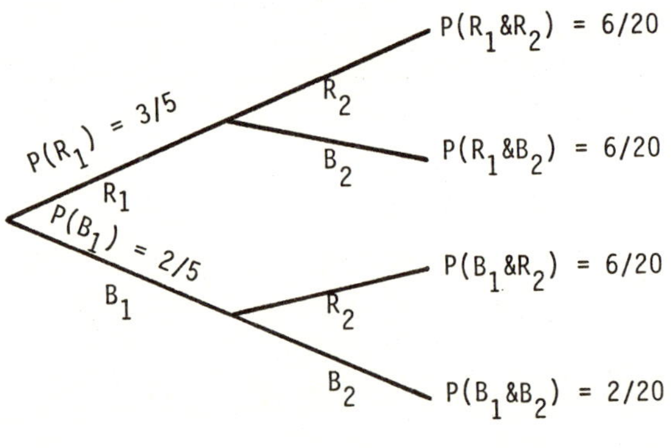

As Hoel points out, the easy way to assign $P(R_2/R_1)$ in the formula $P(R_1\&R_2) = P(R_1)\cdot P(R_2/R_1)$ is to think of the second drawing as a single-stage experiment: in this experiment, since a red ball was drawn on the first drawing, there are only 4 balls left in the urn, whose composition now is 2 red and 2 black balls. Hence, by Rule 4c, $P(R_2/R_1) = 2/4$, a value identical to the $P(R_2/R_1) = 6/12$ we calculated using the 20 entry matrix. Again, this sort of result always will hold.[12]

Note that in the tree (Figure 1.23) we must vary the second-stage urn from which we hypothetically are drawing. For example, if we look at the third path, and wish to find $P(B_1\&R_2) = P(B_1)\cdot P(R_2/B_1)$, then in assigning $P(R_2/B_1)$, the urn is assumed to first have had a black ball removed, leaving 3 red and 1 black ball. Hence, by Rule 4c, $P(R_2/B_1) = 3/4$ (which equals the 6/8 we would calculate using the Venn diagram).

FIGURE 1.23
A Probability Tree Diagram of the Path Probabilities, of the Second-Stage Drawing, and of the Results of the First-Stage Drawing

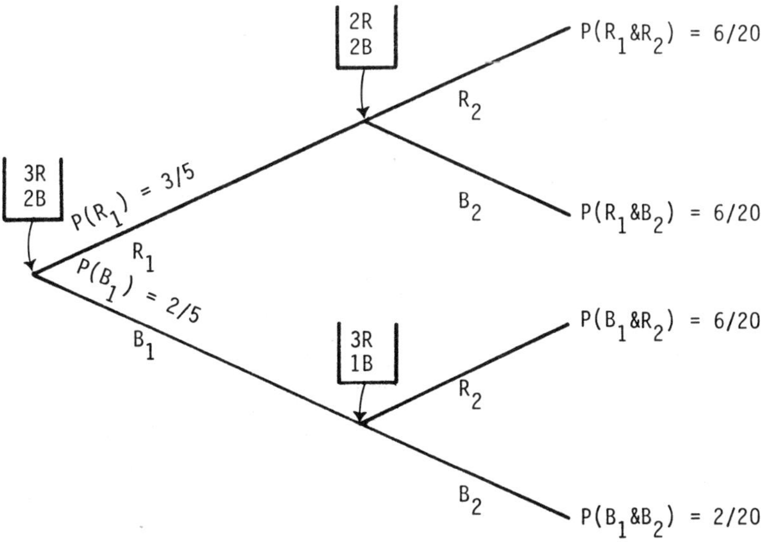

If we want to associate the conditional probabilities assigned at the second stage of the experiment with the second set of branches on the tree, we will get Figure 1.24. Note again, as shown on the tree, that the constituents of the urn to consider for the second-stage drawing vary, depending upon what was drawn first.

Note that, on the top path, reflective of the analysis just completed,

$$P(R_1\&R_2) = P(R_1)\cdot P(R_2/R_1)$$

where $P(R_1) = 3/5$ is calculated on the basis of the first one-stage experiment, and $P(R_2/R_1) = 2/4$ is calculated on the basis of the second one-stage experiment, and these numbers equal the values we get using the 20 entry matrix for the multistage experiment. Moreover, by associating $P(R_1)$ with the first (R_1) branch probability, and $P(R_2/R_1)$ with the second (R_2) branch probability, we get

<u>RULE 14: The probability of a path equals the product of the probabilities of the branches on that path.</u>

This is a terribly important rule, and actually holds no matter how many branches are on a given path.[13]

FIGURE 1.24
A Probability Tree Diagram of the Path Probabilities and of the Results of the First-Stage and Second-Stage Drawings

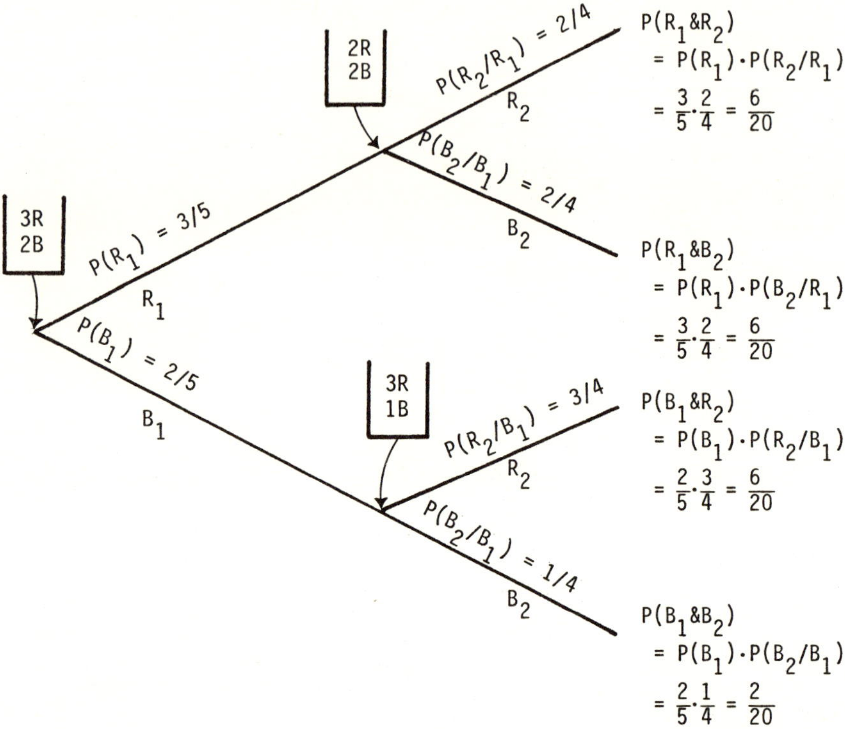

An Example of the Use of a Probability Tree to Solve Problems

Consider the following problem. Suppose we have an urn with three red, two black, and one green ball in it and draw in succession two balls, where we do not replace the first ball drawn. Then,

a) Find the probability of drawing two red balls,
b) Find the probability of drawing a green ball on the second draw,
c) Find the probability of drawing two balls of the same color.

Using our principle of assigning probabilities in multistage experiments and our Rule 14 that the probability of a path equals the product of the probabilities of the branches on that path, we can draw the following tree easily (Figure 1.25).

FIGURE 1.25
A Probability Tree for Drawing (without Replacement) Two Balls from an Urn with Three Red, Two Black, and One Green Ball

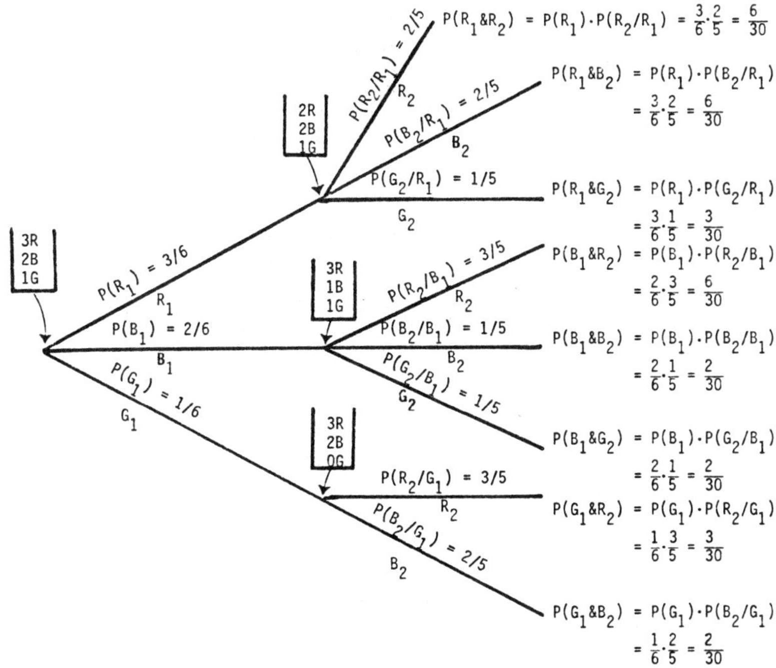

a) Find the probability of drawing two red balls.

Clearly, this condition of drawing two red balls means that we draw a red ball on the first draw and a red ball on the second draw, i.e., the event is $R_1 \& R_2$. So, we want to find $P(R_1 \& R_2)$. But this is just the probability of the top path, so

$$P(R_1 \& R_2) = P(R_1) \cdot P(R_2/R_1) = 3/6 \cdot 2/5 = 6/30 = 0.2.$$

b) Find the probability of drawing a green ball on the second draw.

This is not as obvious. The event of green on the second draw is "G_2", so we want to calculate $P(G_2)$. But, how can we do this? A little thought should convince the reader that underlying our tree is the Venn diagram of Figure 1.26 whose entries (i.e., simple events) represent the results (= color combinations) of the two-stage drawing.

FIGURE 1.26
A Venn Diagram for Drawing Two Balls from the Urn

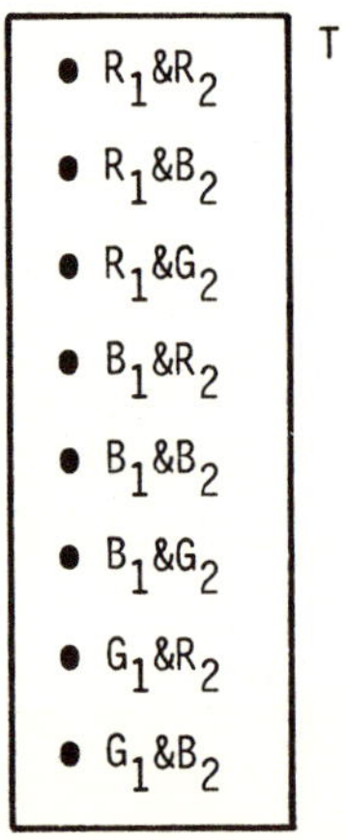

Note that there are eight simple events in this Venn diagram, and that each simple event corresponds to a complete path on the probability tree.

Now, if someone gave you just the Venn diagram and asked you to find $P(G_2)$, you hopefully would say that all you had to do was use Rule 3, which says that the probability of an event equals the sum of the probabilities of the simple events of which it is composed. Hence, to find the probability of the event of drawing green on the second draw (G_2), all you would have to do would be to add up the probabilities of all the simple events where G_2 happens. The event G_2 and the constituent simple events are shown here in Figure 1.27. So,

$$P(G_2) = P(R_1\&G_2) + P(B_1\&G_2) = 3/30 + 2/30 = 5/30 = 0.17.$$

FIGURE 1.27
A Venn Diagram Showing Event "G_2"

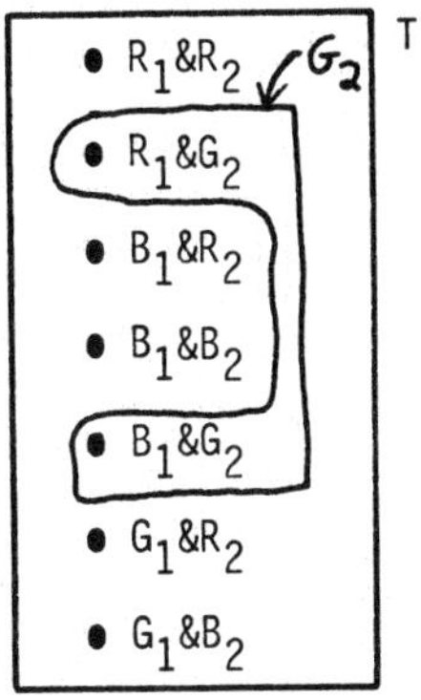

A question that we might ask is this: Can we just use the tree instead of having to go to all the trouble of drawing the Venn diagram? Yes; since each tree path (taken as a whole) corresponds to exactly one simple event in the multistage Venn diagram, the rule is this:

RULE 15: To find the probability of an event, identify the paths on the tree where the event happens and add up the probabilities of those paths. (We will see later that this works as long as we are not finding the conditional probability of an event; Rule 16 below will apply to the later situation.)

This procedure yields exactly the same result as Rule 3 because, as noted, each tree path (taken as a whole), corresponds to exactly one simple event in the multistage Venn diagram. Hence, Rule 15 is the tree equivalent of Rule 3.

c) Find the probability of drawing two balls of the same color.

Repeating our tree, we check off those paths in Figure 1.28 where we have two balls of the same color.

FIGURE 1.28
A Probability Tree for Drawing Two Balls from the Urn

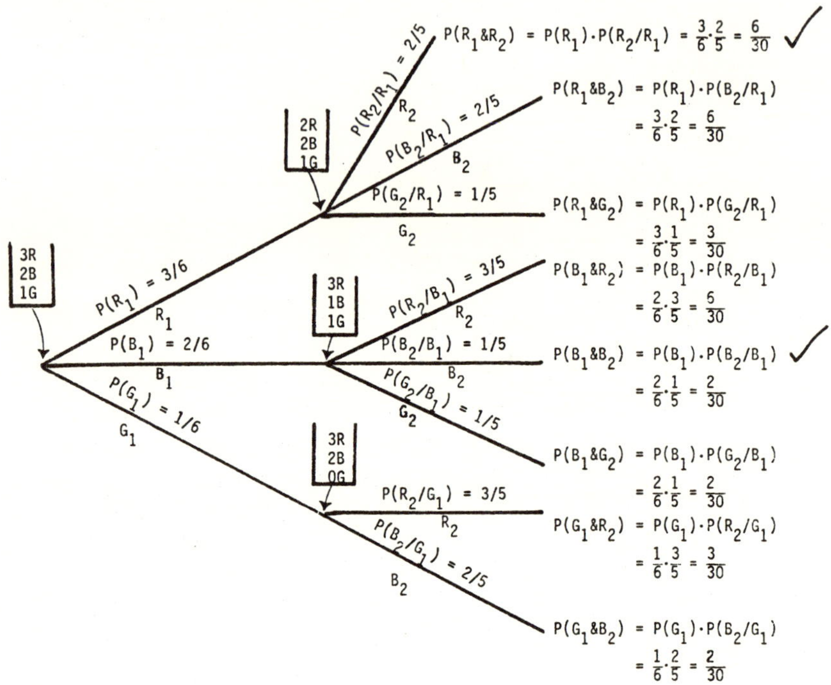

Then: $P(\text{two same color}) = P(R_1\&R_2) + P(B_1\&B_2) = 6/30 + 2/30$
$= 8/30 = 0.27.$

Note carefully that in using Rule 15, we can add only <u>whole</u> path probabilities. Marine student Mike Brock emunicated what we have come to call Brock's law: "We must multiply before we add." That is, we have to multiply out the probabilities of the branches on each path to obtain the path probabilities before we can add up these (entire) path probabilities.

Summary of the Tree Use and Properties

The reader should look carefully at the Figure 1.25 tree used in the example above to verify these properties.

A. From each node, the alternatives listed are mutually exclusive and exhaustive, and therefore, from any node the sum of the probabilities of the alternatives is "1."

As a result of making the alternatives from each node mutually exclusive and exhaustive, when we are done we end up with the next property.

B. Each path on the tree, taken as a whole, is mutually exclusive from every other path and collectively they are exhaustive of all the possible outcomes. Hence, there is a correspondence between the tree and a Venn diagram where each simple event in the Venn diagram corresponds to a complete path on the tree (and the simple events are mutually exclusive and exhaustive as they must be).

Since the paths (simple events) are mutually exclusive and exhaustive, the sum of the path probabilities equals P(T), which is of course "1." So we have the next property.

C. The sum of the probabilities of all the paths in a properly constructed probability tree equals "1."

Several additional critical properties are listed as D, E, and F.

D. The correct mathematical way of representing a path is that it is the intersection of all the events (branches) on the path; e.g., if we have a path as shown in Figure 1.29, then it is denoted as $E_1 \& E_2 \& \cdots \& E_n$.

FIGURE 1.29
The Labeling for One Path on a Probability Tree Diagram

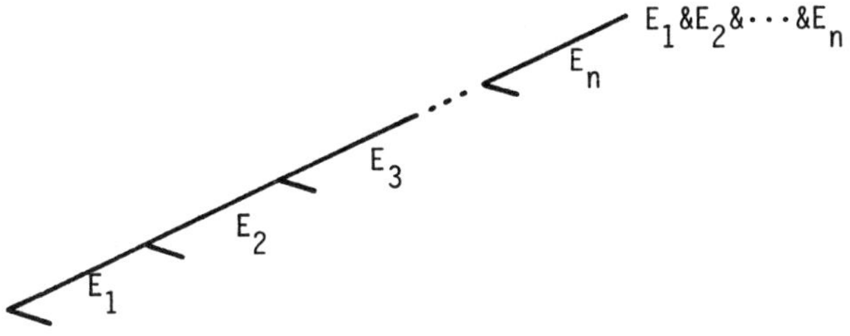

E. Furthermore, the probability of each event (branch) on the path is the conditional probability that the event will occur given that <u>all</u> the preceding branch events have occurred (Figure 1.30).

FIGURE 1.30
The Labeling of Branches and Probabilities of Branches on One Path of a Probability Tree

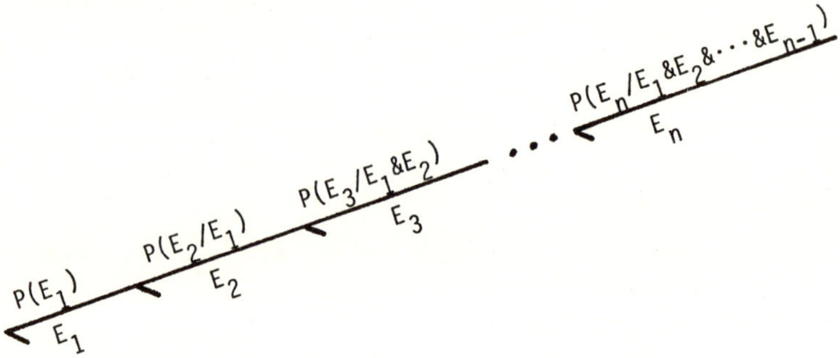

F. Furthermore, as Rule 14 states, the probability of a path equals the product of the probabilities of the branches on that path; i.e.,

$$P(E_1\&E_2\&E_3\&\cdots\&E_n) = P(E_1)\cdot P(E_2/E_1)\cdot P(E_3/E_1\&E_2)\cdots P(E_n/E_1\&E_2\&\cdots\&E_{n-1}).$$

This formula represents the generalization of the formula for P(A&B). Mnemonically, it can be remembered by saying that the probability of the intersection of "n" events equals the probability of the first event times the probability of the second event, given the first event, times the probability of the third event, given the first two events, times the probability of the fourth event, given the first three events, and so forth, with the last term being the probability of the "nth" event given the first "n-1" events.

The notation and reasoning is quite reasonable, and the assignment of probabilities similar to the two-stage tree example above. One comment needs to be made. When we write $P(E_3/E_1\&E_2)$, this is said aloud as "the probability of E_3 given E_1 and E_2." We could use double parentheses, i.e. $P(E_3/(E_1\&E_2))$ to make the point that everything after the slash mark "/" is what is given, but this would make later formulas very messy. So we will write just $P(E_3/E_1\&E_2)$, etc.

The Concepts of Statistical Independence and Conditional Independence

Suppose we consider the experiment of tossing a "fair" coin twice. ("Fairness" implies one side is as likely to come up as the other, and we ignore the possibility of the coin landing on its edge.) Then, the tree will be the one shown in Figure 1.31.

FIGURE 1.31
A Probability Tree for Tossing a Fair Coin Twice

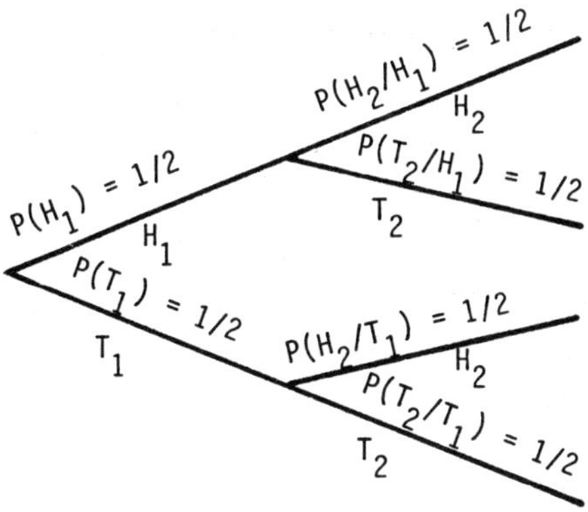

$$P(H_1 \& H_2) = P(H_1) \cdot P(H_2/H_1) = \frac{1}{2} \cdot \frac{1}{2} = \frac{1}{4}$$

$$P(H_1 \& T_2) = P(H_1) \cdot P(T_2/H_1) = \frac{1}{2} \cdot \frac{1}{2} = \frac{1}{4}$$

$$P(T_1 \& H_2) = P(T_1) \cdot P(H_2/T_1) = \frac{1}{2} \cdot \frac{1}{2} = \frac{1}{4}$$

$$P(T_1 \& T_2) = P(T_1) \cdot P(T_2/T_1) = \frac{1}{2} \cdot \frac{1}{2} = \frac{1}{4}$$

Note that when we assigned $P(H_2/H_1) = 1/2$, we were asking the probability of throwing heads on the second toss given that heads had occurred on the first toss. But, did it make any difference that heads was tossed first? The answer is "no." Since the coin was unaffected by the result of the first toss, the probability of getting heads on the second toss was not affected by getting heads on the first toss.

The mathematical way of saying this is to write $P(H_2/H_1) = P(H_2)$, i.e., the probability of getting heads on the second toss, given heads on the first toss, is the same as the probability of getting heads on the second toss (ignoring the occurrence of heads on the first toss). The formal definition is this:

Definition: Events B and A defined on a given universal set T are said to be statistically independent if $P(B/A) = P(B)$.

If we just have points in a Venn diagram, statistical independence is more of an accident than anything else. For example, if the simple events are equiprobable (1/10) as in Figure 1.32, then $P(B/A) = 1/4$, $P(B) = 5/10 = 1/2$, so $P(B/A) \neq P(B)$, but in Figure 1.33 where the events also are equiprobable (1/10), $P(B/A) = 3/5$, while $P(B) = 6/10$, so $P(B/A) = P(B)$.

FIGURE 1.32
A Venn Diagram for Equiprobable Simple Events and Statistical Dependence of Defined Events

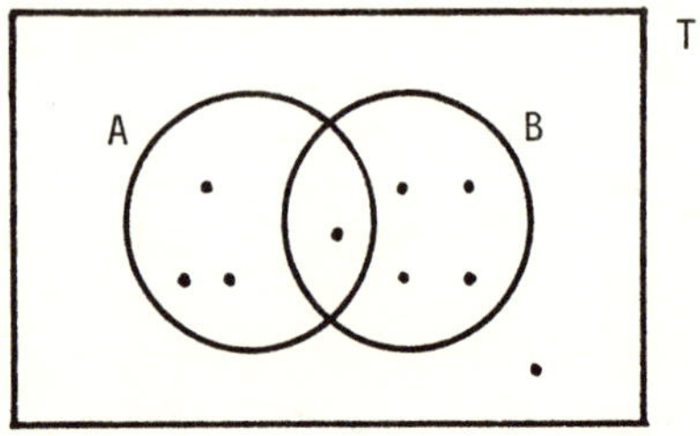

FIGURE 1.33
A Venn Diagram for Equiproable Simple Events and Statistical Independence of Defined Events

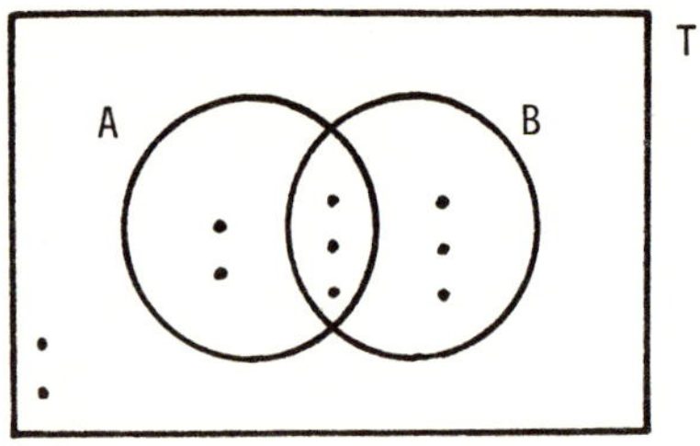

In more concrete cases, such as flipping coins or drawing from urns, statistical independence (or its absence, which we term statistical dependence), may be easy to determine. As Hoel notes, "to determine whether two real life events A_1 and A_2 are independent, it is necessary to ask the question: If A_1 occurs, does that change the chances that A_2 will occur from what they would be if A_1 were completely ignored? If the chances are not changed, the events are independent."[14]

Besides the coin flip example given above, another good illustration of statistical independence is to change the urn problem, where we had 3 red, 2 black, and 1 green ball in the urn by saying that we will put the first ball drawn back in the urn before drawing the second ball. The tree then is as shown in Figure 1.34, and we see that we have statistical independence on all second drawings, i.e.,

FIGURE 1.34
A Probability Tree for Drawing (with Replacement) Two Balls from an Urn with Three Red, Two Black, and One Green Ball

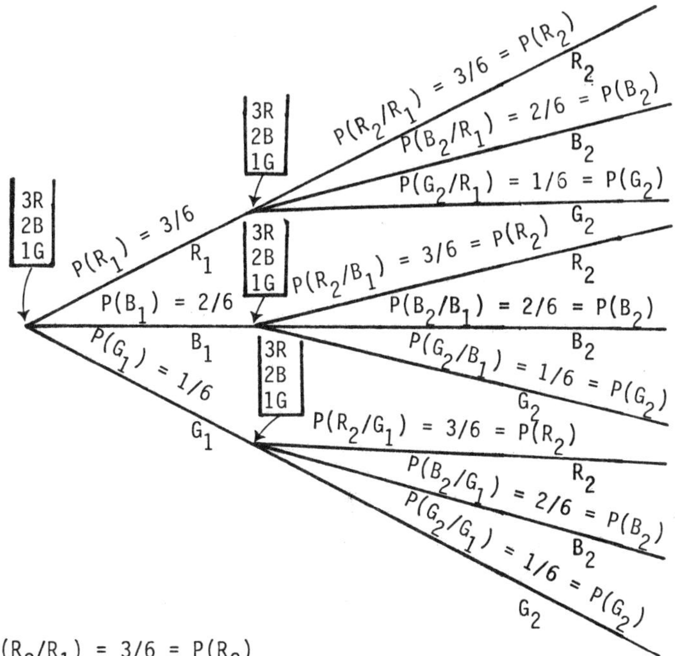

$P(R_2/R_1) = 3/6 = P(R_2)$

$P(B_2/R_1) = 2/6 = P(B_2)$

$P(G_2/R_1) = 1/6 = P(G_2)$

$P(R_2/B_1) = 3/6 = P(R_2)$

(etc.)

It turns out that, if B is statistically independent of A, i.e., if P(B/A)= P(B), it also is true that

(i) P(A&B) = P(A)·P(B) and that

(ii) P(A/B) = P(A).

To show (i), remember that it is always the case that P(A&B) = P(A)·P(B/A). But, if B is statistically independent of A, we can replace the second term, "P(B/A)" in this formula by P(B), thus getting P(A&B) = P(A)·P(B) which is what we wanted to show.

It is only slightly harder to show that, if B is statistically independent of A, we then must have

P(A/B) = P(A).

It is always the case that $P(A\&B) = P(B\&A) = P(B)\cdot P(A/B)$.

So, $P(A\&B) = P(B)\cdot P(A/B)$ always. But we showed in (i) that, if B is statistically independent of A,

$$P(A\&B) = P(A)\cdot P(B).$$

Equating the two things $P(A\&B)$ equals we have

$$P(B)\cdot P(A/B) = P(A)\cdot P(B).$$

Assuming B is not empty, a reasonable assumption, we can cancel out $P(B)$

$$\cancel{P(B)}\cdot P(A/B) = P(A)\cdot \cancel{P(B)}$$

leaving the desired result that $P(A/B) = P(A)$.

But this is saying simply that A is statistically independent of B. We thus have shown that, if B is statistically independent of A, then A is statistically independent of B. This mutual independence makes it permissible simply to say that A and B are statistically independent.

Actually, although we won't show it here, any one of the following three equations can be used as the definition of statistical independence since each one implies the other two:

$$P(B/A) = P(B),$$

$$P(A/B) = P(A),$$

$$P(A\&B) = P(A)\cdot P(B).$$

In words, the last statement can be said this way. "If A and B are statistically independent, the probability of their joint occurrence equals the product of the probabilities of the occurrence of each alone." Since any one of the three equations implies the other two, any one of the three can be used in a specific situation to test for statistical independence.

One common error people make is to confuse the notions of statistical independence and mutual exclusivity. These are totally opposed notions. To see this, note again that two events A and B are mutually exclusive if only one can happen at a time; this is displayed in Figure 1.35.

FIGURE 1.35
A Venn Diagram for Two Mutually Exclusive Events, A and B

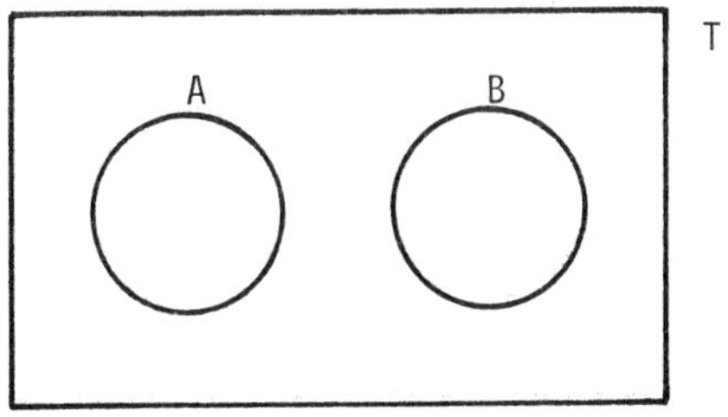

Note that for this situation P(B/A) = 0, i.e., if A happens, the probability that B also can happen is "0," since only one can happen at a time. Also, as long as B is not the null set, we must have $P(B) > 0$, and hence $P(B) \neq P(B/A) = 0$, i.e., B and A can't be statistically independent.

So, the picture for statistical independence of sets A and B has to be one of overlap (Figure 1.36).

FIGURE 1.36
A Venn Diagram for Statistically Independent Sets A and B

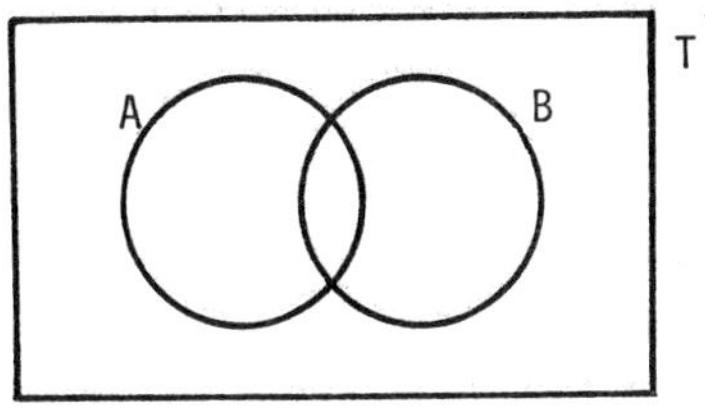

On the other hand, as our discussion of Figures 1.32 and 1.33 showed, sometimes with overlap we have statistical independence, and sometimes we don't have statistical independence (i.e., we have statistical dependence).

In summary, then, where "$\Rightarrow$" means "implies," we have three distinct types of situations (Figure 1.37).

FIGURE 1.37
Three Distinct Types of Situations

(a) mutually exclusive $\Rightarrow$ 

(b) statistical independence $\Rightarrow$ 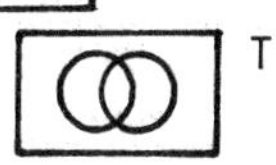

(c) 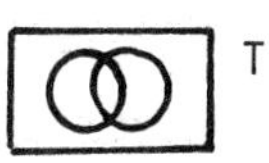

$\Rightarrow$ we can have either statistical independence or statistical dependence.

At this point, a reasonable question is the following: How does all this relate to trees? The answer is relatively straightforward. In a properly constructed tree, each path, taken as a whole, is mutually exclusive from every other path and collectively the paths are exhaustive; on any given path, the events on the path will either be statistically independent or dependent.

Often, books drop the term "statistical" from "statistical independence" and just speak of independence. This is an extremely unfortunate practice because there is a notion of "conditional independence" (and, of course, "conditional dependence") that is of primary importance in Bayesian analysis and must be distinguished from statistical independence and dependence. This notion can be shown rather clearly in this example: suppose we have two urns, indistinguishable on the outside, the first containing 6 red and 4 black balls, and the second containing 5 red and 7 black balls. Suppose we select an urn at random, draw out a ball, record its color, and then replace it before drawing a second ball whose color we also record. Suppose we want to find the probability of drawing two red balls. The tree for this three-stage experiment is Figure 1.38.

FIGURE 1.38
A Probability Tree Diagram for Selecting One of Two Urns and Drawing (with Replacement) Two Balls

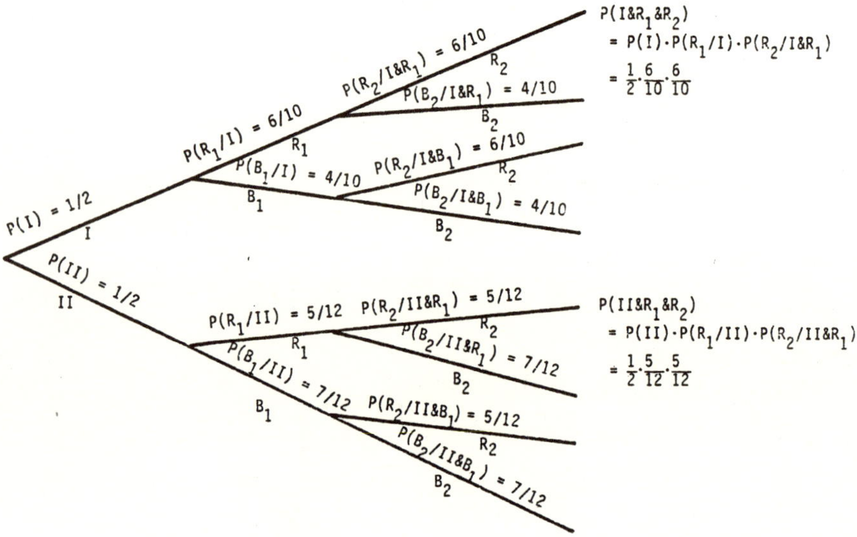

We want to find the probability of drawing two red balls. This is the event of drawing red the first time and red the second time, i.e., $R_1\&R_2$. Thus we want to find $P(R_1\&R_2)$. $P(R_1\&R_2)$ will be found by adding up the probabilities of the paths where the event $R_1\&R_2$ happens. There are two such paths: $I\&R_1\&R_2$ and $II\&R_1\&R_2$. Thus,

$$\begin{aligned} P(R_1\&R_2) &= P(I\&R_1\&R_2) + P(II\&R_1\&R_2) \\ &= P(I)\cdot P(R_1/I)\cdot P(R_2/I\&R_1) \\ &\quad + P(II)\cdot P(R_1/II)\cdot P(R_2/II\&R_1) \\ &= \frac{1}{2}\cdot\frac{6}{10}\cdot\frac{6}{10} + \frac{1}{2}\cdot\frac{5}{12}\cdot\frac{5}{12} = 0.18 + 0.087 = 0.267. \end{aligned}$$

Now, one interesting thing about the tree is the assignment of probabilities to the second drawing outcomes. Looking at the top path, we found that $P(R_2/I\&R_1) = 6/10$. In assigning this probability, the fact that R_1 had occurred was irrelevant since the first ball drawn was replaced; the fact that it was assumed we were drawing from urn I was the critical factor in assigning the probability that R_2 would occur. Thus, the probability of R_2 was conditional on drawing from urn I, but independent of the fact R_1 was assumed to have been drawn first, and we can write

$$P(R_2/I\&R_1) = P(R_2/I).$$

This is the notion, then, of <u>conditional independence</u>.[15] Note that, if we left the first ball drawn out of the urn before drawing the second ball, the probability of the occurrence of R_2 would be conditional not only on which urn we assumed we were drawing from but also dependent on which ball was assumed to have been drawn first. We would have <u>conditional dependence</u> and would not be able to simplify $P(R_2/I\&R_1)$, i.e, it would <u>not</u> be true that $P(R_2/I\&R_1) = P(R_2/I)$. The tree for this case would be Figure 1.39.

In this new situation,

$$\begin{aligned} P(R_1\&R_2) &= P(I\&R_1\&R_2) + P(II\&R_1\&R_2) \\ &= P(I)\cdot P(R_1/I)\cdot P(R_2/I\&R_1) \\ &\quad + P(II)\cdot P(R_1/II)\cdot P(R_2/II\&R_1) \\ &= \frac{1}{2}\cdot\frac{6}{10}\cdot\frac{5}{9} + \frac{1}{2}\cdot\frac{5}{12}\cdot\frac{4}{11} = \frac{1}{6} + \frac{5}{66} = \frac{11}{66} + \frac{5}{66} = \frac{16}{66} = 0.24. \end{aligned}$$

FIGURE 1.39
A Probability Tree Diagram for Selecting One of Two Urns and Drawing (without Replacement) Two Balls

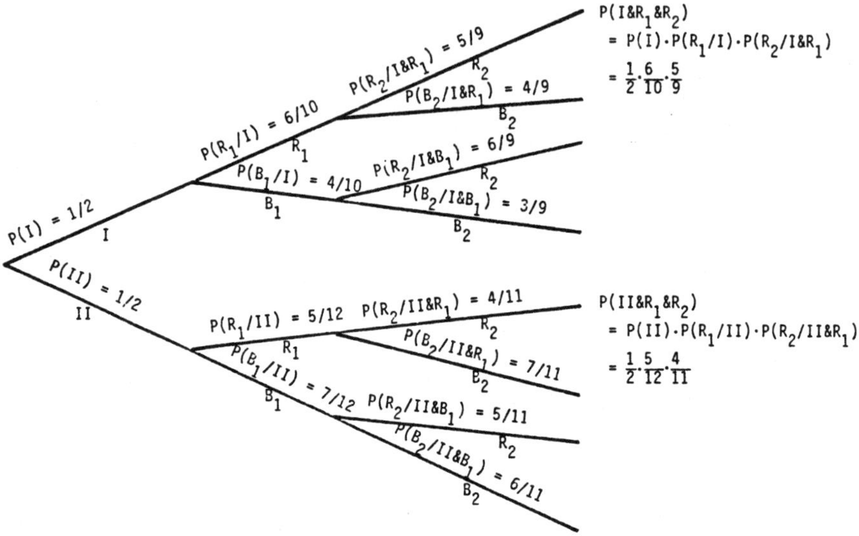

Note, of course, that if we have conditional dependence, the answer is different from the one where we have conditional independence.

Given the tree in Figure 1.38 and rule 15 which says that the probability of an event equals the sum of the path probabilities where the event happens, we can ask such questions as: What is the probability of drawing a red ball on the second draw? What is the probability of drawing two balls of the same color? and so on.

A trickier question, however, is to ask the probability of drawing a red ball on the second draw if we have drawn a red ball on the first draw. Mathematically, this is asking $P(R_2/R_1)$. Our rule has to be modified if we are asked to find a conditional probability. What we must do is make use of the formula $P(B/A) =$

$\frac{P(B\&A)}{P(A)}$; if we take $P(R_2/R_1)$, use of this formula yields

$$P(R_2/R_1) = \frac{P(R_2\&R_1)}{P(R_1)}.$$

We now can use the tree and rule 15 to calculate the numerator $P(R_2\&R_1)$ and also to calculate the denominator $P(R_1)$. That is, using the tree we have

$$P(R_2\&R_1) = P(R_1\&R_2) = P(I\&R_1\&R_2) + P(II\&R_1\&R_2),$$

while

$$P(R_1) = P(I\&R_1\&R_2) + P(I\&R_1\&B_2) + P(II\&R_1\&R_2) + P(II\&R_1\&B_2). \text{ So}$$

$$P(R_2/R_1) = \frac{P(I\&R_1\&R_2) + P(II\&R_1\&R_2)}{P(I\&R_1\&R_2) + P(I\&R_1\&B_2) + P(II\&R_1\&R_2) + P(II\&R_1\&B_2)}$$

$$= \frac{P(I)\cdot P(R_1/I)\cdot P(R_2/I\&R_1) + P(II)\cdot P(R_1/II)\cdot P(R_2/II\&R_1)}{P(I)\cdot P(R_1/I)\cdot P(R_2/I\&R_1) + P(I)\cdot P(R_1/I)\cdot P(B_2/I\&R_1) + P(II)\cdot P(R_1/II)\cdot P(R_2/II\&R_1) + P(II)\cdot P(R_1/II)\cdot P(B_2/II\&R_1)}$$

$$= \frac{\frac{1}{2}\cdot\frac{6}{10}\cdot\frac{6}{10} + \frac{1}{2}\cdot\frac{5}{12}\cdot\frac{5}{12}}{\frac{1}{2}\cdot\frac{6}{10}\cdot\frac{6}{10} + \frac{1}{2}\cdot\frac{6}{10}\cdot\frac{4}{10} + \frac{1}{2}\cdot\frac{5}{12}\cdot\frac{5}{12} + \frac{1}{2}\cdot\frac{5}{12}\cdot\frac{7}{12}}$$

$$= \frac{0.1800 + 0.0868}{0.1800 + 0.1200 + 0.0868 + 0.1215} = \frac{0.2668}{0.5083} = 0.525.$$

Another way of looking at all this is to go to the approach we used to explain what $P(B/A)$ really was getting at. If we take the tree and put down the corresponding Venn diagram we get Figure 1.40.

FIGURE 1.40
A Venn Diagram for Selecting an Urn and Drawing Two Balls

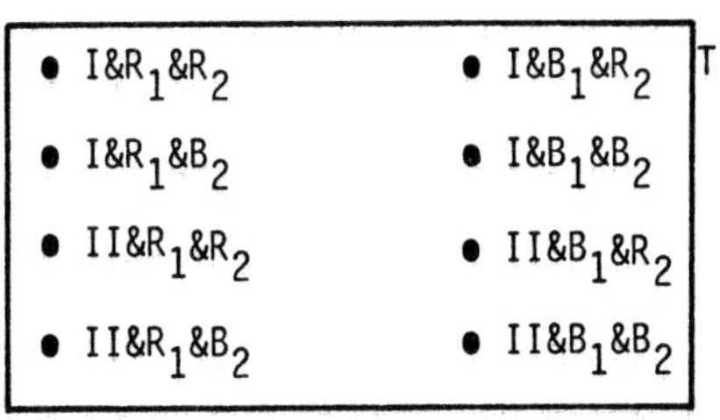

If we ask for $P(R_2/R_1)$, we can see that event R_1 consists of the four events circled in Figure 1.41 and event R_2 (given R_1) consists of the two events circled in Figure 1.41.

FIGURE 1.41
A Venn Diagram for Selecting an Urn and Drawing Two Balls and Showing the Events of R_1 and R_2 (given R_1)

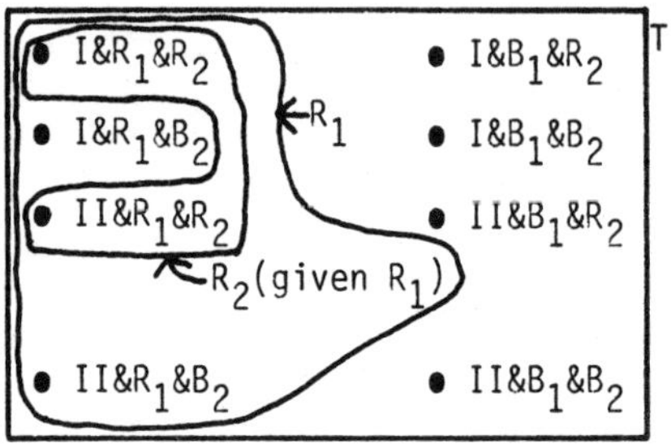

Given event R_1, we want to find what proportion of the "time" R_2 happens; event R_2 (given R_1) is shown within event R_1. But, to find the proportion of the time R_2 happens, we want to add up the probabilities of the two simple events where R_2 happens, i.e., simple events $I\&R_1\&R_2$ and $II\&R_1\&R_2$, and then divide by the probabilities of the four simple events where R_1 happens, i.e., simple events $I\&R_1\&R_2$, $I\&R_1\&B_2$, $II\&R_1\&R_2$, and $II\&R_1\&B_2$. The probabilities of each of these simple events can be read off the tree that the Venn diagram matches, and we get

$$P(R_2/R_1) = \frac{P(I\&R_1\&R_2) + P(II\&R_1\&R_2)}{P(I\&R_1\&R_2) + P(I\&R_1\&B_2) + P(II\&R_1\&R_2) + P(II\&R_1\&B_2)}$$

which is just what we had before, of course.

Thus, we have

RULE 16: To find the conditional probability of an event, convert the conditional probability (using $P(B/A) = \frac{P(B\&A)}{P(A)}$) and then read the numerator off the tree and the denominator off the tree (by using Rule 15 to calculate the numerator and then using Rule 15 again to calculate the denominator).

A Bayes Problem

Suppose we use the second, conditional dependence version of the two urns problem as shown in Figure 1.39, and consider this question. Prior to drawing any balls, the probability that by chance we have selected urn I is of course 1/2. The results of the two drawings, however, should change our mind somewhat about which urn we have selected. Note that in urn I we have 6 red and 4 black balls (a ratio favorable to red) so that it is fairly easy to draw two red balls, but in urn II we have 5 red and 7 black balls (a ratio favorable to black) so that it is easier to draw black balls than red balls. So, if we actually do draw two red balls, intuitively we may feel this should increase the probability that we are drawing from urn I. Mathematically, what we would want to calculate is $P(I/R_1\&R_2)$, i.e., the (revised) probability that urn I has been selected if two red balls have been chosen.

Now, by the rule we just developed (i.e., Rule 16), we should convert $P(I/R_1\&R_2)$. Doing so yields

$$P(I/R_1\&R_2) = \frac{P(I\&(R_1\&R_2))}{P(R_1\&R_2)}$$

We can eliminate the inner parentheses in the numerator, and thus have

$$P(I/R_1\&R_2) = \frac{P(I\&R_1\&R_2)}{P(R_1\&R_2)}.$$

Now, using Rule 15, we can read the numerator off the tree and the denominator off the tree and we have

$$\begin{aligned}
P(I/R_1\&R_2) &= \frac{P(I\&R_1\&R_2)}{P(R_1\&R_2)} \\
&= \frac{P(I\&R_1\&R_2)}{P(I\&R_1\&R_2) + P(II\&R_1\&R_2)} \\
&= \frac{P(I)\cdot P(R_1/I)\cdot P(R_2/I\&R_1)}{P(I)\cdot P(R_1/I)\cdot P(R_2/I\&R_1) + P(II)\cdot P(R_1/II)\cdot P(R_2/II\&R_1)} \\
&= \frac{\frac{1}{2}\cdot\frac{6}{10}\cdot\frac{5}{9}}{\frac{1}{2}\cdot\frac{6}{10}\cdot\frac{5}{9} + \frac{1}{2}\cdot\frac{5}{12}\cdot\frac{4}{11}} \\
&= \frac{\frac{1}{6}}{\frac{1}{6} + \frac{5}{66}} = \frac{\frac{11}{66}}{\frac{11}{66} + \frac{5}{66}} = \frac{11}{16} \\
&= 0.69.
\end{aligned}$$

Note that the probability that we are dealing with urn I has gone from the original value of 1/2 up to 0.69, an increase as we had intuitively reasoned it should be.

The reason for calling this a Bayes problem is that we are revising the initial probabilities of the events from the first node in the light of evidence. The events from the first node can be viewed as hypotheses (in this case, hypotheses regarding which urn we have selected). We are, in effect, revising the probabilities of hypotheses in the light of evidence, and this is what Bayes's theorem is all about, as we will find in the next chapter. Note that in the example we gave, the numerator was a repeat of one of the terms of the denominator. This always will be the case in Bayes problems.

NOTES

1. Note three things. First, A and B are subsets of T, i.e., they are defined on the same universal set T. Second, the "&" sign is <u>not</u> a "plus" sign. It means "and," and, as noted, can be said aloud as "A and B" or "A intersection B." Third, the standard notation has seemed less useful than the employment of the "&" sign. (That is, for students with a minimal mathematics background, the "&" sign usually is much easier to learn than the "$\cap$" symbol.)

2. Sample spaces are not always unique. Another sample space for the experiment of throwing a fair die once is

T^1 = (even number shows, odd number shows).

3. By "fair" it is meant that it can be <u>assumed</u>, for reasons of symmetry, that all the outcomes of the experiment are equally likely.

4. The term often used here is "composite event," not just "event," where a composite event is defined as a collection of simple events. Thus some authors discuss the topic of assigning probabilities to both simple events and composite events. Only in the discussion of problem 12 in Chapter 5 do I feel the need to use the term composite event for clarity.

5. See GOLDBERG, SAMUEL (1960) Probability: An Introduction. Englewood Cliffs, N.J.: Prentice Hall, Inc., pp. 56 and 61 for a nice (and more rigorous) discussion of "assigning" probabilities.

6. We really are going in circles here. The assignment of P(1) = P(2) = P(3) = P(4) = P(5) = P(6) = 1/6 was based on an implicit <u>assumption</u> that the sum of the probabilities of the simple events would be the number "1." Elementary treatments of probability which do not use an axiomatic approach run the risk of ending up going in circles like this. Indeed, essentially any approach to probability that is based on relative frequency eventually ends up

going in circles and, actually, really "defines" the concept of probability in terms of itself - which is invalid. For most elementary treatments of probability, however, the relative frequency approach still provides the most satisfactory explanations of the concepts of probability, and a great many books adopt this approach.

7. That is, since we defined an event as a set defined on a given sample space T, we theoretically can have the situation where there are no simple events in A, i.e., the collection has no elements, and is, therefore, the null set.

8. Rule 9 provides, however, a more rigorous way of proving Rule 7. We want to show that P(E) + P(not E) = 1. Note that "E" and "not E" are mutually exclusive, since only one can happen at a time. So we can use Rule 9 to say P(E or not E) = P(E) + P(not E). Since "E or not E" = T (see the figure),

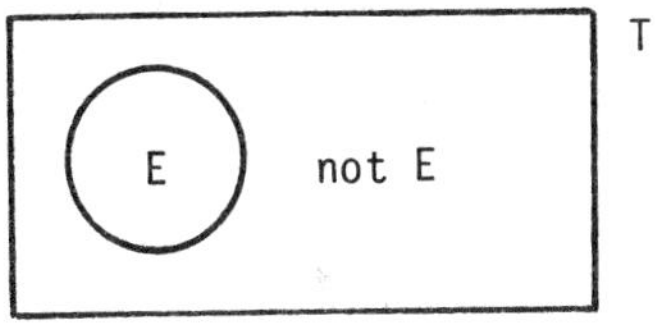

we have

P(E or not E) = P(T) = 1.

Since we noted above that P(E or not E) = P(E)+P(not E), we have P(E)+P(not E) = P(T) = 1.

9. Purists will note that I do not use "if and only if." Being this precise did not seem necessary at this point.

10. HOEL, PAUL G. (1976) Elementary Statistics, 4th edition. New York: John Wiley & Sons, Inc., p. 51.

11. Ibid., pp. 52-53.

12. Ibid.

13. There still probably are two questions in the minds of some readers.

1. Why is $P(R_1) = 3/5$, assigned by considering a one-stage experiment, the same as $P(R_1) = 12/20$ assigned on the basis of the two-stage experiment? The answer is that, if we draw a complete tree, as shown, we find that if we focus just on the branches coming from the first node, the proportion of branches that are red is 3 out of 5, so that assigning a value of 3/5 to $P(R_1)$ is

most reasonable indeed. On the other hand, looking

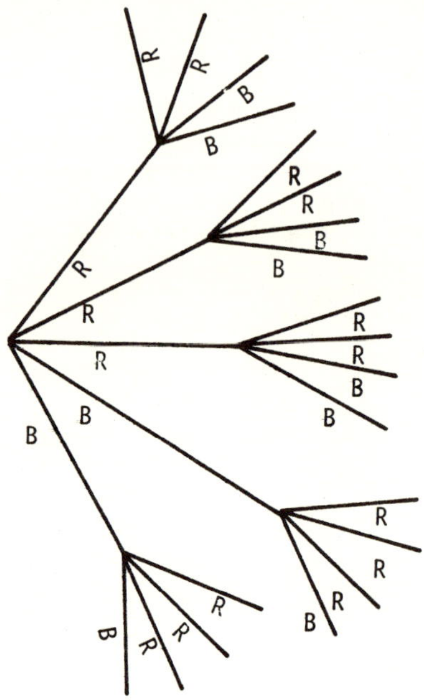

at the entire two-stage tree, the proportion of paths starting with red is a multiple of 4 of all the branches from the first node--that is, each and every node has added onto it at its second node the same number -4- of branches. In essence, the fraction 3/5 is multiplied above and below by 4 to yield

$$\frac{12 \text{ (number of paths starting with red)}}{20 \text{ (total number of paths)}}$$

Since we assume that each of the 20 paths is just as likely as any other, the only reasonable assignment of $P(R_1)$ based on the whole, 20 path tree, is 12/20. We thus end up with a consistent assignment of $P(R_1)$ whether we consider a single-stage experiment, or the whole, two-stage experiment.

2. Why is $P(R_2/R_1) = 2/4$, assigned considering the second, one-stage experiment, the same as $P(R_2/R_1) = 6/12$ assigned on the basis of the entire two-stage experiment?

The answer is much more involved than the answer to question 1. If we look at the drawing of the complete tree, and then first limit our analysis to the 3 red branches coming from the first node, (so that R_1 is given), and then look only at the proportion of the second red branches coming off those 3 initial red branches, we find that the proportion of red branches from the second nodes is always the same - namely, 2 out of 4 branches (where each of the branches is assumed as likely as another). So, an assignment of $P(R_2/R_1) = 2/4$ based solely on the second one-stage experiment is reasonable.

On the other hand, looking once more at the complete tree, if we limit the analysis to those paths starting with the 3 red branches coming from the first node, the proportion of entire paths with red branches on the second draw is a multiple of 3 (the number of red branches from the initial node). That is, each of the 3 sets of 4 branches from the second nodes on the paths starting with red has added onto it the same number - 1 - of initial red branches, so that we end up with $1 \cdot 4 + 1 \cdot 4 + 1 \cdot 4 = 12$ paths, but the proportion of those 12 paths with red is just 6/12. In essence, the fraction 2/4 is multiplied above and below by 3 to yield

$$\frac{6 \text{ (number of paths ending with red)}}{12 \text{ (total number of paths starting with red)}}.$$

14. HOEL (1976), p. 55.

15. When I first started to read the literature on Bayesian analysis, I ran across the concept of conditional independence. Unfortunately, I could find no formal definition of the term. I was quite frustrated to find that none of my probability or statistics texts even mentioned it. Even today, most such texts fail to mention the concept. The major exception to this rule is the fine treatment found in PFEIFFER, PAUL E. and SCHUM, DAVID A. (1973) Introduction to Applied Probability. New York, New York: Academic Press, Inc. A more advanced treatment is to be found in PFEIFFER, PAUL E. (1979) Conditional Independence in Applied Probability. Newton, Mass: Education Development Center, Inc. (The UMAP Expository Monograph Series.)

2
Introduction to Bayesian Analysis: Development of Bayesian Formulas and the Tabular Array Solution Format

As noted in the preface, Bayesian analysis will be treated in this book as a technique of statistical inference used to revise the analyst's probability assessments of hypotheses about reality in the light of new evidence or events. This revision of probability assessments involves application of Bayes's theorem (i.e., formula), named after Rev. Thomas Bayes, who first formulated it in the 18th century. The formula reads:

$$P(H/E) = \frac{P(H) \cdot P(E/H)}{P(E)}$$

where P(H) is the initial (or "prior") probability that a particular hypothesis H about the state of nature (reality) is true, P(E/H) is the (conditional) probability that an event E will occur if the particular hypothesis H is assumed to be true, P(E) is the probability that the event E will occur (whether hypothesis H or one of a number of alternative hypotheses about the state of nature is true), and P(H/E) is the revised probability that the particular hypothesis H is true assuming that event E has occurred.[1] (See Footnote 4 for alternative wording.)

Note that P(H/E) and P(E/H) are the same sort of conditional probabilities that we have been dealing with, except that P(E/H), as we will find, is usually a subjective (i.e., personal) conditional probability. Note also that there is nothing really new about Bayes's formula. From the definition of P(B/A), i.e., from the definition that $P(B/A) = \frac{P(B\&A)}{P(A)}$, we can say that

$$P(H/E) = \frac{P(H\&E)}{P(E)}.$$

But we also know that P(H&E) = P(H)·P(E/H) (because P(B&A) = P(B)·P(A/B)). Hence

$$P(H/E) = \frac{P(H) \cdot P(E/H)}{P(E)}$$

which is just Bayes's formula. The only thing that will be new is the interpretation we will give to the H and E in the formula.

Development of Bayes's Formula for Two Hypotheses and Any Number of Events

We will develop Bayes's formula first for the case where we have two hypotheses. We will discuss the cases where there are one, two, and then three events and indicate how the two-hypotheses formula can be generalized for any number of events.

Suppose that we have two hypotheses, call them H_1 and H_2, about the state of reality. H_1 might be, for example, the hypothesis that country A is planning to attack country B within three weeks. Now, when we use Bayesian analysis one rule is that the hypotheses must be mutually exclusive and exhaustive. If there are only two hypotheses, it follows that H_2 = not H_1. Thus, if H_1 is the hypothesis that country A is planning to attack country B within three weeks, then H_2 must be the hypothesis that country A is not planning to attack country B within three weeks.[2]

Having decided on the hypotheses, H_1 and H_2, the next step is to assign a probability to H_1, i.e., assign $P(H_1)$. ($P(H_2)$ is then determined as $1-P(H_1)$.) $P(H_1)$ and $P(H_2)$ are referred to as prior probabilities, meaning prior to additional evidence being received.

If we had more than two hypotheses, we would have to assign prior probabilities to all the hypotheses except the last one, whose probability would be determined as "1-sum of probabilities of all the previous hypotheses." In making these assessments, we generally will not have a symmetrical urn, die, or card situation, or much frequency count data. Therefore, the assignments generally will be subjective in nature, and the subjective, personal probabilities assigned will represent the assessor's degree of belief concerning the truth of the various hypotheses. Different analysts probably will assign different prior probabilities to the hypotheses.

In this case, suppose we assess $P(H_1) = 0.8$. Then $P(H_2) = 1-P(H_1) = 0.2$. We can diagram the situation on a tree as follows.

FIGURE 2.1
Probability Tree Diagram for Two Hypotheses

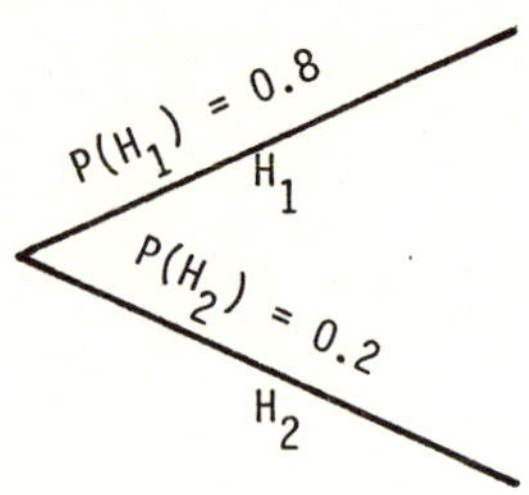

Next, suppose that an event E_1 occurs. If, for example, country A and country B are separated by a river, event E_1 might be that

country A brings up bridging equipment. The question is, then, what are the revised probabilities of H_1 and H_2 in light of the evidence, E_1? That is, what are $P(H_1/E_1)$ and $P(H_2/E_1)$?

In answering the question, we will not directly modify our belief in the probability of the hypotheses, i.e., we will not directly assess $P(H_1/E_1)$ and $P(H_2/E_1)$. Instead, we will in turn assume each hypothesis to be true (i.e., to be given) and will assess the conditional probability of the event in light of this assumption. In other words, we will provide our subjective, personal probability assessments to the conditional probabilities $P(E_1/H_1)$, and $P(E_1/H_2)$. Having done this, Bayes's formula will provide us mechanically with a revised estimate for the probabilities of H_1 and H_2.

To see how this is done, note that if we draw the usual sort of tree diagram we have Figure 2.2.

FIGURE 2.2
Probability Tree Diagram for Two Hypotheses and One Event

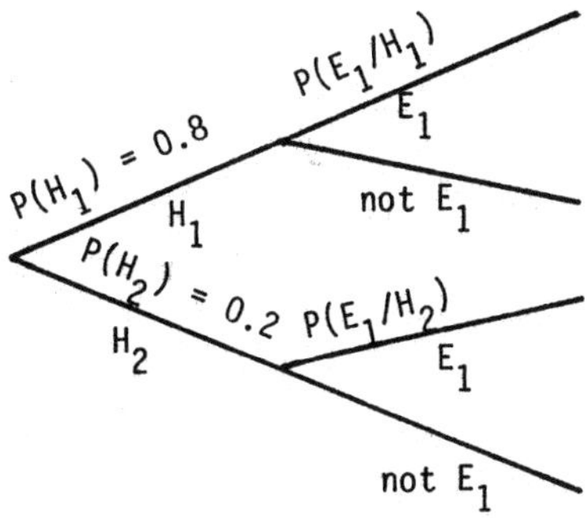

The tree itself suggests that we must assign probabilities to $P(E_1/H_1)$, and $P(E_1/H_2)$. Suppose we feel that $P(E_1/H_1) = 0.9$ (reflecting a feeling that it is highly likely that bridging equipment would be used if war was going to occur), and $P(E_1/H_2) = 0.8$ (perhaps reflecting country A's monthly practice with bridging equipment). Then our tree becomes Figure 2.3.

At this point, by the way, a common question is this: "Why are we assigning probabilities when we know that E_1 has happened?"; E_1 is "certain," the argument goes, so what sense does it make to assign $P(E_1/H_1)$, $P(E_1/H_2)$? It may be easier to answer the question if we look back at the urn problem of Figure 1.39 at the end of the last chapter. If we had let

FIGURE 2.3
Probability Tree Diagram for Two Hypotheses and One Event with Assigned Probabilities

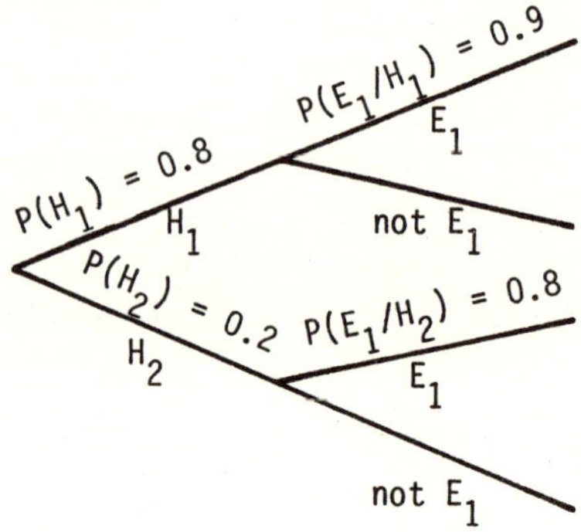

H_1 = urn 1 with 6R,4B

H_2 = urn 2 with 5R,7B

E_1 = red ball is drawn first,

then our tree would be the one shown as Figure 2.4.

FIGURE 2.4
Probability Tree Diagram of Urn Problem with Two Hypotheses and One Event

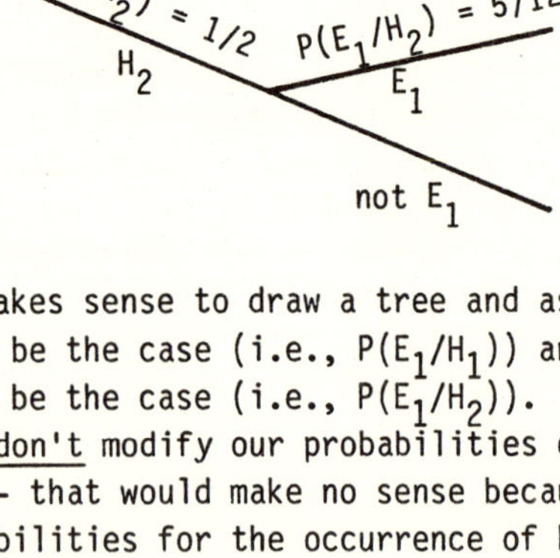

That is, it makes sense to draw a tree and ask how likely E_1 is assuming H_1 to be the case (i.e., $P(E_1/H_1)$) and how likely E_1 is assuming H_2 to be the case (i.e., $P(E_1/H_2)$). If E_1 actually happens, then we <u>don't</u> modify our probabilities of $P(E_1/H_1)$ and $P(E_1/H_2)$ to "1" -- that would make no sense because the (objective) probabilities for the occurrence of E_1 are not affected by E_1 happening. Similarly, in this case, it is reasonable to ask how likely E_1 is if H_1 is the case (i.e., $P(E_1/H_1)$) and how likely E_1 is if H_2 is the case (i.e., $P(E_1/H_2)$), and realize that we don't change our probabilities just because E_1 actually happens.

Another way of looking at this assignment is to say that if H_1 is the case, we expect E_1 to happen 90 percent of the time (i.e., $P(E_1/H_1) = 0.9$), while if H_2 is the case, we expect E_1 to happen 80 percent of the time (i.e., $P(E_1/H_2) = 0.8$). Then if E_1 actually does happen, we don't want to change these probabilities. What we do want to do is find out the implication of the actual occurrence of E_1 for the likelihoods of H_1 and H_2. Continuing on with our example, then, the tree becomes the previously noted Figure 2.3. We wish first to find $P(H_1/E_1)$, i.e., to see how the probability that H_1 is the case is affected by E_1 having happened. Then, we have, since $P(B/A) = \frac{P(B\&A)}{P(A)}$,

$$P(H_1/E_1) = \frac{P(H_1\&E_1)}{P(E_1)}.$$ Since, by Rule 15,

$P(E_1) = P(H_1\&E_1) + P(H_2\&E_1)$, it follows that

$$P(H_1/E_1) = \frac{P(H_1\&E_1)}{P(H_1\&E_1) + P(H_2\&E_1)}$$

$$= \frac{P(H_1)\cdot P(E_1/H_1)}{P(H_1)\cdot P(E_1/H_1) + P(H_2)\cdot P(E_1/H_2)}.$$

At this point, we simply can plug in our values for the probabilities and solve for $P(H_1/E_1)$.

$$P(H_1/E_1) = \frac{(0.8)(0.9)}{(0.8)(0.9) + (0.2)(0.8)} = \frac{0.72}{0.72 + 0.16} = \frac{0.72}{0.88} = 0.82.$$

This result says that the probability that H_1 is indeed the "true" state of reality has increased slightly because E_1 has happened. This result is intuitively correct, it can be argued. This is because if we look back at the tree, what we were saying was that E_1 was 90 percent likely to happen if H_1 was true, but less likely -- 80 percent -- to happen if H_2 was true. So E_1 was slightly more likely to happen if H_1 was true than if H_2 was true. The actual occurrence of E_1 then slightly favors the truth of H_1 vs H_2. So we would expect the revised probability of H_1 to be slightly larger than before, which is what we found.[3]

Turning back to the formula, note that in the formula

$$P(H_1/E_1) = \frac{P(H_1)\cdot P(E_1/H_1)}{P(H_1)\cdot P(E_1/H_1) + P(H_2)\cdot P(E_1/H_2)}$$

the numerator, $P(H_1)\cdot P(E_1/H_1)$, is simply a repeat of the first row of the denominator, and that this row involves H_1 everywhere.

Note also that, if we had instead wanted to calculate $P(H_2/E_1)$, we would have had

$$P(H_2/E_1) = \frac{P(H_2\&E_1)}{P(E_1)}$$

$$= \frac{P(H_2\&E_1)}{P(H_1\&E_1) + P(H_2\&E_1)}$$

$$= \frac{P(H_2)\cdot P(E_1/H_2)}{P(H_1)\cdot P(E_1/H_1) + P(H_2)\cdot P(E_1/H_2)}$$

In this latter case, we observe that the numerator, $P(H_2)\cdot P(E_1/H_2)$, is simply a repeat of the second row of the denominator and that this row involves H_2 everywhere. If we compare

$$P(H_1/E_1) = \frac{P(H_1)\cdot P(E_1/H_1)}{P(H_1)\cdot P(E_1/H_1) + P(H_2)\cdot P(E_1/H_2)}$$

and $$P(H_2/E_1) = \frac{P(H_2)\cdot P(E_1/H_2)}{P(H_1)\cdot P(E_1/H_1) + P(H_2)\cdot P(E_1/H_2)}$$

we can observe that the <u>denominators are identical</u>, with one row for H_1 and one row for H_2, while the numerator in each case is a repeat of the row from the denominator corresponding to the hypothesis whose revised probability we wish to calculate. That is, if we wish to calculate the revised probability of H_1, we simply make our numerator a repeat of the row from the denominator involving H_1, i.e., the first row of the denominator. On the other hand, if we want to calculate the revised probability of H_2, we simply make our numerator a repeat of the row from the denominator involving H_2, i.e., the second row of the denominator.

A slightly different and intuitive approach to what is involved in calculating $P(H_1/E_1)$ may be illuminating. The underlying Venn diagram for the tree is Figure 2.5.

FIGURE 2.5
Venn Diagram for Two Hypotheses and One Event

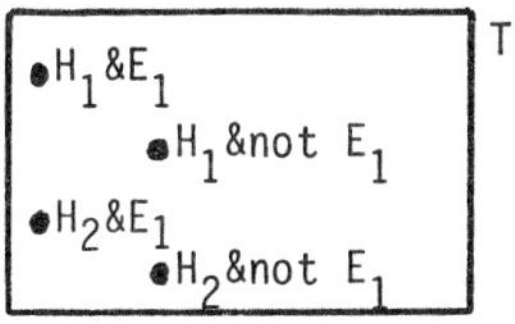

Then, if we wish to find $P(H_1/E_1)$, since E_1 is given, we see that the new Venn diagram T' is Figure 2.6 because either $H_1\&E_1$

FIGURE 2.6
Venn Diagram for Two Hypotheses when One Event is Given

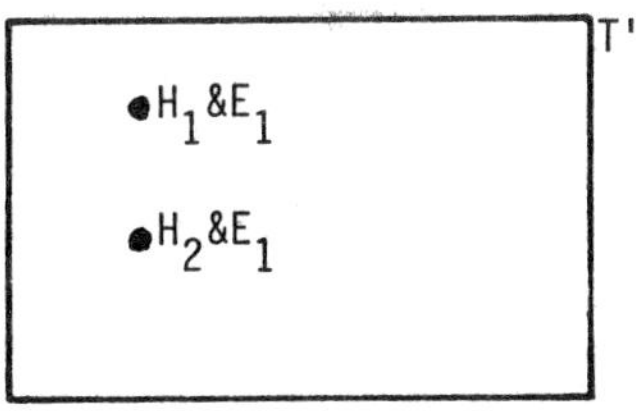

has "happened" or $H_2\&E_1$ has "happened." We actually are asking what proportion of the "time" $H_1\&E_1$ happens, i.e., what proportion $H_1\&E_1$ is of the total of $H_1\&E_1$ plus $H_2\&E_1$. So $P(H_1/E_1)$ intuitively should be equal to $\frac{P(H_1\&E_1)}{P(H_1\&E_1) + P(H_2\&E_1)}$, which it is. If we want to find $P(H_2/E_1)$, we really are asking what proportion of the "time" $H_2\&E_1$ happens, i.e., what proportion $H_2\&E_1$ is of the total which is $H_1\&E_1$ plus $H_2\&E_1$; so $P(H_2/E_1)$ intuitively should be equal to $\frac{P(H_2\&E_1)}{P(H_1\&E_1) + P(H_2\&E_1)}$, which it is.

So far, in developing our formula for two hypotheses we have found that, if we have one event then

$$P(H_1/E_1) = \frac{P(H_1)\cdot P(E_1/H_1)}{P(H_1)\cdot P(E_1/H_1) + P(H_2)\cdot P(E_1/H_2)}$$

and $$P(H_2/E_1) = \frac{P(H_2)\cdot P(E_1/H_2)}{P(H_1)\cdot P(E_1/H_1) + P(H_2)\cdot P(E_1/H_2)}.$$

A consequence of this is that

$$P(H_1/E_1) + P(H_2/E_1) = \frac{P(H_1)\cdot P(E_1/H_1)}{P(H_1)\cdot P(E_1/H_1) + P(H_2)\cdot P(E_1/H_2)} + \frac{P(H_2)\cdot P(E_1/H_2)}{P(H_1)\cdot P(E_1/H_1) + P(H_2)\cdot P(E_1/H_2)}$$

$$= \frac{P(H_1)\cdot P(E_1/H_2)+P(H_2)\cdot P(E_1/H_2)}{P(H_1)\cdot P(E_1/H_1) + P(H_2)\cdot P(E_1/H_2)} = 1$$

since the numerator and the denominator of the big fraction are the same. What this is saying is that the revised probabilities of the hypotheses must add to "1," just as the original probabilities summed to "1."

Next, suppose that a second event, E_2, occurs. Suppose E_2 = A mobilizes an elite marine battalion. Then our tree becomes the one shown here as Figure 2.7.

FIGURE 2.7
Probability Tree Diagram for Two Hypotheses and Two Events

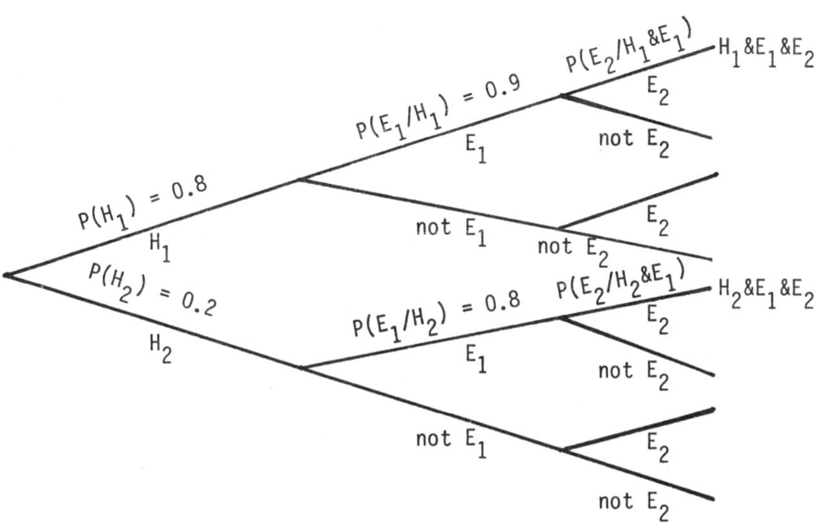

The question is, what do we want to calculate and what values do we need to assign? Since two events have occurred, we want to calculate $P(H_1/E_1\&E_2)$ and $P(H_2/E_1\&E_2)$. Applying Rule 16 for dealing with such probabilities,

$$P(H_1/E_1\&E_2) = \frac{P(H_1\&E_1\&E_2)}{P(E_1\&E_2)}.$$

The numerator here is simply the probability of the top path on the tree. The denominator is the sum of probabilities of paths on the tree where the event "$E_1\&E_2$" happens; there are two such paths, $H_1\&E_1\&E_2$ and $H_2\&E_1\&E_2$. So

$$P(H_1/E_1\&E_2) = \frac{P(H_1\&E_1\&E_2)}{P(H_1\&E_1\&E_2) + P(H_2\&E_1\&E_2)}.$$

Remembering that, by Rule 14, the probability of a path equals the product of the probabilities of the branches on the path, we see that we need to assign $P(E_2/H_1\&E_1)$ on the top path and $P(E_2/H_2\&E_1)$ on the fifth path. Suppose that the values assigned are $P(E_2/H_1\&E_1) = 0.95$, $P(E_2/H_2\&E_1) = 0.8$. Then, the tree becomes the one displayed as Figure 2.8.

FIGURE 2.8
Probability Tree Diagram for Two Hypotheses and Two Events with Assigned Probabilities

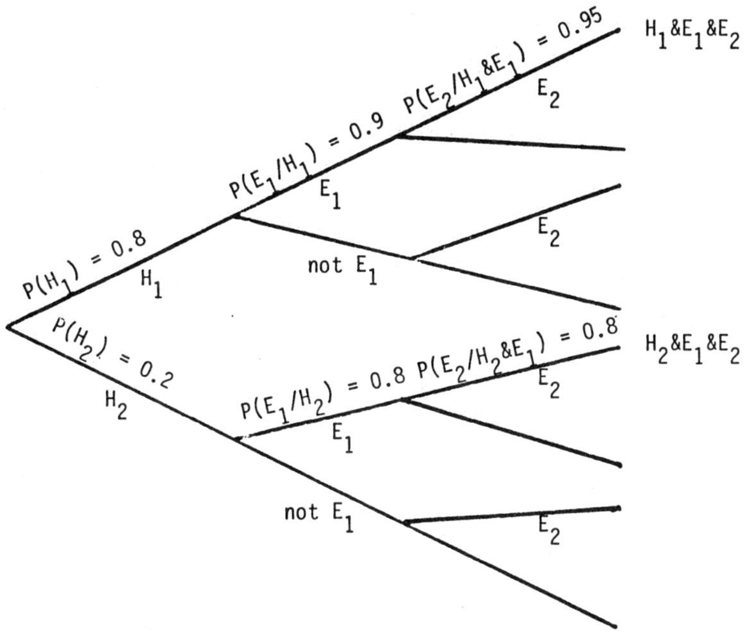

Note that we only need to assign values on the paths where both $E_1\&E_2$ have happened and that we can ignore the other paths on the tree. Then

$$P(H_1/E_1\&E_2) = \frac{P(H_1)\cdot P(E_1/H_1)\cdot P(E_2/H_1\&E_1)}{P(H_1)\cdot P(E_1/H_1)\cdot P(E_2/H_1\&E_1) + P(H_2)\cdot P(E_1/H_2)\cdot P(E_2/H_2\&E_1)}$$

$$= \frac{(0.8)(0.9)(0.95)}{(0.8)(0.9)(0.95) + (0.2)(0.8)(0.8)} = \frac{0.684}{0.684 + 0.128} = \frac{0.684}{0.812} = 0.84$$

and, once more, the probability of H_2 has been revised slightly upward.

There is, by the way, what is called the recursive form of Bayes, which takes the revised probability of H_1 after the first event and uses it as the prior probability of the hypothesis for the calculation for the second event. I do not like this form of Bayes because I like to use a tree format (with which it is not very compatible) and because it does not easily lend itself to the consideration of conditional dependence issues. Therefore, the recursive form of Bayes will not be used.

Note that in the tree the probabilities that we assigned for E_2 were $P(E_2/H_1\&E_1)$ and $P(E_2/H_2\&E_1)$. In making these assessments, we were assuming that conditional dependence was the case, that is, that in each case the probability was conditional not only upon the hypothesis but dependent also on the preceding event, E_1. So, when the judgments were made, they had to be phrased like this: "$P(E_2/H_1\&E_1)$ means the (subjective) probability that E_2 would happen given the H_1 assumption that A is planning to attack B within three weeks and given that E_1 has happened (i.e., that A brought up bridging equipment);" "$P(E_2/H_2\&E_1)$ means the (subjective) probability that E_2 would happen given the H_2 assumption that A is not planning to attack B within three weeks and given that E_1 has happened (i.e., that A brought up bridging equipment)." These are indeed subjective probabilities and need to be justified and explained, not just assigned arbitrarily. Political/military factors should be considered in making the assessments.

The reader might wonder if conditional independence might be the case, i.e., if we might just assign probabilities to E_2 based only on the hypotheses and ignoring the previous item of evidence, E_1. Occasionally, conditional independence may hold, but in Chapter Four it will be argued at length that conditional dependence probably is more realistic.

Turning to another topic, we can note that in the formula for two hypotheses, the numerator again is a repeat of the row of the denominator corresponding to the hypothesis whose revised probability we wish to calculate. Thus

$$P(H_1/E_1\&E_2) = \frac{P(H_1)\cdot P(E_1/H_1)\cdot P(E_2/H_1\&E_1)}{P(H_1)\cdot P(E_1/H_1)\cdot P(E_2/H_1\&E_1) + P(H_2)\cdot P(E_1/H_2)\cdot P(E_2/H_2\&E_1)}$$

and $$P(H_2/E_1\&E_2) = \frac{P(H_2)\cdot P(E_1/H_2)\cdot P(E_2/H_2\&E_1)}{P(H_1)\cdot P(E_1/H_1)\cdot P(E_2/H_1\&E_1) + P(H_2)\cdot P(E_1/H_2)\cdot P(E_2/H_2\&E_1)}.$$

(The reader can confirm that these revised probabilities also add up to "1.")

If we compare these last formulas for two events to the formulas where only one event E_1 has occurred, i.e., to these formulas:

$$P(H_1/E_1) = \frac{P(H_1)\cdot P(E_1/H_1)}{P(H_1)\cdot P(E_1/H_1) + P(H_2)\cdot P(E_1/H_2)}$$

and $$P(H_2/E_1) = \frac{P(H_2)\cdot P(E_1/H_2)}{P(H_1)\cdot P(E_1/H_1) + P(H_2)\cdot P(E_1/H_2)}$$

the reader may note that the set of formulas for two events in essence just results from adding on "a column" for E_2 to the formulas for one event. That is, when E_2 occurred, a new "column" corresponding to the conditional probability of E_2 was "added onto" the formulas which we had developed for the first event E_1. This same sort of "pattern" will be repeated for the formulas for the third event, as we will see now.

Suppose then that an event E_3 occurs. Then, the truncated tree becomes Figure 2.9 (where we have left out uninteresting paths).

FIGURE 2.9
Probability Tree Diagram for Two Hypotheses and Three Events

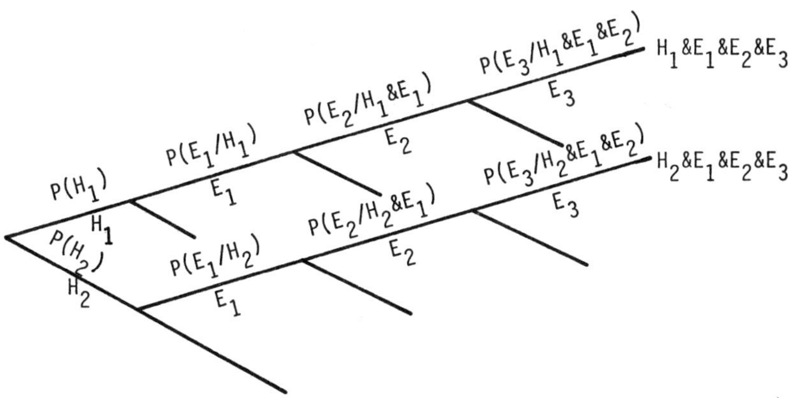

Then we want to calculate

$$P(H_1/E_1\&E_2\&E_3) = \frac{P(H_1\&E_1\&E_2\&E_3)}{P(E_1\&E_2\&E_3)}$$

$$= \frac{P(H_1\&E_1\&E_2\&E_3)}{P(H_1\&E_1\&E_2\&E_3) + P(H_2\&E_1\&E_2\&E_3)}$$

$$= \frac{P(H_1)\cdot P(E_1/H_1)\cdot P(E_2/H_1\&E_1)\cdot P(E_3/H_1\&E_1\&E_2)}{P(H_1)\cdot P(E_1/H_1)\cdot P(E_2/H_1\&E_1)\cdot P(E_3/H_1\&E_1\&E_2) + P(H_2)\cdot P(E_1/H_2)\cdot P(E_2/H_2\&E_1)\cdot P(E_3/H_2\&E_1\&E_2)}.$$

Clearly, the formula for H_2 is

$$P(H_2/E_1\&E_2\&E_3) = \frac{P(H_2)\cdot P(E_1/H_2)\cdot P(E_2/H_2\&E_1)\cdot P(E_3/H_2\&E_1\&E_2)}{P(H_1)\cdot P(E_1/H_1)\cdot P(E_2/H_1\&E_1)\cdot P(E_3/H_1\&E_1\&E_2) + P(H_2)\cdot P(E_1/H_2)\cdot P(E_2/H_2\&E_1)\cdot P(E_3/H_2\&E_1\&E_2)}.$$

We note again that (1) the numerator is a repeat of the row of the denominator corresponding to the hypothesis whose revised probability we wish to calculate, (2) the revised probabilities of H_1 and H_2 add up to "1," and (3) that we in essence "added" on a "column" for E_3 to the formulas for two events.

Given this pattern, the reader should be able to guess the general formula for "n" events (where "n" is any positive integer, e.g., 1 or 2 or 3 or 4, etc.) for the tree in Figure 2.10 where

FIGURE 2.10
Probability Tree Diagram for Two Hypotheses and "n" Events

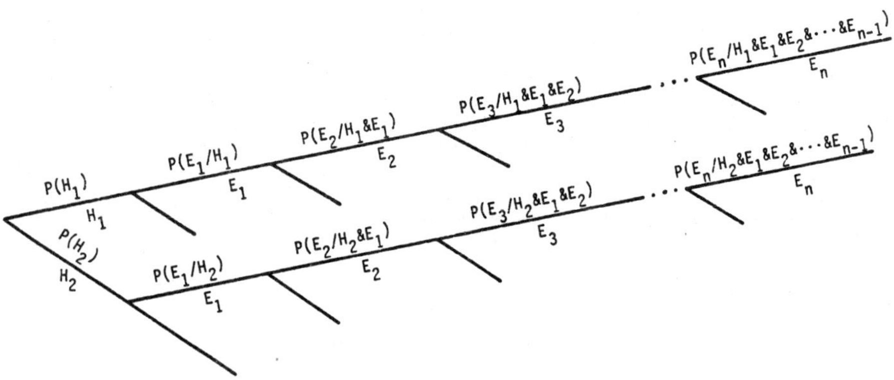

the 3 dots "···" indicate we just keep on adding branches. Then

$$P(H_1/E_1\&E_2\&\cdots\&E_n)$$
$$= \frac{P(H_1)\cdot P(E_1/H_1)\cdot P(E_2/H_1\&E_1)\cdots P(E_n/H_1\&E_1\&E_2\&\cdots\&E_{n-1})}{P(H_1)\cdot P(E_1/H_1)\cdot P(E_2/H_1\&E_1)\cdots P(E_n/H_1\&E_1\&E_2\&\cdots\&E_{n-1}) + P(H_2)\cdot P(E_1/H_2)\cdot P(E_2/H_2\&E_1)\cdots P(E_n/H_2\&E_1\&E_2\&\cdots\&E_{n-1})}.$$

Looking at the left hand side of the equation, when we write $P(H_1/E_1\&E_2\&\cdots\&E_n)$ the three dots in "$E_1\&E_2\cdots\&E_n$" simply mean that for ease of writing we have not listed the events after E_2 up to the last event (which is E_n), but that it is to be understood that all the events from the first (E_1) through the last (E_n) are being considered. So, the revised probability of H_1 is being calculated given all the events E_1 through E_n.

Similarly, looking at the right side of the equation, we use three dots between the "column" for E_2 and the last "column" for the last event E_n, where the dots indicate that for ease of writing we have not listed the "columns" for events after the second event up to the column for the event before the last column (which is for event E_n), but that it is understood that columns for all the events from the first (E_1) through the last (E_n) are being used in the calculations.

In the last column, the column for E_n, the term $P(E_n/H_1\&E_1\&E_2\&\cdots\&E_{n-1})$ asks for the probability of E_n given the assumption that H_1 is the case and given all the events through E_{n-1}, meaning events E_1, E_2, E_3, etc, up through the event preceding E_n. If E_n is, say, E_{10} then the last event before E_{10} is $E_{10-1} = E_9$. Hence, the "E_{n-1}" in the formula.

Looking at this formula, we see that we have a first "column" for the hypotheses followed by a "column" for each event. The revised probabilities of $P(H_1/E_1\&E_2\&\cdots\&E_n)$ and $P(H_2/E_1\&E_2\&\cdots\&E_n)$ will add to 1, and the formula for $P(H_2/E_1\&E_2\&\cdots\&E_n)$ differs only from the formula for $P(H_1/E_1\&E_2\&\cdots\&E_n)$ by having a repeat of the second row of the denominator as the numerator.

Therefore, we can say that for n events and two hypotheses, the revised probability of hypothesis H_i (where i = 1 or 2) is

$$P(H_i/E_1\&E_2\&\cdots\&E_n)$$
$$= \frac{P(H_i)\cdot P(E_1/H_i)\cdot P(E_2/H_i\&E_1)\cdots P(E_n/H_i\&E_1\&E_2\&\cdots\&E_{n-1})}{P(H_1)\cdot P(E_1/H_1)\cdot P(E_2/H_1\&E_1)\cdots P(E_n/H_1\&E_1\&E_2\&\cdots\&E_{n-1}) + P(H_2)\cdot P(E_1/H_2)\cdot P(E_2/H_2\&E_1)\cdots P(E_n/H_2\&E_1\&E_2\&\cdots\&E_{n-1})}.$$

<u>Development of the General Formula for "n" Events and "m" Hypotheses</u>.

Suppose we have three hypotheses, instead of two. Then our tree becomes the one shown in Figure 2.11

FIGURE 2.11
Probability Tree Diagram for Three Hypotheses and One Event

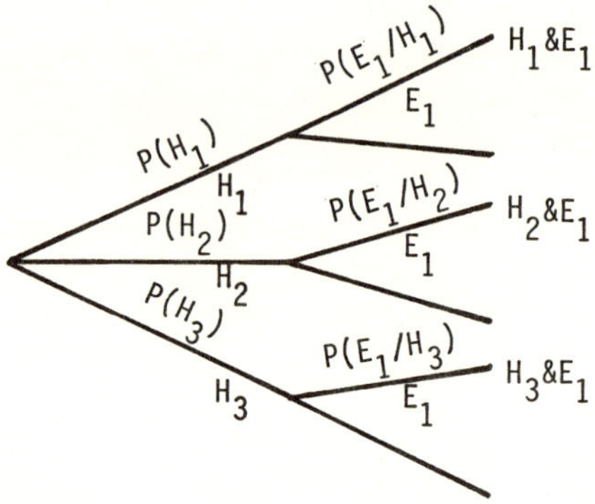

and $P(H_1/E_1) = \frac{P(H_1\&E_1)}{P(E_1)}$

$$= \frac{P(H_1\&E_1)}{P(H_1\&E_1) + P(H_2\&E_1) + P(H_3\&E_1)}$$

$$= \frac{P(H_1)\cdot P(E_1/H_1)}{P(H_1)\cdot P(E_1/H_1) + P(H_2)\cdot P(E_1/H_2) + P(H_3)\cdot P(E_1/H_3)}.$$

If the reader will compare this to the formula for $P(H_1/E_1)$ for two hypotheses:

$$P(H_1/E_1) = \frac{P(H_1)\cdot P(E_1/H_1)}{P(H_1)\cdot P(E_1/H_1) + P(H_2)\cdot P(E_1/H_2)}$$

it is clear that the only difference is that, when we go to three hypotheses, we add on a third row in the denominator for that third hypotheses. Similarly, if we had four hypotheses, the tree would be as shown in Figure 2.12 and the formula would be

FIGURE 2.12
Probability Tree Diagram for Four Hypotheses and One Event

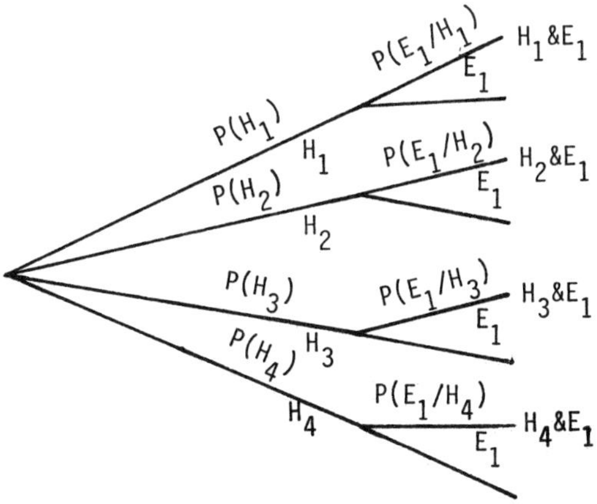

$$P(H_1/E_1) = \frac{P(H_1\&E_1)}{P(E_1)}$$

$$= \frac{P(H_1\&E_1)}{P(H_1\&E_1) + P(H_2\&E_1) + P(H_3\&E_1) + P(H_4\&E_1)} = \frac{P(H_1)\cdot P(E_1/H_1)}{P(H_1)\cdot P(E_1/H_1) + P(H_2)\cdot P(E_1/H_2) + P(H_3)\cdot P(E_1/H_3) + P(H_4)\cdot P(E_1/H_4)}.$$

Thus, the only thing about the formula that changes as we add hypotheses is the denominator - for each hypothesis we add on we also add on a row in the denominator. We have as many rows in the denominator as we have hypotheses. Moreover, as the reader can easily verify, the numerator in the formula for $P(H_i/E_1)$ is a repeat of the row of the denominator corresponding to the hypothesis H_i whose revised probability we want.

As we add on events, we continue to add on a column for each event.

The general formula, then, for the revised probability of any hypothesis H_i, when there are "n" events and "m" hypotheses is[4]

$$P(H_i/E_1\&E_2\&\cdots\&E_n)$$

$$= \frac{P(H_i)\cdot P(E_1/H_i)\cdot P(E_2/H_i\&E_1)\cdots P(E_n/H_i\&E_1\&E_2\&\cdots\&E_{n-1})}{\begin{array}{l} P(H_1)\cdot P(E_1/H_1)\cdot P(E_2/H_1\&E_1)\cdots P(E_n/H_1\&E_1\&E_2\&\cdots\&E_{n-1}) \\ +P(H_2)\cdot P(E_1/H_2)\cdot P(E_2/H_2\&E_1)\cdots P(E_n/H_2\&E_1\&E_2\&\cdots\&E_{n-1}) \\ \vdots \\ +P(H_m)\cdot P(E_1/H_m)\cdot P(E_2/H_m\&E_1)\cdots P(E_n/H_m\&E_1\&E_2\&\cdots\&E_{n-1}) \end{array}}.$$

When we put three vertical dots,"$\vdots$", in the denominator between the second and last rows, it means simply that we keep listing rows after the second one until we get to the last one. In the expression $P(H_i/E_1\&E_2\&\cdots\&E_n)$, "$H_i$" stands for any one of the hypotheses. That is, we are giving a general formula that will tell us how to calculate the revised probability for any one of the hypotheses. Thus "i" can be any number between "1" and "m." So, if i=1, we use the formula to calculate H_1. We do this by replacing all the "i"s everywhere they appear to the left and the right of the "=" sign by the number "1." Then in the numerator of the right hand side of the equation, we would replace the "i" in $P(H_i)\cdot P(E_1/H_i)\cdot P(E_2/H_i\&E_1)\cdots P(E_n/H_i\&E_1\&E_2\&\cdots\&E_{n-1})$ by "1" and then have $P(H_1)\cdot P(E_1/H_1)\cdot P(E_2/H_1\&E_1)\cdots P(E_n/H_1\&E_1\&E_2\&\cdots\&E_{n-1})$.

If i = 4, to give another example, we would be calculating the revised probability of H_4, i.e., $P(H_4/E_1\&E_2\&\cdots\&E_n)$ and the numerator would be
$P(H_4)\cdot P(E_1/H_4)\cdot P(E_2/H_4\&E_1)\cdots P(E_n/H_4\&E_1\&E_2\&\cdots\&E_{n-1})$.

The Tabular Array Approach to Bayes.

The general formula just presented hopefully makes good sense to the reader. Note again that the numerator is a repeat of the row of the denominator corresponding to the hypothesis H_i, whose revised probability we wish to calculate.

It would be quite troublesome in a time urgent situation with "n" pieces of data to write out this lengthy equation each time we want to use it. A quicker approach is easily discovered when we remember our talk about the "column" for the hypotheses followed by the "columns" for the first event E_1 through the last E_n. If we look then at the denominator

$$\begin{array}{l} P(H_1)\cdot P(E_1/H_1)\cdot P(E_2/H_1\&E_1)\cdots P(E_n/H_1\&E_1\&E_2\&\cdots\&E_{n-1}) \\ +P(H_2)\cdot P(E_1/H_2)\cdot P(E_2/H_2\&E_1)\cdots P(E_n/H_2\&E_1\&E_2\&\cdots\&E_{n-1}) \\ \quad\vdots \qquad\quad \vdots \qquad\qquad \vdots \\ +P(H_m)\cdot P(E_1/H_m)\cdot P(E_2/H_m\&E_1)\cdots P(E_n/H_m\&E_1\&E_2\&\cdots\&E_{n-1}) \end{array}$$

instead of writing all individual terms we can use column headings of $P(H_j)$, $P(E_1/H_j)$, $P(E_2/H_j\&E_1),\cdots$, $P(E_n/H_j\&E_1\&E_2\&\cdots\&E_{n-1})$.

To arrive at this conclusion, consider the following example. Suppose that we have four hypotheses, call them H_1, H_2, H_3, and H_4 and assume that the hypotheses are mutually exclusive and exhaustive. As noted earlier, if we want to run a Bayesian analysis we start off by assessing subjective probabilities for all but the last hypothesis. In this case, then, we assess $P(H_1)$, $P(H_2)$, $P(H_3)$. Then, $P(H_4) = 1 -(P(H_1)+P(H_2)+P(H_3))$, since the hypotheses are mutually exclusive and exhaustive.

For this example, suppose we decide that $P(H_1) = 0.4$, $P(H_2) =$

0.2, $P(H_3) = 0.1$. Then $P(H_4) = 1-0.7 = 0.3$. Next, suppose we use these initial values in our Bayesian analysis and that we wish to revise the probabilities in light of an event E_1. According to the formula the following is true:

$$P(H_1/E_1) = \frac{P(H_1)\cdot P(E_1/H_1)}{P(H_1)\cdot P(E_1/H_1) + P(H_2)\cdot P(E_1/H_2) + P(H_3)\cdot P(E_1/H_3) + P(H_4)\cdot P(E_1/H_4)}$$

$$P(H_3/E_1) = \frac{P(H_3)\cdot P(E_1/H_3)}{P(H_1)\cdot P(E_1/H_1) + P(H_2)\cdot P(E_1/H_2) + P(H_3)\cdot P(E_1/H_3) + P(H_4)\cdot P(E_1/H_4)}$$

$$P(H_2/E_1) = \frac{P(H_2)\cdot P(E_1/H_2)}{P(H_1)\cdot P(E_1/H_1) + P(H_2)\cdot P(E_1/H_2) + P(H_3)\cdot P(E_1/H_3) + P(H_4)\cdot P(E_1/H_4)}$$

$$P(H_4/E_1) = \frac{P(H_4)\cdot P(E_1/H_4)}{P(H_1)\cdot P(E_1/H_1) + P(H_2)\cdot P(E_1/H_2) + P(H_3)\cdot P(E_1/H_3) + P(H_4)\cdot P(E_1/H_4)}.$$

Strictly speaking, once we calculate $P(H_1/E_1)$, $P(H_2/E_1)$, and $P(H_3/E_1)$, we can calculate $P(H_4/E_1)$ by the formula $P(H_4/E_1) = 1 - (P(H_1/E_1) + P(H_2/E_1) + P(H_3/E_1))$ (since the revised probabilities must sum to "1"). However, as a check on all our calculations, we really should go ahead and calculate $P(H_4/E_1)$ just to make sure the numbers come out right.

Note again that in the formulas for $P(H_1/E_1)$, $P(H_2/E_1)$, $P(H_3/E_1)$, and $P(H_4/E_1)$, the denominators are all the same. Only the numerators vary.

In order to use the formulas, we now must assign these subjective conditional probabilities:

$P(E_1/H_1)$
$P(E_1/H_2)$
$P(E_1/H_3)$
$P(E_1/H_4)$.

In order to assess $P(E_1/H_1)$, we ask ourselves, "What is the probability that event E_1 would occur if we assume (i.e. given) hypothesis H_1 to be true." In order to assess $P(E_1/H_2)$ we ask ourselves, "What is the probability that event E_1 would occur if we assume (i.e., given) hypothesis H_2 to be true." We continue to ask the same kinds of questions, assuming H_3 to be true (i.e., given) in order to assess $P(E_1/H_3)$ and then H_4 to be true (i.e., given) in order to assess $P(E_1/H_4)$.

Suppose we decide on the following probabilities:

$P(E_1/H_1) = 0.8$
$P(E_1/H_2) = 0.9$
$P(E_1/H_3) = 0.4$
$P(E_1/H_4) = 0.5.$

Note that there is no need for these probabilities to add up to "1," since they are assigned to branches coming from different nodes in the tree.

Once these values are assigned, we simply plug them into our equations. In this case, we get the following results:

$$P(H_1/E_1) = \frac{P(H_1)\cdot P(E_1/H_1)}{\begin{array}{l}P(H_1)\cdot P(E_1/H_1)\\+P(H_2)\cdot P(E_1/H_2)\\+P(H_3)\cdot P(E_1/H_3)\\+P(H_4)\cdot P(E_1/H_4)\end{array}} = \frac{(0.4)(0.8)}{\begin{array}{l}(0.4)(0.8)\\+(0.2)(0.9)\\+(0.1)(0.4)\\+(0.3)(0.5)\end{array}} = \frac{0.32}{\begin{array}{l}0.32\\+0.18\\+0.04\\+0.15\end{array}}$$

$$= \frac{0.32}{0.69} = 0.464.$$

Since, as noted before, the denominators are all the same for $P(H_1/E_1)$, $P(H_2/E_1)$, $P(H_3/E_1)$, $P(H_4/E_1)$, we simply can write:

$$P(H_2/E_1) = \frac{P(H_2)\cdot P(E_1/H_2)}{0.69} = \frac{(0.2)(0.9)}{0.69} = \frac{0.18}{0.69} = 0.261$$

$$P(H_3/E_1) = \frac{P(H_3)\cdot P(E_1/H_3)}{0.69} = \frac{(0.1)(0.4)}{0.69} = \frac{0.04}{0.69} = 0.058$$

$$P(H_4/E_1) = \frac{P(H_4)\cdot P(E_1/H_4)}{0.69} = \frac{(0.3)(0.5)}{0.69} = \frac{0.15}{0.69} = 0.217.$$

So, the results are:

$P(H_1/E_1) = 0.464$
$P(H_2/E_1) = 0.261$
$P(H_3/E_1) = 0.058$
$P(H_4/E_1) = \underline{0.217}$
$\qquad\qquad\quad 1.000$

and the results, that is, the revised probabilities, add to "1" as they should.

Thus far, we have the results shown in Table 2.1.

TABLE 2.1
Initial Probabilities, Conditional Probabilities of Events, and Revised Probabilities of Hypotheses after One Event

$P(H_1) = 0.4$	$P(E_1/H_1) = 0.8$	$P(H_1/E_1) = 0.464$
$P(H_2) = 0.2$	$P(E_1/H_2) = 0.9$	$P(H_2/E_1) = 0.261$
$P(H_3) = 0.1$	$P(E_1/H_3) = 0.4$	$P(H_2/E_1) = 0.058$
$P(H_4) = 0.3$	$P(E_1/H_3) = 0.5$	$P(H_4/E_1) = 0.217$

Suppose now that E_2 occurs, and we wish to calculate our revised probabilities in light of the latest evidence. According to the formula, we wish to calculate:

$$P(H_1/E_1\&E_2) = \frac{P(H_1)\cdot P(E_1/H_1)\cdot P(E_2/H_1\&E_1)}{\begin{array}{l} P(H_1)\cdot P(E_1/H_1)\cdot P(E_2/H_1\&E_1) \\ +P(H_2)\cdot P(E_1/H_2)\cdot P(E_2/H_2\&E_1) \\ +P(H_3)\cdot P(E_1/H_3)\cdot P(E_2/H_3\&E_1) \\ +P(H_4)\cdot P(E_1/H_4)\cdot P(E_2/H_4\&E_1) \end{array}}$$

$$P(H_2/E_1\&E_2) = \frac{P(H_2)\cdot P(E_1/H_2)\cdot P(E_2/H_2\&E_1)}{\begin{array}{l} P(H_1)\cdot P(E_1/H_1)\cdot P(E_2/H_1\&E_1) \\ +P(H_2)\cdot P(E_1/H_2)\cdot P(E_2/H_2\&E_1) \\ +P(H_3)\cdot P(E_1/H_3)\cdot P(E_2/H_3\&E_1) \\ +P(H_4)\cdot P(E_1/H_4)\cdot P(E_2/H_4\&E_1) \end{array}}$$

$$P(H_3/E_1\&E_2) = \frac{P(H_3)\cdot P(E_1/H_3)\cdot P(E_2/H_3\&E_1)}{\begin{array}{l} P(H_1)\cdot P(E_1/H_1)\cdot P(E_2/H_1\&E_1) \\ +P(H_2)\cdot P(E_1/H_2)\cdot P(E_2/H_2\&E_1) \\ +P(H_3)\cdot P(E_1/H_3)\cdot P(E_2/H_3\&E_1) \\ +P(H_4)\cdot P(E_1/H_4)\cdot P(E_2/H_4\&E_1) \end{array}}$$

$$P(H_4/E_1\&E_2) = \frac{P(H_4)\cdot P(E_1/H_4)\cdot P(E_2/H_4\&E_1)}{\begin{array}{l} P(H_1)\cdot P(E_1/H_1)\cdot P(E_2/H_1\&E_1) \\ +P(H_2)\cdot P(E_1/H_2)\cdot P(E_2/H_2\&E_1) \\ +P(H_3)\cdot P(E_1/H_3)\cdot P(E_2/H_3\&E_1) \\ +P(H_4)\cdot P(E_1/H_4)\cdot P(E_2/H_4\&E_1) \end{array}}.$$

As a check on our calculations, we should calculate $P(H_4/E_1\&E_2)$ as well as the other probabilities. Note again that all the denominators in the above equations are the same. Only the numerators differ.

In order to use the formulas, we now must assign these subjective conditional probabilities:

$P(E_2/H_1\&E_1)$
$P(E_2/H_2\&E_1)$
$P(E_2/H_3\&E_1)$
$P(E_2/H_4\&E_1)$.

In order to assess (assign) the conditional probability $P(E_2/H_1\&E_1)$, we ask ourselves "What is the probability that E_2 would occur if we assume (i.e., given) hypothesis H_1 to be true and if (i.e., given) event E_1 has occurred." In order to assess $P(E_2/H_2\&E_1)$, we ask ourselves, "What is the probability event E_2 would occur if we assume (i.e., given) hypothesis H_2 to be true and if (i.e., given) event E_1 has occurred." We continue on to ask the same sorts of questions, first assuming H_3 to be true and E_1 to have occurred in order to assess $P(E_2/H_3\&E_1)$ and next assuming H_4 to be true and E_1 to have occurred in order to assess $P(E_2/H_4\&E_1)$.

Suppose we decide on these values:

$P(E_2/H_1\&E_1) = 0.2$
$P(E_2/H_2\&E_1) = 0.8$
$P(E_2/H_3\&E_1) = 0.6$
$P(E_2/H_4\&E_1) = 0.3$.

Note again that there is no need for these values to add up to "1." Once these values are assigned we simply plug them into our equations. To make our calculations easier to follow, we will first repeat all the values we have assessed so far.

TABLE 2.2
Assessments of Prior Probabilities and Probabilities of Events

$P(H_1) = 0.4$	$P(E_1/H_1) = 0.8$	$P(E_2/H_1\&E_1) = 0.2$
$P(H_2) = 0.2$	$P(E_1/H_2) = 0.9$	$P(E_2/H_2\&E_1) = 0.8$
$P(H_3) = 0.1$	$P(E_1/H_3) = 0.4$	$P(E_2/H_3\&E_1) = 0.6$
$P(H_4) = 0.3$	$P(E_1/H_4) = 0.5$	$P(E_2/H_4\&E_1) = 0.3$

Taking these values, then, we can plug the numbers into our formulas. We get:

$$P(H_1/E_1\&E_2) = \frac{P(H_1)\cdot P(E_1/H_1)\cdot P(E_2/H_1\&E_1)}{\begin{array}{l}P(H_1)\cdot P(E_1/H_1)\cdot P(E_2/H_1\&E_1)\\+P(H_2)\cdot P(E_1/H_2)\cdot P(E_2/H_2\&E_1)\\+P(H_3)\cdot P(E_1/H_3)\cdot P(E_2/H_3\&E_1)\\+P(H_4)\cdot P(E_1/H_4)\cdot P(E_2/H_4\&E_1)\end{array}}$$

$$= \frac{(0.4)(0.8)(0.2)}{\begin{array}{l}(0.4)(0.8)(0.2)\\+(0.2)(0.9)(0.8)\\+(0.1)(0.4)(0.6)\\+(0.3)(0.5)(0.3)\end{array}}$$

$$= \frac{0.064}{0.064 + 0.144 + 0.024 + 0.045}$$

$$= \frac{0.064}{0.277} = 0.231.$$

Noting again that since the denominators in the equations for calculating $P(H_1/E_1\&E_2)$, $P(H_2/E_1\&E_2)$, $P(H_3/E_1\&E_2)$, $P(H_4/E_1\&E_2)$ are the same, we simply can write:

$$P(H_2/E_1\&E_2) = \frac{P(H_2)\cdot P(E_1/H_2)\cdot P(E_2/H_2\&E_1)}{0.277}$$

$$= \frac{(0.2)(0.9)(0.8)}{0.277} = \frac{0.144}{0.277} = 0.520$$

$$P(H_3/E_1\&E_2) = \frac{P(H_3)\cdot P(E_1/H_3)\cdot P(E_2/H_3\&E_1)}{0.277}$$

$$= \frac{(0.1)(0.4)(0.6)}{0.277} = \frac{0.024}{0.277} = 0.087$$

$$P(H_4/E_1\&E_2) = \frac{P(H_4)\cdot P(E_1/H_4)\cdot P(E_2/H_4\&E_1)}{0.27}$$

$$= \frac{(0.3)(0.5)(0.3)}{0.277} = \frac{0.045}{0.277} = 0.162.$$

So the results are as follows:

$$\begin{array}{l}P(H_1/E_1\&E_2) = 0.231\\P(H_2/E_1\&E_2) = 0.520\\P(H_3/E_1\&E_2) = 0.087\\P(H_4/E_1\&E_2) = \underline{0.162}\\\qquad\qquad\qquad\quad 1.000 = \text{sum}\end{array}$$

and the results, that is, the revised probabilities, add to "1" as they should.

At this point, the reader may have noted something very interesting. If you look back at Table 2.2, you can see that the denominators we just calculated were just what you would get if you multiplied the entries together for each row in Table 2.2 and added up the results. Look at the denominator in the equation for $P(H_1/E_1\&E_2)$ following Table 2.2. It matches the array we set out in Table 2.2, i.e., each term (i.e., row) in the denominator matches a corresponding row in the table.

As for the numerators we just calculated, if we want the revised probability of any hypothesis, the numerator comes from multiplying the entries together for the row in Table 2.2 that corresponds to the hypothesis of interest. For example, we get the numerator for $P(H_3/E_1\&E_2)$ from the third row entries of Table 2.2 - compare to the actual calculation we made for $P(H_3/E_1\&E_2)$ to verify this.

All this is very important because it means that we simply can plug our entries into a table and multiply out the rows and add up the results in order to calculate the denominators. Any numerator for a given hypothesis comes from the corresponding row for that hypothesis. Let us do this for the following example.

Suppose we keep on with the previous case, but now have four events. We have to assess the entries shown in Table 2.3.

TABLE 2.3
Assessments of Prior Probabilities and Probabilities of Events

					Row Product
$P(H_1)=0.4$	$P(E_1/H_1)=0.8$	$P(E_2/H_1\&E_1)=0.2$	$P(E_3/H_1\&E_1\&E_2)=0.6$	$P(E_4/H_1\&E_1\&E_2\&E_3)=0.3$	0.01152
$P(H_2)=0.2$	$P(E_1/H_2)=0.9$	$P(E_2/H_2\&E_1)=0.8$	$P(E_3/H_2\&E_1\&E_2)=0.7$	$P(E_4/H_2\&E_1\&E_2\&E_3)=0.6$	0.06048
$P(H_3)=0.1$	$P(E_1/H_3)=0.4$	$P(E_2/H_3\&E_1)=0.6$	$P(E_3/H_3\&E_1\&E_2)=0.1$	$P(E_4/H_3\&E_1\&E_2\&E_3)=0.2$	0.00048
$P(H_4)=0.3$	$P(E_1/H_4)=0.5$	$P(E_2/H_4\&E_1)=0.3$	$P(E_3/H_4\&E_1\&E_2)=0.5$	$P(E_4/H_4\&E_1\&E_2\&E_3)=0.9$	0.02025
					0.09273

To verify that this array really corresponds to the entries in the formula, look at that equation and substitute in m=4, n=4. It should be clear that each row of Table 2.2 corresponds to a row of the denominator of the formula. The row products in Table 2.3 correspond to the values of the terms (i.e., rows) in the denominator of the formula.

Using the table, since the sum of the row products is 0.09273, we have:

$$P(H_1/E_1\&E_2\&E_3\&E_4) = \frac{0.01152}{0.09273} = 0.124$$

$$P(H_2/E_1\&E_2\&E_3\&E_4) = \frac{0.06048}{0.09273} = 0.652$$

$$P(H_3/E_1\&E_2\&E_3\&E_4) = \frac{0.00048}{0.09273} = 0.005$$

$$P(H_4/E_1\&E_2\&E_3\&E_4) = \frac{0.02025}{0.09273} = \underline{0.218}$$

$$0.999 = \text{sum}$$

Our sum for the revised probabilities is essentially "1," as it should be.

This sort of tabular approach to Bayes greatly simplifies our calculations. We simply list as many rows as there are hypotheses. The first column will be the initial, or prior, probabilities assessed for the probabilities of the hypotheses. The following columns correspond to the entries assessed as the events occur. If n=4, as in the above example, we simply fill in the appropriate entries - what is appropriate is quickly clear from an examination of the denominator of the formula.

Further simplification is possible and desirable. If we look at Table 2.3, the reader might ask why we should use anything but column headings for all the entries? We can write as shown in Table 2.4 where the intersection of the "j" number and the column heading tells what entry we write into the table. For example, if j=3 and the column heading is $P(E_2/H_j\&E_1)$, then the entry to put into the table at the intersection of the row and column is $P(E_2/H_3\&E_1)$, which is 0.6. Use of column headings means the calculating process can be much more rapid than if we were to write out a lengthy equation each time we want to revise the probabilities of our hypotheses.

TABLE 2.4
Tabular Array for Prior Probabilities of Hypotheses, Probabilities of Events, and Revised Probabilities of Hypotheses

j	$P(H_j)$	$P(E_1/H_j)$	$P(E_2/H_j\&E_1)$	$P(E_3/H_j\&E_1\&E_2)$	$P(E_4/H_j\&E_1\&E_2\&E_3)$	Row Product	Revised Prob of H_j
1	0.4	0.8	0.2	0.6	0.3	0.01152	$\frac{0.01152}{0.09273}$ = 0.124
2	0.2	0.9	0.8	0.7	0.6	0.06048	$\frac{0.06048}{0.09273}$ = 0.652
3	0.1	0.4	0.6	0.1	0.2	0.00048	$\frac{0.00048}{0.09273}$ = 0.005
4	0.3	0.5	0.3	0.5	0.9	0.02025	$\frac{0.02025}{0.09273}$ = 0.218
						0.09273	0.999

Using the above example, suppose two new items of evidence, call them E_5 and E_6, are received. Then, the format to use is as follows (Table 2.5).

TABLE 2.5
Tabular Array for Dealing with New Items of Evidence

j	Previous Row Product	$P(E_5/H_j\&E_1\cdots\&E_4)$	$P(E_6/H_j\&E_1\&\cdots\&E_5)$	Latest Row Product	Latest Revised Probability of H_j
1	0.01152	0.8	0.1	0.0009216	0.035
2	0.06048	0.3	0.8	0.0145152	0.548
3	0.00048	0.4	0.7	0.0001344	0.005
4	0.02025	0.6	0.9	0.010935	0.413
				0.0265062	1.001

This is the approach recommended for quick, hand calculator approaches to using Bayes. (If EMP is a worry, slide rules are advised.) That is, each time a new set of items of evidence is received, simply take the previous row products, set up a column for each new item of evidence, make the appropriate assessments, and calculate the new row products and latest revised probabilities of hypotheses.

The "Odds-Likelihood" Form of Bayes

What will be presented next is a derivation of the so-called "odds-likelihood" form of Bayes. Before deriving the desired equation, let us look at a simple numerical problem which will be analogous to what we will do. Consider the equations

$$3 = \frac{6}{2}$$

$$7 = \frac{14}{2}.$$

We can divide the second equation into the first by dividing the left side of the second into the left side of the first, and the right side of the second into the right side of the first. We get

$$\frac{3}{7} = \frac{\frac{6}{2}}{\frac{14}{2}}.$$

Next, we want to simplify the right hand side of the above equation. We can do this by multiplying above and below by 2. We have

$$\frac{3}{7} = \frac{\frac{6}{2} \cdot 2}{\frac{14}{2} \cdot 2} = \frac{6}{14}.$$

We will follow an analogous procedure to divide the equation for $P(H_1/E_1\&E_2\&\cdots\&E_n)$ by the equation for $P(H_2/E_1\&E_2\&\cdots\&E_n)$.

Suppose, then, that we have only 2 hypotheses, H_1 & H_2. We have, by what we did above,

$$P(H_1/E_1\&E_2\&\cdots\&E_n)$$

$$= \frac{P(H_1)\cdot P(E_1/H_1)\cdot P(E_2/H_1\&E_1)\cdots P(E_n/H_1\&E_1\&E_2\&\cdots\&E_{n-1})}{P(H_1)\cdot P(E_1/H_1)\cdot P(E_2/H_1\&E_1)\cdots P(E_n/H_1\&E_1\&E_2\&\cdots\&E_{n-1}) + P(H_2)\cdot P(E_1/H_2)\cdot P(E_2/H_2\&E_1)\cdots P(E_n/H_2\&E_1\&E_2\&\cdots\&E_{n-1})}$$

and

$$P(H_2/E_1\&E_2\&\cdots\&E_n)$$

$$= \frac{P(H_2)\cdot P(E_1/H_2)\cdot P(E_2/H_2\&E_1)\cdots P(E_n/H_2\&E_1\&E_2\&\cdots\&E_{n-1})}{P(H_1)\cdot P(E_1/H_1)\cdot P(E_2/H_1\&E_1)\cdots P(E_n/H_1\&E_1\&E_2\&\cdots\&E_{n-1}) + P(H_2)\cdot P(E_1/H_2)\cdot P(E_2/H_2\&E_1)\cdots P(E_n/H_2\&E_1\&E_2\&\cdots\&E_{n-1})}.$$

Dividing the second equation into the first we get

$$\frac{P(H_1/E_1\&E_2\&\cdots\&E_n)}{P(H_2/E_1\&E_2\&\cdots\&E_n)}$$

$$= \frac{\dfrac{P(H_1)\cdot P(E_1/H_1)\cdot P(E_2/H_1\&E_1)\cdots P(E_n/H_1\&E_1\&E_2\&\cdots\&E_{n-1})}{P(H_1)\cdot P(E_1/H_1)\cdot P(E_2/H_1\&E_1)\cdots P(E_n/H_1\&E_1\&E_2\&\cdots\&E_{n-1}) + P(H_2)\cdot P(E_1/H_2)\cdot P(E_2/H_2\&E_1)\cdots P(E_n/H_2\&E_1\&E_2\&\cdots\&E_{n-1})}}{\dfrac{P(H_2)\cdot P(E_1/H_2)\cdot P(E_2/H_2\&E_1)\cdots P(E_n/H_2\&E_1\&E_2\&\cdots\&E_{n-1})}{P(H_1)\cdot P(E_1/H_1)\cdot P(E_2/H_1\&E_1)\cdots P(E_n/H_1\&E_1\&E_2\&\cdots\&E_{n-1}) + P(H_2)\cdot P(E_1/H_2)\cdot P(E_2/H_2\&E_1)\cdots P(E_n/H_2\&E_1\&E_2\&\cdots\&E_{n-1})}}$$

$$= \frac{P(H_1)\cdot P(E_1/H_1)\cdot P(E_2/H_1\&E_1)\cdots P(E_n/H_1\&E_1\&E_2\&\cdots\&E_{n-1})}{P(H_2)\cdot P(E_1/H_2)\cdot P(E_2/H_2\&E_1)\cdots P(E_n/H_2\&E_1\&E_2\&\cdots\&E_{n-1})}.$$

Splitting apart the factors in the above equation we get

Revised odds

$$\frac{P(H_1/E_1\&E_2\&\cdots\&E_n)}{P(H_2/E_1\&E_2\&\cdots\&E_n)}$$

Prior odds	1st Likelihood Ratio	2nd Likelihood Ratio	...	Last (n th) Likelihood Ratio

$$= \frac{P(H_1)}{P(H_2)} \cdot \frac{P(E_1/H_1)}{P(E_1/H_2)} \cdot \frac{P(E_2/H_1\&E_1)}{P(E_2/H_2\&E_1)} \cdots \frac{P(E_n/H_1\&E_1\&E_2\&\cdots\&E_{n-1})}{P(E_n/H_2\&E_1\&E_2\&\cdots\&E_{n-1})}.$$

This is called the <u>odds-likelihood form of Bayes</u>. I personally don't particularly like it, but some <u>very</u> influential Bayesian specialists do, so it is presented here. One good thing about it is that it is a fairly easy formula to remember. We have on the left side of the equation the so-called "revised odds," where odds means a ratio of probabilities and where revised simply means revised in light of the evidence of E_1 through E_n.

The first term on the right hand side is called reasonably the "prior odds," where we mean the (initial) odds prior to the evidence. The next term is the so-called first likelihood ratio, which is reasonable language because it is a ratio of two likelihoods (= probabilities), and it is for the first event. There is a likelihood ratio for each of the "n" events.

By the way, instead of assessing, say, $P(H_1) = 0.8$, $P(H_2) = 1-P(H_1) = 0.2$ and saying that the prior odds equals

$$\frac{P(H_1)}{P(H_2)} = \frac{0.8}{0.2} = \frac{8}{2} = \frac{4}{1},$$

some analysts think <u>directly</u> in terms of odds, and might say that the prior odds in favor of H_1 are 4 to 1. The same procedure might be used in assigning $P(E_1/H_1)$ and $P(E_1/H_2)$. That is, when figuring the first likelihood ratio instead of, say, assessing $P(E_1/H_1) = 0.9$, $P(E_1/H_2) = 0.8$, the analyst might say that the odds favoring E_1 (given H_1) over E_1 (given H_2) are 9 to 8.

One real problem with the odds-likelihood form simply is how to calculate the revised probabilities of H_1 and H_2 on the basis of the odds. Analysts seem to have trouble remembering how to do it correctly. Suppose, for example, the odds-likelihood formula has been used in a particular situation, and we have

$$\frac{P(H_1/E_1\&E_2\&\cdots\&E_n)}{P(H_2/E_1\&E_2\&\cdots\&E_n)}$$

$$= \frac{P(H_1)}{P(H_2)} \cdot \frac{P(E_1/H_1)}{P(E_1/H_2)} \cdot \frac{P(E_2/H_1\&E_1)}{P(E_2/H_2\&E_1)} \cdots \frac{P(E_n/H_1\&E_1\&E_2\&\cdots\&E_{n-1})}{P(E_n/H_2\&E_1\&E_2\&\cdots\&E_{n-1})}$$

$$= \frac{0.8}{0.2} \cdot \frac{0.9}{0.8} \cdot \frac{0.95}{0.8} \cdot \frac{0.4}{0.9} \cdot \frac{0.7}{0.6}$$

$$= \frac{0.19152}{0.06912} = \frac{19152}{6912} = \frac{9576}{3456} = \frac{4788}{1728} = \frac{2394}{864}$$

$$= \frac{1197}{432} = \frac{399}{144} = \frac{133}{48}.$$

The question is what to do with this result in order to calculate $P(H_1/E_1\&E_2\&...E_n)$? The answer is that since

$$\frac{P(H_1/E_1\&E_2\&\cdots\&E_n)}{P(H_2/E_1\&E_2\&\cdots\&E_n)} = \frac{133}{48}, \text{ we then have}$$

$P(H_1/E_1\&E_2\&\cdots\&E_n) = \frac{133}{133+48} = \frac{133}{181} = 0.735$, i.e., the numerator on the right, divided by itself + the denominor on the right equals what we want. It sounds like black magic. It does make sense, but it is not too easy to show. To try to see why, remember that before we split the factors out, we had

$$\frac{P(H_1/E_1\&E_2\&\cdots\&E_n)}{P(H_2/E_1\&E_2\&\cdots\&E_n)}$$

$$= \frac{P(H_1)\cdot P(E_1/H_1)\cdot P(E_2/H_1\&E_1)\cdots P(E_n/H_1\&E_1\&E_2\&\cdots\&E_{n-1})}{P(H_2)\cdot P(E_1/H_2)\cdot P(E_2/H_2\&E_1)\cdots P(E_n/H_2\&E_1\&E_2\&\cdots\&E_{n-1})}.$$

If we take the numerator on the right, and divide it by itself + the denominator on the right, then we get as a result

$$\frac{P(H_1)\cdot P(E_1/H_1)\cdot P(E_2/H_1\&E_1)\cdots P(E_n/H_1\&E_1\&E_2\&\cdots\&E_{n-1})}{P(H_1)\cdot P(E_1/H_1)\cdot P(E_2/H_1\&E_1)\cdots P(E_n/H_1\&E_1\&E_2\&\cdots\&E_{n-1}) + P(H_2)\cdot P(E_1/H_2)\cdot P(E_2/H_2\&E_1)\cdots P(E_n/H_2\&E_1\&E_2\&\cdots\&E_{n-1})}$$

which is just the formula for 2 hypotheses for $P(H_1/E_1\&E_2\&\cdots\&E_n)$! So, that is why the "trick" of "numerator over numerator + denominator" works to yield $P(H_1/E_1\&E_2\&\cdots\&E_n)$.

By the way, as noted in the example, we can get rid of decimal points and also reduce fractions by dividing above and below by common factors. To convince yourself, just note that if we take the original equation

$$\frac{P(H_1/E_1\&E_2\&\cdots\&E_n)}{P(H_2/E_1\&E_2\&\cdots\&E_n)} = \frac{0.19152}{0.06912}$$

and calculate

$$P(H_1/E_1\&E_2\&\cdots\&E_n) = \frac{0.19152}{0.19152 + 0.06912} = \frac{0.19152}{0.26064} = 0.735$$

we get the same answer.

What we can't do on the right is <u>ever</u> get rid of a fraction. For example, if the revised odds in favor of H_1 end up as $\frac{4}{1}$, we <u>can not</u> just write "4" - it makes no sense! The revised odds are 4 to 1, written $\frac{"4"}{1}$, and the revised probability of H_1 will be $\frac{4}{4+1}$ = 0.8.

By the way, if we want the revised probability of H_2, it is, of course, 1- revised probability of H_1. Instead of trying to do anything clever, we should calculate the revised probability of H_2 by using this relation.

<u>The Issues of Conditional Independence and Dependence</u>

Near the end of chapter one, we introduced the notions of conditional independence and conditional dependence. In terms of the notation introduced in this chapter, we can say that we have conditional independence for a given hypothesis H_j if it is true for each item of evidence E_i that

$$P(E_i/H_j\&E_1\&E_2\&\cdots\&E_{i-1}) = P(E_i/H_j).$$

This says that the conditional probability of the occurrence of a particular item of evidence E_i, given a particular hypothesis H_j and all the items of evidence preceding E_i, equals the conditional probability of the occurrence of E_i given just the hypothesis H_j. In short, the probability of the occurrence of E_i depends just on H_j -- all the previous items of evidence can be ignored in calculating the probability of the occurrence of E_i. The previous items of evidence don't matter. The probability of the occurrence of E_i is conditional upon the hypothesis H_j, but independent of the previous items of evidence. (As we saw at the end of chapter one, an urn example would be where we had several different urns with different compositions, then selected one urn and drew repeatedly. If we drew with replacement, the probabilities of the possible results would be conditionally independent because they would be conditional upon the assumption of which urn we were drawing from but independent of the results of previous drawings (because the replacement each time of what was drawn left the composition of the urn unchanged).)

If the relation

$$P(E_i/H_j \& E_1 \& E_2 \& \cdots \& E_{i-1}) = P(E_i/H_j)$$

does not hold for a given hypothesis H_j, then we have conditional dependence. In this case, the probability of the occurrence of E_i is conditional upon the hypothesis H_j and also dependent upon the previous items of evidence. (In terms of the urn analogy introduced above, if we did not replace what we drew from the urn, we would have conditional dependence because the probabilities of the possible results would be conditional upon the assumption of which urn we were drawing from and dependent on the results of the previous drawings (because leaving out each time what was drawn changed the urn's composition).)

In developing the formulas of this chapter, we assumed that we had conditional dependence of the evidence for all the hypotheses. This is shown, for example, in Figure 2.10 where the probability of the branch for the "n th" item of evidence, E_n, was $P(E_n/H_1 \& E_1 \& E_2 \& \cdots \& E_{n-1})$ for the top path (the path for hypothesis H_1) and was $P(E_n/H_2 \& E_1 \& E_2 \& \cdots \& E_{n-1})$ for the path for hypothesis H_2. This assumption of conditional dependence was retained when we developed our general formula for "n" events and "m" hypotheses and also retained in our development of the tabular array format and of the odds-likelihood formula.

It should be noted at this point that the conditionally independent formulas normally are the ones developed in the literature on Bayes. What, then, would these formulas be? The answer can be given rather simply if we notice that the branch probabilities of Figure 2.10 would simplify to those shown here in Figure 2.13 if we have conditional independence of the evidence for all the hypotheses.

FIGURE 2.13
Probability Tree Diagram for Two Hypotheses and "n" Events, when we have Conditional Independence

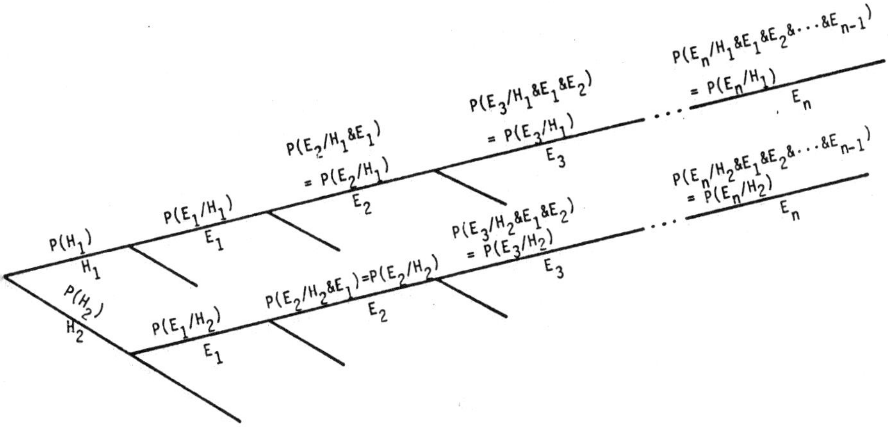

As a result of this, the general formula for "n" events and two hypotheses would simplify to

$$P(H_i/E_1\&E_2\&\cdots\&E_n) = \frac{P(H_i)\cdot P(E_1/H_i)\cdot P(E_2/H_i)\cdots P(E_n/H_i)}{P(H_1)\cdot P(E_1/H_1)\cdot P(E_2/H_1)\cdots P(E_n/H_1) + P(H_2)\cdot P(E_1/H_2)\cdot P(E_2/H_2)\cdots P(E_n/H_2)}.$$

Further, the general formula for "n" events and "m" hypotheses becomes, if we have conditional independence,

$$P(H_i/E_1\&E_2\&\cdots\&E_n) = \frac{P(H_i)\cdot P(E_1/H_i)\cdot P(E_2/H_i)\cdots P(E_n/H_i)}{\begin{array}{l} P(H_1)\cdot P(E_1/H_1)\cdot P(E_2/H_1)\cdots P(E_n/H_1) \\ +P(H_2)\cdot P(E_1/H_2)\cdot P(E_2/H_2)\cdots P(E_n/H_2) \\ \vdots \\ +P(H_m)\cdot P(E_1/H_m)\cdot P(E_2/H_m)\cdots P(E_n/H_m) \end{array}}.$$

Also, the headings in the tabular array approach become, if we have conditional independence, the following:

j	$P(H_j)$	$P(E_1/H_j)$	$\cdots$	$P(E_n/H_j)$	Row Product	Revised Probability of H_j

Finally, the odds-likelihood form of Bayes becomes, if we have conditional independence,

$$\frac{P(H_1/E_1\&E_2\&\cdots\&E_n)}{P(H_2/E_1\&E_2\&\cdots\&E_n)} = \frac{P(H_1)}{P(H_2)} \cdot \frac{P(E_1/H_1)}{P(E_1/H_2)} \cdot \frac{P(E_2/H_1)}{P(E_2/H_2)} \cdots \frac{P(E_n/H_1)}{P(E_n/H_2)}.$$

We will see in the fifth chapter that there are very important problems involved in the choice of whether to use the conditionally independent or dependent formulas (or tabular arrays).

Presenting the Results of a Bayesian Analysis

Various sorts of tabular presentations can be used to present the results of a Bayesian analysis, i.e., to present the revised probabilities of hypotheses. One way of doing this for the example of Tables 2.1 through 2.5 is shown in Table 2.6.

TABLE 2.6 Initial or Revised Probabilities of Hypotheses after "n" Items of Evidence

Number of Items of Evidence	Initial or Revised Probabilities of: H_1	H_2	H_3	H_4
none	0.4	0.2	0.1	0.3
E_1	0.464	0.261	0.058	0.217
$E_1\&E_2$	0.231	0.520	0.087	0.162
$E_1\&E_2\&\cdots\&E_4$	0.124	0.652	0.005	0.218
$E_1\&E_2\&\cdots\&E_6$	0.035	0.548	0.005	0.413

In addition, the analyst <u>always</u> should present the following type of graph to the decision makers (Figure 2.14). The latter should find the graph to be extremely helpful in seeing what the trends are.

FIGURE 2.14
Graphical Presentation of Initial and Revised Probabilities of Hypotheses

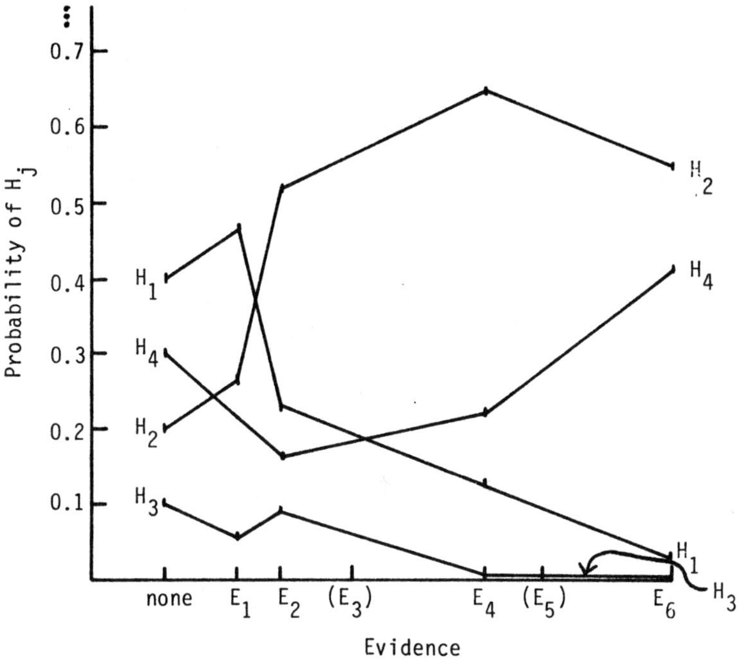

NOTES

1. The "state of nature," as used here, is a standard term in decision theory and simply means the true state of reality with respect to some issue. Used in the plural, the term "states of nature" refers to the mutually exclusive and exhaustive set of possibilities with respect to some issue. As we will note below, the hypotheses in a particular Bayesian analysis are mutually exclusive and exhaustive and actually specify the possible "states of nature" concerning some issue. Using this language, the point of a Bayesian analysis is to discover (through the processing of items of evidence) which one of the specified "states of nature" (i.e., hypotheses) really is the (one and only true) "state of nature," i.e., which represents the truth concerning the particular issue.

2. As we will see in the discussion of problem 3 in chapter four, there are some "buried" subhypotheses in H_2. For our purposes here, however, we can ignore this issue.

3. Dr. Hopple has pointed out to me in his written comments on this manuscript that in many cases an analyst's intuitive revision

probably would not equal the Bayesian revision, and that this is exactly what we want - i.e., comparative probabilities. We want, he notes, to use Bayesian analysis as an independent and systematic check on our usual analysis.

4. Suppose we have just one item of evidence, E. Then the formula would be

$$P(H_i/E) = \frac{P(H_i)\cdot P(E/H_i)}{\begin{array}{l} P(H_1)\cdot P(E/H_1) \\ + P(H_2)\cdot P(E/H_2) \\ \vdots \\ + P(H_m)\cdot P(E/H_m) \end{array}} = \frac{P(H_i)\cdot P(E/H_i)}{\sum_{i=1}^{m} P(H_i)\cdot P(E/H_i)}$$

where we use the sigma "$\sum$" to mean that we add up all the terms in the denominator. Nicholas Schweitzer has an extremely lucid interpretation of this equation, which is that

> E is an event, an "item" of intelligence;
> H is an hypothesis, a hypothetical cause of events;
> Hi is one of a set of [m] mutually exclusive hypotheses;
> P(Hi) is the starting, or "prior," probability of an hypothesis;
> P(E/Hi) is the probability of an event given Hi, [i.e.,] of an event occurring, given a particular underlying cause;
> P(Hi/E) is the probability of an hypothesis given E, [i.e.,] the "revised" probability of an hypothesis, given that a particular event has occurred.
>
> In words, it says that,
>
> given an analyst's starting probabilities P(Hi) -- his feeling for the likelihoods of a set of more or less mutually exclusive hypotheses, and
> given his assessments P(E/Hi) of how likely an event would be if each of the hypotheses were true, then
> the updated version of the hypotheses themselves P(Hi/E) can be calculated in a straightforward fashion.

SCHWEITZER, NICHOLAS (1976) "Bayesian Analyses for Intelligence: Some Focus on the Middle East." A paper presented at the 17th Annual Convention of the International Studies Association, 28 February, Appendix 1, p. i. This important paper has appeared completely or in parts in several other publications. See SCHWEITZER, NICHOLAS (1978) "Bayesian Analysis: Estimating the Probability of Middle East Conflict," pp. 11-30 in Richards J. Heuer, Jr., (ed.) Quantitative Approaches to Political Intelligence: The CIA Experience. Boulder, Colorado: Westview Press.

3
Major Advantages of Bayesian Analysis

In this chapter we will present a number of the major advantages of using Bayesian analysis. The treatment is not exhaustive of all the advantages, but does touch on most of the major ones.

1. The first advantage is that the use of Bayesian analysis overcomes the conservative bias it is said that we supposedly have regarding our revision of initial probabilities in the light of evidence. Many laboratory experiments, where there is a mathematically correct solution, have shown that we do not revise our initial estimates of probabilities as much as we should - up or down - given the evidence we have received.[1] If Bayesian analysis is used, however, the revisions are much closer to what they should be. That is, when Bayesian analysis is used, the results are non-conservative and ahead of the intuitively (subjectively) derived results of the experiments in the sense that if $P(H_i/\text{evidence})$ is increasing, the increase that Bayes will yield will be greater than what we intuitively assign. On the other hand, if $P(H_i/\text{evidence})$ is decreasing, the decrease that Bayes will yield will be greater than what we intuitively assign. The Bayesian revisions in either direction are larger in magnitude and more correct than those we make intuitively. Many researchers feel that the Bayesian procedure, which yields these more correct results, also will yield more correct results if applied to political/military analyses (where there is no mathematically correct solution, however). It is felt that we (analysts) are (incorrectly) conservative, and that Bayes will overcome this conservatism. This view is not universally held.[2] Dr. Hopple notes that it may be that in political/military analyses, where there is no known, mathematically correct solution, the Bayesian results are just different. Richards J. Heuer argues cogently that

> we cannot say, *a priori*, that the Bayesian result is "better" than the intuitive conclusion; there are good and sound reasons for using Bayes and the odds favor the Bayesian conclusion being more accurate than the unaided inituitive

judgment, but there are some circumstances under which the Bayesian conclusion will be worse -- not better -- than unaided intuitive judgment.[3]

2. A second and closely related advantage is that "more information can be extracted from the [body of] available data because the technique allows [calls for] each piece of evidence, central or marginal [in terms of diagnostic probabilities] to add its weight to the final assessment in a systematic way."[4] Thus, a number of items with low diagnostic value can outweigh one item with a large diagnostic value.[5] (An item is diagnostic if it is considered more likely to occur for some hypotheses than for others. The greater the ratio between the probabilities of occurrence for different hypotheses, the greater the degree of diagnosticity.)

3. A third advantage is that use of the technique compels us to employ an improved system of accounting for the evidence used. That is, we are forced to list the events, the evidence, that we are considering.[6]

The accounting aspect helps reduce arguments among analysts that they have not considered all the evidence. Indeed, common, expanded lists of events can be constructed from the separate lists of different analysts and considered. If such a master list is prepared, arguments by analysts that each has overlooked relevant events tend to disappear. Thus, a significant class of arguments is eliminated.

4. A fourth, and related advantage, is that if different analysts reach different conclusions, the source of the disagreements can be determined much more easily than in normal, verbal analysis.[7] This is because the analysts have had to assess how important the listed events are by their assignment of conditional probabilities, and should have identified the key factors influencing their conditional probability judgments. The reasoning behind the assessment of the probabilities should be available for critical review by others approaching the question from varying angles. We easily can determine the relative importance given by different analysts to any given event by examining the conditional probabilities that they assign to that event.

5. The fifth advantage is that using Bayes forces us to consider alternative hypotheses. The formulation of the questions forces us to consider alternative explanations of the facts we see, "thus loosening the bonds of established opinions."[8] In other words, the analyst "is asked to look at how well the evidence explains [possible] hypotheses [or scenarios] other than the one he [may, indeed probably, already] has decided is most likely."[9]

The example, if an analyst is convinced that country A is not planning on attacking country B in the near future, the use of Bayes will force that analyst to consider the hypothesis that an attack is planned, and to evaluate the evidence in light of this hypothesis, too.

The intelligence community in particular has been criticized severely in the past for not considering alternative hypotheses or explanations for observed evidence. Intelligence analysts simply cannot use Bayes without considering alternative hypotheses.

6. A sixth and closely related advantage is the decoupling of the analyst's ego from the probability figures assigned. After our initial assignment of probabilities to the hypotheses, we are not defending the hypotheses, but are assessing the probability of events given (that is, assuming) that the hypotheses are true. We do not take the available evidence as given, and draw from it our conclusions about the relative merits of the opposing hypotheses, as is done in normal verbal analyses.

That is, we do not directly assess $P(H_i/\text{evidence})$, an assessment which is subject to psychological biases which tend to lead to the selection and interpretation of evidence to support the hypothesis in which we believe. Instead, when the Bayesian procedure is used, we postulate by turn the truth of each hypothesis, and address only the likelihood that each item of evidence would appear, first under the assumption that one hypothesis about the state of nature is true, and then under the assumption that another hypothesis about the state of nature is true, etc. We are free of any ego-supporting need to hold to positions previously taken on the merits of the respective hypotheses. That is, since we are not directly concerned with assessing the probability of the hypotheses in the light of new evidence, we do not feel called upon to reinforce our self-esteem by reaffirmation of opinions about the probability of the hypotheses initially put on the record. Since the procedures for estimation require us to postulate, not debate, the truth of opposing hypotheses when assessing $P(E_i/H_j\&E_1\&E_2\&\cdots\&E_{i-1})$, Bayes's theorem helps us to get around one of our most troublesome pitfalls - the human tendency to hold fast to prior estimates when uncommitted opinion would go along with a change.[10]

Let me say at this time that I don't completely agree with this argument. The analyst is aware of the direction of influence of the estimates (s)he makes regarding $P(E_1/H_1)$, $P(E_1/H_2)$, etc. on $P(H_1/E_1)$. If (s)he supports H_1, and subconsciously wants to discount a piece of data E_1, (s)he simply has to say that $P(E_1/H_1) = P(E_1/H_2) = \cdots$ and the data has no effect. If (s)he supports H_1 and feels that the evidence does, (s)he can make the process reflect this belief simply by giving a higher value to $P(E_1/H_1)$ than to $P(E_1/H_2)$, $P(E_1/H_3)$, etc. That is, if (s)he subconsciously wants to manipulate the process to support his (her) previous belief in the probability of an hypotheses, there is a way to try to do so. Hopefully, (s)he will be aware of his (her) biases. Even if (s)he tries to be, if (s)he feels $P(H_1)$ is high, (s)he is more likely to feel that the E_1 that has occurred was indeed likely to occur, and to say that $P(E_1/H_1)$ is higher than $P(E_1/H_2)$,

$P(E_1/H_3)$, etc.

Fortunately, the fact that an analyst's judgments are open to the questions of other analysts will tend to somewhat undermine subconscious manipulations. Perhaps someday, it will be the case that interaction or decision making models will become more sophisticated and provide us with correlation coefficients (which can be converted to probabilities), particularly for such situations as the processes leading to wars. We then would have more objective rather than subjective, estimates for $P(E_1/H_1)$, $P(E_1/H_2)$, etc.

7. The seventh advantage is that we are making deductive judgments when we assess $P(E_i/H_j \& E_1 \& E_2 \& \cdots \& E_{i-1})$, and deductive judgments are <u>easier</u> to make than inductive judgments like $P(H_j/E_1 \& E_2 \& \cdots \& E_n)$. The Bayesian procedure, in other words, does not ask us to make a direct inductive judgment as to the revised probability of a hypothesis in the light of the evidence, i.e., $P(H_j/E_1 \& E_2 \& \cdots \& E_n)$ but instead to make a series of deductive judgments as to the likelihood of a given datum assuming, in turn, various hypotheses to be true. We are not, that is, asked to assess $P(H_j/E_1 \& E_2 \& \cdots \& E_n)$ directly in an inductive fashion. Instead, we deductively assess the elements of the right hand side of the Bayes equation, i.e., we deductively assess $P(E_i/H_j \& E_1 \& E_2 \& \cdots \& E_{i-1})$ for each E_i and all H_j. The mathematics of the Bayes equation correctly combines all the deductive judgments to yield a (mathematically) correct inductive judgment.[11] (The judgment is still subjective, of course, if the deductive judgments are subjective, as they usually are.)

(An interesting question to research would be this: if very good analysts are very good at induction (which almost certainly is true), will they reach good judgments using a Bayesian procedure which calls them to be good deductive analysts? Perhaps there is a negative correlation between being good at induction and good at deduction. Perhaps poor analysts (if they are poor at induction) would be better served by Bayes if they are good at deduction. Reader comments are welcome on this issue.)[12]

8. An eighth advantage is that we are required to quantify, i.e., make explicit, and get away from the ambiguity of words, judgments which we do not ordinarily express in numerical terms.[13] This requirement to quantify draws fire from some critics who say that analysts will disagree in their opinion of the exact figures that should represent the diagnostic value of an item of evidence, i.e., figures for $P(E_i/H_j \& E_1 \& E_2 \& \cdots \& E_{i-1})$. The Bayesian rebuttal is that disagreement among analysts is just as much a characteristic of the traditional method and is no less serious for being traditionally implicit rather than explicit in the Bayesian analysis. Words simply are terribly ambiguous. There is no uniform agreement as to what is meant by such words as "likely," "probable," etc. When we assign our probabilities we are, as Dr. Gordon T. Shahin has put it, being precise about the

degree of our uncertainty.[14]

9. The ninth advantage is that the revised probabilities of the hypotheses can change very quickly. This advantage is a major one. What is being said here is that, not only is Bayes not conservative with regard to the revision of probabilities (i.e., not only is it the case that using Bayes results in larger increases or decreases in probabilities than intuitive revisions yield), it also is the case that Bayes can lead to radically revised probabilities on the basis of relatively little data. That is, the revised probabilities can "quickly" change substantially - quickly in the sense of few items of data. Bayes is, in short, highly sensitive to the input data. (We are, then, talking about how "fast" - in terms of input data - the conservative bias can be overcome and we are saying that it can be overcome very quickly. The "rate of change" can be high.)

Several numerical examples may make this advantage of Bayes clear. We will do two examples, each of which involves two hypotheses. We will employ the odds-likelihood form of Bayes.

Example 1. Suppose our prior odds are $\frac{0.5}{0.5}$ and we have six events. Assume further that for each of the six events the likelihood ratio is $\frac{0.55}{0.45}$. Then, our revised odds are

$$\frac{P(H_1/E_1\&E_2\&\cdots\&E_6)}{P(H_2/E_1\&E_2\&\cdots\&E_6)} = \frac{0.5}{0.5}\cdot\frac{0.55}{0.45}\cdot\frac{0.55}{0.45}\cdot\frac{0.55}{0.45}\cdot\frac{0.55}{0.45}\cdot\frac{0.55}{0.45}\cdot\frac{0.55}{0.45}$$

$$= \frac{0.0138403}{0.0041518}.$$

$$\text{So } P(H_1/E_1\&E_2\&\cdots\&E_6) = \frac{0.0138403}{0.0138403+0.0041518} = \frac{0.0138403}{0.0179921}$$

$$= 0.77.$$

Thus, on the basis of relatively little input data (that is, six items of evidence), we see a rather substantial revision of the probability of H_1. The probability has gone from 0.5 to almost 0.8. Note also that each of the few items of evidence that led to this substantial change was in itself not particularly diagnostic. That is, the likelihood ratios were $\frac{0.55}{0.45}$ each, which means that each item of evidence was only slightly more likely (i.e., 0.1) to occur if H_1 was assumed to be true than if H_2 was assumed to be true.

Example 2. A Middle East scenario. Suppose that we consider the situation prior to the 1973 Arab-Israeli war. From the viewpoint of an Israeli intelligence officer, the prior probabilities of war and no war might have been 0.1 and 0.9 respectively. Suppose an event E_1 occurred and was assessed as follows: $P(E_1/War) = 0.95$, $P(E_1/no\ War) = 0.05$. E_1 is a highly diagnostic event, then, in the sense that it is considered to be much more likely to occur assuming one of the hypotheses to be true than it is to occur assuming the other hypothesis to be true. Using these assessments the revised odds are

$$\frac{P(War/E_1)}{P(no\ War/E_1)} = \frac{0.1}{0.9} \cdot \frac{0.95}{0.05} = \frac{0.095}{0.045}.$$

$$\text{So } P(War/E_1) = \frac{0.095}{0.095+0.045} = \frac{0.095}{0.140} = 0.68.$$

Thus, the probability of war has been revised from 0.1 to almost 0.7 on the basis of only one item of evidence.

The fact that Bayes can "blow up" on us very quickly, i.e., that the revised probability of a hypothesis can go up or down quickly, can be seen to be a very major advantage. That is, if we have good input data in the sense that we have evidence that is diagnostic (even if it is not highly diagnostic) then we can expect to find only a little data leading to substantially revised probabilities -- well in advance of our intuitive revisions.

However, given this high sensitivity to diagnostic evidence, it cannot be overemphasized that we must have very good reasons for assigning different conditional probabilities for a given event. That is, when we consider $P(E_i/H_1\&E_1\&E_2\&\cdots\&E_{i-1})$ and $P(E_i/H_2\&E_1\&E_2\&\cdots\&E_{i-1})$ for a particular event E_i, which is just considering respectively how likely E_i is to occur assuming H_1 and the previous items of evidence to be true, and how likely E_i is to occur assuming H_2 and the previous items of evidence to be true, we must be able to justify assigning different numbers for these conditional probabilities. As we have seen, even minor differences in the assigned numbers can have a dramatic impact. We must have very good reasons indeed if we are to assign dramatically different values such as those in the second example.

While the examples specifically, and the previous paragraph implicitly, deal with the odds-likelihood form of Bayes, all of our remarks apply to the multiple hypotheses case and we could give numerical examples to show how, say, hypothesis 3 in a multiple-hypotheses case could quickly rise or fall.

The moral of all this is as follows: If we can back up our assignment of our conditional probabilities, diagnostic evidence can provide reliable early warning by <u>quickly</u> yielding revised probabilities that are much larger or smaller than we would assess intuitively.[15]

Other advantages could be listed,[16] but I think that most of the major ones have been discussed.

NOTES

1. EDWARDS, WARD (1966) "Nonconservative Probabilistic Information Processing Systems." Decision Sciences Laboratory, Electronic Systems Division, Air Force Systems Command Technical Documentary Report No. ESD-TR-66-404, Dec., Bedford, Mass. (This is only one of many reports and articles by Professor Edwards.) See also SCHWEITZER (1978), p. 24.

2. WINKLER, ROBERT L. and MURPHY, ALLAN H. (1973) "Experiments in the Laboratory and the Real World." Organizational Behavior and Human Performance 10:252-270.

3. HEUER, RICHARDS J., JR. (1981, p. 101) "Applications of Bayesian Inference in Political Intelligence," pp. 93-121 in Gerald W. Hopple and James A. Kuhlman Expert-Generated Data: Applications in International Affairs. Boulder, Colorado: Westview Press. (Indeed, the reader is urged to examine his entire discussion in pp. 98-103 and his footnote 4.)

4. SCHWEITZER (1978), p. 24. Underlining added.

5. _Ibid_.

6. _Ibid_.

7. _Ibid_.

8. _Ibid_.

9. _Ibid_.

10. Anonymous (1974) "Bayes for Political Analysts." Central Intelligence Agency, Office of Current Intelligence, OCI Notes on Methodology No. 24, March 22, Langley, Va. See also JERVIS, ROBERT (1976) Perception and Misperception in International Politics. Princeton, N.J.: Princeton University Press.

11. SCHWEITZER (1978), p. 24.

12. Dr. Hopple notes that Bayesian advocates/users do not appear to have researched questions like this empirically. He also notes that some contractor's "decision aids seem to show no awareness that analysts A and B could come up with very different conclusions with the same aid - like the failure to even compare analysts' intuitive judgments - via interjudge reliability coefficients."

13. SCHWEITZER (1978), p. 24.

14. A point usually made by him in his lectures for the government on probability theory and/or Bayesian analysis.

Dr. Hopple notes that Sherman Kent prepared a normative chart which was intended to provide numerical probability equivalents for words like "almost certain," "probable," etc. (Reported on in Platt, Washington (1962) Strategic Intelligence Production. New York: Frederick A. Praeger, pp 207-208.)

This idea was an excellent one. As Dr. Hopple points out, the essence of much intelligence analysis is estimates, and estimates almost invariably are probabilistic in nature. The probablistic predictions, however, still normally are delivered with words, rather than with numbers. There has been no effective way of compelling intelligence writer's to use Kent's or anyone else's scale. The meaning of words such as "probably" or "likely" normally therefore is known only to the authors. (Maybe the authors don't know themselves. A not-so-funny case in the U.S. intelligence community involved two co-authors of a report who had used the same (probabilistic) word, but later found their numerical probability estimates were greatly at variance. This use of words, rather than numbers, could be quite dangerous, really, because the consumer will interpret words according to his own scale of meanings and probably (P 0.9) will have different numerical equivalents than the authors.)

15. Dr. Hopple points out that the results of the Bayesian analysis are "useful even if 'wrong' - as a contrast to intuitive judgments." He also notes that two methods implying a conclusion always are better than one method implying a conclusion.

16. See HEUER (1981), pp. 104-107 for more discussion of advantages of normal (non-hierarchical that is) Bayes. Dr. Hopple points out that we might consider the fact that computerized aids are reliable, cheap, and readily available to be another advantage. "Basic Bayes can be programmed on any system as a trival matter," he notes. I feel that with a hand calculator, any individual should be able to use the tabular array approach quickly.

4
Problems Which Arise at the Beginning of a Bayesian Analysis and Which Involve Hypotheses

From this point on, the chapters of this book will deal with many of the methodological problems involved with Bayesian analysis. Whenever possible, solutions will be offered for the problems presented.

This chapter will discuss problems involving hypotheses which arise at the beginning, when we start to carry out a Bayesian analysis.

<u>PROBLEM 1</u>. Hypotheses not mutually exclusive and/or not exhaustive: the problem of hypotheses which are defective because they are not mutually exclusive and/or not exhaustive.

<u>The problem</u>: Although the hypotheses that are used in Bayesian analysis must be mutually exclusive and collectively exhaustive, it is quite easy to find actual analyses where one or both of these requirements is violated. The resulting analyses are therefore quite incorrect - the evidence will not be considered in light of correct hypotheses.

<u>The solution</u>: The use of a tree diagram in developing the hypotheses generally should eliminate this problem.

When reading someone else's analysis, check to see if his or her hypotheses are mutually exclusive by asking if more than one hypothesis logically can be true at a time. If such is the case, then the hypotheses are <u>not</u> mutually exclusive.

To check for exhaustivity, ask if there is anything that logically could happen that is not included in the list of hypotheses. If such is the case, the hypotheses are not exhaustive.

If the hypotheses are complex, try putting them into a tree diagram format as an aid in checking on whether or not they are mutually exclusive and collectively exhaustive.

<u>Discussion</u>: As has been noted before, the hypotheses in Bayesian analyses must be both mutually exclusive and collectively exhaustive. A set of hypotheses will be defective if one or both of the above conditions is violated. Perhaps the more common error of the two is to have hypotheses that are not mutually

exclusive. It is quite easy to find actual analyses where one or both of these requirements is violated.

An example of this sort of problem is the following. Suppose we have developed these hypotheses:

H_1 = country A is using nerve agent GB in country B,

H_2 = country A is using nerve agent VX in country B,

H_3 = country A is using mustard gas in country B,

H_4 = country A is not using poisonous gases in country B.

After specifying the set of hypotheses, the first question to ask is this: "Are these hypotheses mutually exclusive? That is, can only one be true at a time?" Clearly, looking at the above set of hypotheses, the answer is "No!" H_1 and H_2 can be true at the same time, H_2 and H_3 can be true at the same time, and H_1 and H_3 can be true at the same time. So, the list is <u>not</u> made up of mutually exclusive hypotheses.[1] If we want to consider the possibility of various combinations of gases being used, the list can be fixed up by replacing it with the following list:

H_1 = country A is using GB and VX and mustard in country B,

H_2 = country A is using GB and VX and <u>no</u> mustard in country B,

H_3 = country A is using GB and <u>no</u> VX and mustard in country B,

H_4 = country A is using GB and <u>no</u> VX and <u>no</u> mustard in country B,

H_5 = country A is using <u>no</u> GB, but VX and mustard in country B,

H_6 = country A is using <u>no</u> GB, but VX and <u>no</u> mustard in country B,

H_7 = country A is using <u>no</u> GB and <u>no</u> VX, but mustard in country B,

H_8 = country A is using <u>no</u> GB and <u>no</u> VX and <u>no</u> mustard in country B.

We get such a list of mutually exclusive hypotheses from a simple tree like the one in Figure 4.1. The list still is not as complete as possible because, although it is exhaustive with respect to the usage of GB, VX, and mustard, it is not exhaustive with respect to the use of gases in country B. They could be using other poisonous gas(es) (call this "OP"), non-poisonous gas(es) (call this non-P), or no gases at all. To make the hypotheses on the list both mutually exclusive and exhaustive we first should put the possibilities on a tree (Figure 4.2). The last, bottom pathway (number 32) in the tree is the one of no gases being used at all. Pathway 31 is the pathway of only non-poisonous gas being used. On any other pathway, at least one sort of poisonous gas is being used.

FIGURE 4.1
Tree Diagram for Logical Possibilities of Country A's Use of Gases in Country B.

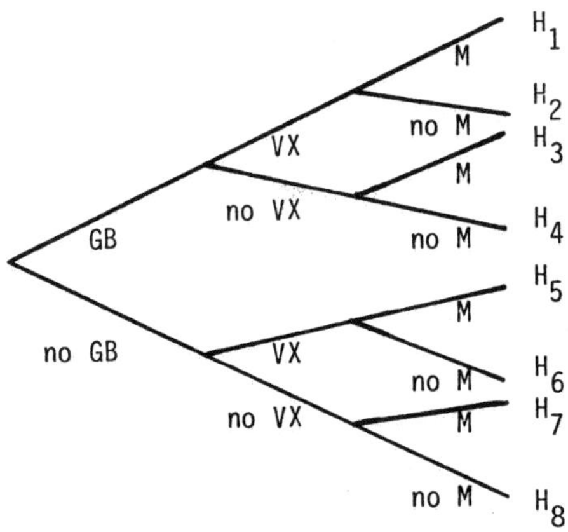

The tree has gotten terribly complicated, and yields one hypothesis for each pathway taken as a whole. The hypotheses, however, now are mutually exclusive and exhaustive with respect to the possibility of gases being used. The price we pay for such a large tree is having to ask the conditionally independent probabilities $P(E_i/H_j)$, j = 1-32 for each E_i (or the conditionally dependent probabilities $P(E_i/H_j\&E_1\&E_2\&\cdots\&E_{i-1})$, j = 1-32 for each E_i). We can simplify somewhat, depending on the questions we want to answer. The huge tree might be useful in pinning down the type of poisonous gas(es), if any, being used. If it is a question of (a) poisonous gas(es) or (b) non-poisonous gas(es) or (c) no gases, a simpler tree (to be quickly modified) like Figure 4.3 would be more useful.

FIGURE 4.2
Tree Showing Possible Combinations of Use of Gases

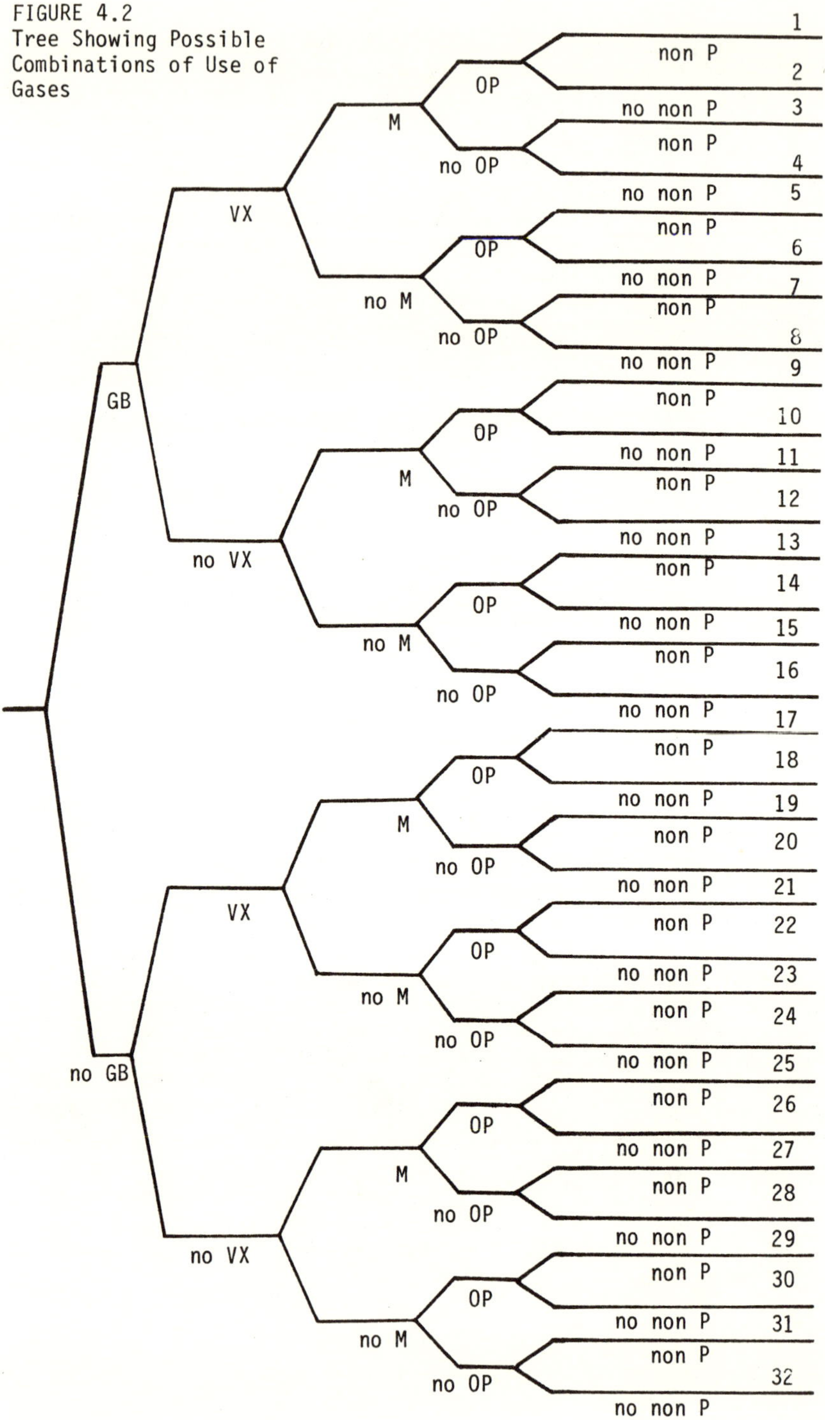

FIGURE 4.3
"Working" Tree of Logical Possibilities of Country A's Use of Poisonous, Non-Poisonous or No Gases in Country B.

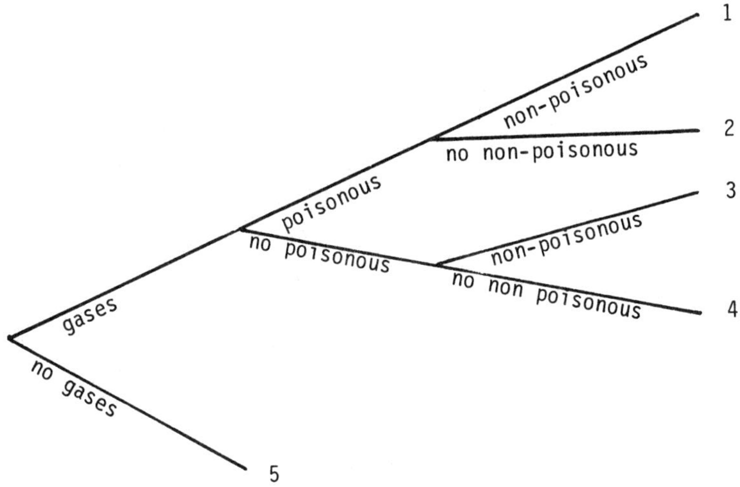

Clearly, this "working" tree needs to be revised because the fourth path of gases and no poisonous and no non-poisonous gases is contradictory. Properly constructed, the tree is the one shown as Figure 4.4.

FIGURE 4.4
Final Tree for Logical Possibilities of Country A's Use of Poisonous, Non-Poisonous or No Gases in Country B.

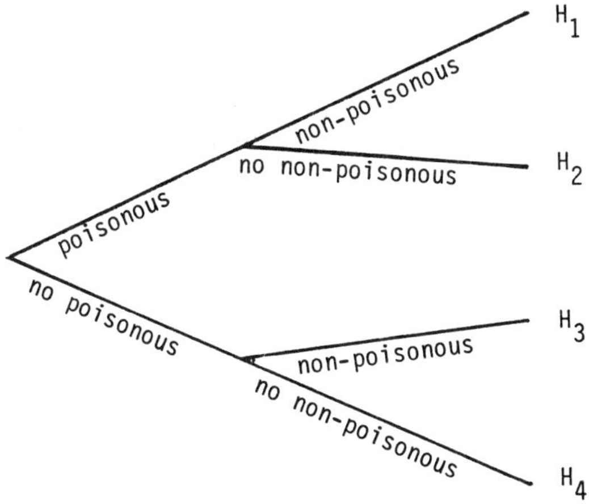

Thus, we get 4 mutually exclusive and collectively exhaustive hypotheses, instead of 32, which are

H_1 = both poisonous and non-poisonous,

H_2 = poisonous, but no non-poisonous,

H_3 = no poisonous, but non-poisonous,

H_4 = no gas at all (no poisonous, no non-poisonous).

If we use trees to formulate the hypotheses in Bayes, we are far more likely to end up with a properly constructed list. Using a tree to check on the hypotheses developed by others who employ Bayes can be useful as an aid in checking on their being mutually exclusive and collectively exhaustive.

PROBLEM 2. A recognizably defective residual hypothesis: the problem of a final hypothesis which is recognizably too inclusive, meaningless, and "catch-all."

The problem: The last, i.e., residual hypothesis, in a list of hypotheses may clearly (to us) be too inclusive, and hence meaningless, thus making it difficult, if not impossible, for us to assign the conditional probabilities $P(E_i/H_{residual}$ and $E_1\&E_2\&E_3\&\cdots\&E_{i-1})$ or $P(E_i/H_{residual})$.

The solution: Replace the residual hypothesis by a number of more specific hypotheses so that the breadth and diversity of the contents of the new residual hypothesis are greatly reduced. (Subdividing the residual hypothesis technically does the same thing.)

Discussion: The last, or residual, hypothesis in our listing may clearly (to us) be so broad in scope as to be meaningless, making it difficult, if not impossible, to assign the conditional probabilities to $P(E_i/H_{residual}$ & $E_1\&E_2\&E_3\&\cdots\&E_{i-1})$ (or $P(E_i/H_{residual})$). This is likely to be particularly true in the case of two hypotheses. Suppose, for example, that the hypotheses are formulated as "threat" and "no threat." The latter hypothesis encompasses such a wide variety of possible options or policies or scenarios that it is difficult, if not impossible, for a given event E_i to assess the conditionally dependent probability $P(E_i$/"no threat" & $E_1\&E_2\&E_3\&\cdots\&E_{i-1})$(or the conditionally independent probability $P(E_i$/"no threat")).

There are several possible ways of alleviating this problem. The first is to replace the residual hypothesis by a number of more specific hypotheses so that the breadth and diversity of the contents of the new residual hypothesis are greatly reduced. We can go from two to five hypotheses, for example.

Another solution, which technically does the same thing, but yields a result for the original residual hypothesis, is to subdivide the final hypothesis into subhypotheses.

As an example of the two solutions, suppose we have three mutually exclusive and exhaustive hypotheses, H_1, H_2, and H_3, and we decide that H_3 is just too inclusive and meaningless to be used. The first choice of what to do is to split up H_3. Suppose we decide to split it into three mutually exclusive (and exhaustive) parts: (new) H_3, H_4, and H_5. If the resulting H_5 is now a meaningful (new) residual hypothesis, our efforts are worthwhile. Then we end up with five mutually exclusive and exhaustive hypotheses:

(old) H_1,
(old) H_2,
(new) H_3,
H_4,
H_5.

Alternatively, if we want to keep H_3, but still recognize its lack of meaningfulness, we can consider splitting it into mutually exclusive and exhaustive subhypotheses H_{31}, H_{32}, H_{33}, where H_{31} corresponds to the (new) H_3, H_{32} corresponds to H_4, and H_{33} corresponds to H_5.

To make the above more concrete, suppose the initial values assigned were $P(H_1) = 0.2$, $P(H_2) = 0.1$, $P(H_3) = 0.7$. Suppose further that, when H_3 is felt to be too inclusive and meaningless and is split into three mutually exclusive and exhaustive parts (new) H_3, H_4, and H_5, the 0.7 probability for H_3 is allocated as follows: $P((\text{new})\ H_3) = 0.4$, $P(H_4) = 0.1$, $P(H_5) = 0.2$. Alternatively, as noted, we can think of H_3 as being split into subhypotheses of H_{31}, H_{32}, H_{33}, where H_{31} corresponds to (new) H_3, H_{32} corresponds to H_4, and H_{33} corresponds to H_5. Then we have

$P(H_1) = 0.2$	or	$P(H_1) = 0.2$
$P(H_2) = 0.1$		$P(H_2) = 0.1$
$P((\text{new})\ H_3) = 0.4$		$P(H_{31}) = 0.4$
$P(H_4) = 0.1$		$P(H_{32}) = 0.1$
$P(H_5) = 0.2$		$P(H_{33}) = 0.2$.

Suppose next that E_1 occurs, and that these assignments are made

$P(E_1/H_1) = 0.8$	or	$P(E_1/H_1) = 0.8$
$P(E_1/H_2) = 0.6$		$P(E_1/H_2) = 0.6$
$P(E_1/(\text{new})\ H_3) = 0.9$		$P(E_1/H_{31}) = 0.9$
$P(E_1/H_4) = 0.3$		$P(E_1/H_{32}) = 0.3$

$P(E_1/H_5) = 0.5$ $\qquad$ $P(E_1/H_{33}) = 0.5.$

Using a tabular array for the new set of 5 hypotheses we have the following (Table 4.1).

TABLE 4.1
Assignment of Probabilities (to Hypotheses and for One Event) and Revised Probabilities of Hypotheses

j	$P(H_j)$	$P(E_1/H_j)$	Row Product	Revised Probability of H_j
1	0.2	0.8	0.16	0.23
2	0.1	0.6	0.06	0.08
3	0.4	0.9	0.36	0.51
4	0.1	0.3	0.03	0.04
5	0.2	0.5	0.10	0.14
	1.0		0.71	1.00

If we are thinking in terms of subhypotheses, we would have instead Table 4.2.

TABLE 4.2
Assignment of Probabilities (to Hypotheses and Subhypotheses and for One Event) and Revised Probabilities of Hypotheses

j	$P(H_j)$	$P(E_1/H_j)$	Row Product	Revised Probability of H_j	
1	0.2	0.8	0.16	0.23	
2	0.1	0.6	0.06	0.08	
31	0.4	0.9	0.36	0.51	
32	0.1	0.3	0.03	0.04	0.69 for H_3 (31–33)
33	0.2	0.5	0.10	0.14	
	1.0		0.71	1.00	

So, the revised probability for the residual hypothesis H_3 is 0.69, a result in which we can have some faith since it was calculated on the basis of analyzing its meaningful subhypotheses.

PROBLEM 3. "Buried" subhypotheses: the problem of unrecognized subhypotheses in a hypothesis (often, but not always, the residual hypothesis), which render the hypothesis somewhat meaningless.

The problem: This problem is closely related to the previous problem, but differs in that (1) the hypothesis in question is not always the residual hypothesis, and (2) the hypothesis, because its "buried" subhypotheses are unrecognized, is itself not recognized as being somewhat meaningless. Use of such a hypothesis then results in probability judgments which are somewhat misleading and revised probability judgments which are not recognized as being somewhat inaccurate. This problem often is associated with hypotheses referring to the possible decisions which have been reached by an actor. It also is found, however, in other types of hypotheses.

The solution: When our hypotheses deal with the possible decisions which have been reached by an actor, we should consider using hypothesis sets (without hedging options) of five or four hypotheses. These expanded hypothesis sets will alleviate the "buried" hypothesis problem to a great extent by making explicit the major, previously "buried," subhypotheses. However, there always will be "buried" subhypotheses in the sense of additional conditions which might or might not be the case for any given hypothesis, so the problem ultimately is insolvable. All we can do is alleviate the difficulty.

When our hypotheses deal with future decisions (what an actor, in the future, will decide to do), or with non-decisional situations, a tree diagram can be used to develop expanded hypotheses which will make explicit previously buried subhypotheses. This process will alleviate the problem.

Discussion: This problem is closely related to the previous problem, but differs, as noted, in that (1) the hypothesis in question is not always the residual hypothesis, and (2) the hypothesis, because its "buried" subhypotheses are unrecognized, is itself not recognized as being somewhat meaningless. Use of such a hypothesis then results in probability judgments which are somewhat misleading and revised probability judgments which are not recognized as being somewhat inaccurate. This appears to be one of the most common problems in Bayesian analysis.

This problem often is associated with hypotheses referring to the possible decisions which have been reached by an actor.

A good example is the following:

H_1 = the Soviet Union has "decided to invade Poland within one month" (where within means up to and including one month),

H_2 = the Soviet Union has not "decided to invade Poland within one month."

Since the hypotheses are of the form "H" and "not H," they are fine technically in that they are mutually exclusive and exhaustive.

However, there are at least four subhypotheses buried in H_2 and they are very different in content. They are as follows:

(1) "the Soviet Union has decided to invade Poland sometime after one month;"
(2) "The Soviet Union has decided to not invade Poland at all;"
(3) "The Soviet Union has decided to not invade Poland within one month, but has not made up its mind whether it will or will

not invade sometime after one month;"
(4) "The Soviet Union has not made up its mind[2] whether it will or will not invade Poland within one month."

The problem here is that we may charge ahead and make the judgment $P(E_i/H_2\&E_1\&E_2\&\cdots\&E_{i-1})$ without recognizing that (1) H_2 really contains so many "buried" subhypotheses that it is somewhat meaningless as worded and that (2) therefore the judgments of $P(E_i/H_2\&E_1\&E_2\&\cdots\&E_{i-1})$ actually are somewhat meaningless. By a somewhat meaningless judgment we mean simply (1) that the probability judgment made is not uniquely determined but instead depends on which of the multiple buried subhypotheses we happen to use implicitly (or, less often, explicitly) in assessing $P(E_i/H_2\&E_1\&E_2\&\cdots\&E_{i-1})$, and (2) that the probability judgment will be invalid mathematically because the implicit (or explicit) choice of one of the buried subhypotheses means that there no longer is exhaustivity of hypotheses. That is, we are not asking $P(E_i/H_j\&E_1\&E_2\&\cdots\&E_{i-1})$ for all of the buried subhypotheses but rather for just one, thus biasing the whole Bayesian process which requires that the hypotheses be not only mutually exclusive but exhaustive as well. Unfortunately, the less justification we provide for the assessment of $P(E_i/H_2\&E_1\&E_2\&\cdots\&E_{i-1})$, the more likely it is that both we and the consumer will not realize that an implicit assumption has been made about which subhypothesis (buried in H_2) is being used to assess $P(E_i/H_2\&E_1\&E_2\&\cdots\&E_{i-1})$.

The implicit assumption most likely to be made by us and the consumer is that H_2 = the Soviet Union has "decided to not invade Poland at all." Note that this is not the original wording of H_2, but probably is the implicit way we will interpret the originally worded H_2. This implicit assumption will cause us to ignore the other buried subhypotheses that were listed above.

The approach to alleviating this problem that is most likely to work is to use a tree diagram to come up with the possible alternatives. For the example being used, the working tree would look like Figure 4.5.

FIGURE 4.5
Tree Diagram for Soviet Decisions Regarding a Polish Invasion

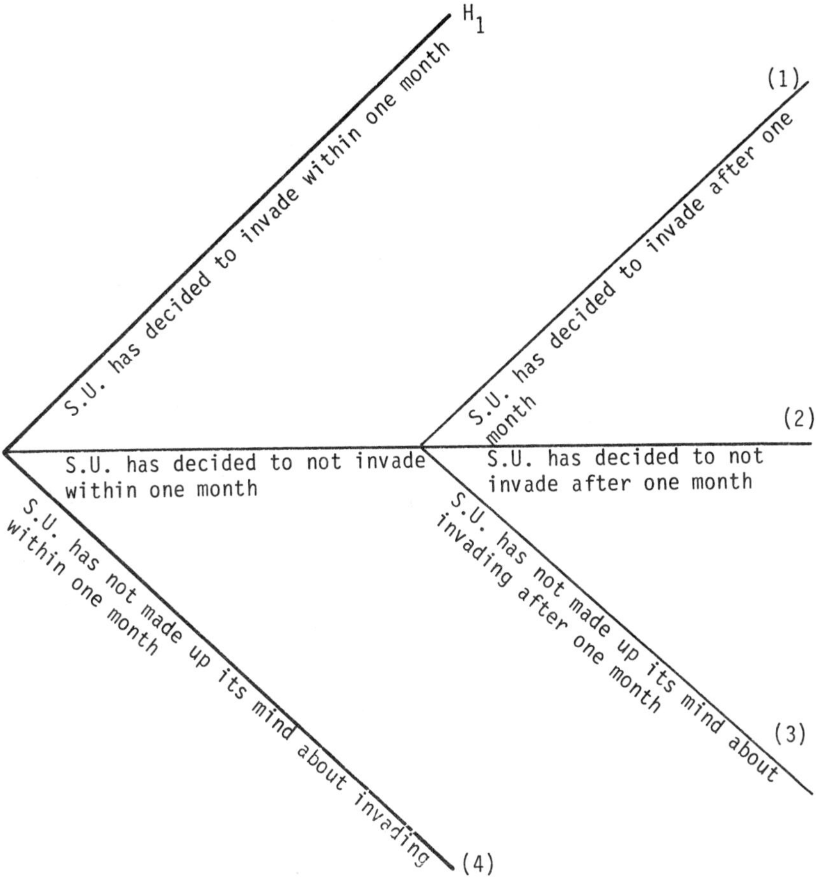

The moral of this example would seem to be that, if we are talking about the possible decisions which have been reached by an actor, then two alternatives simply will not work. There always will be the logical decisional alternative of "no final decision." Indeed, the more general tree would appear to be Figure 4.6.

FIGURE 4.6
A General Tree for Possible Decisions by an Actor with Respect to Doing "x" within Time Period "y"

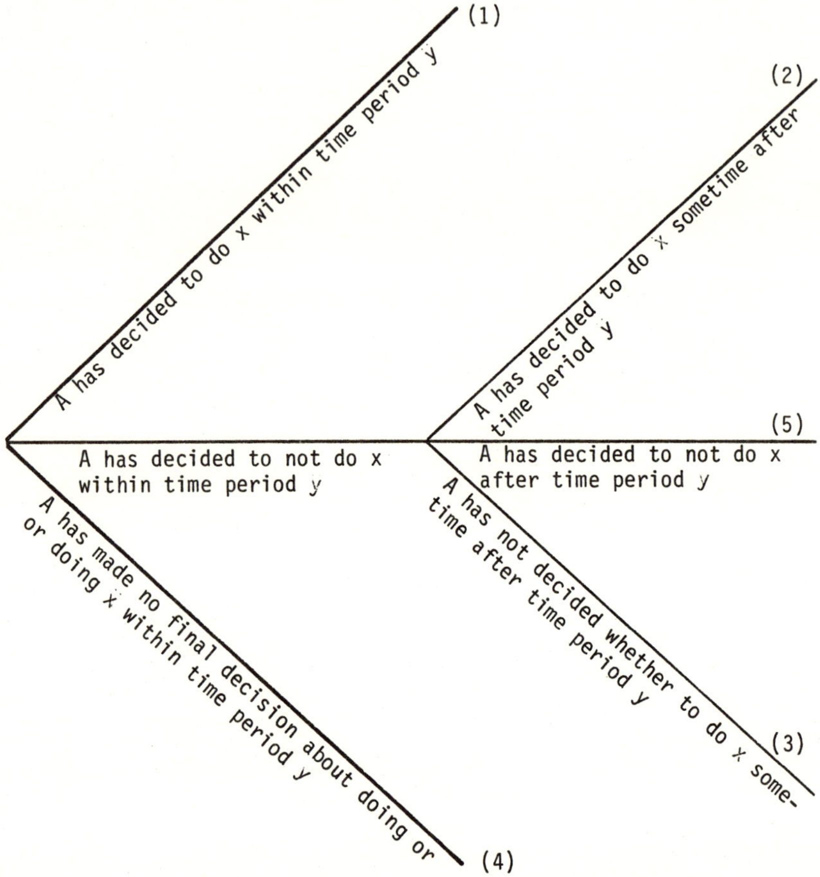

Thus, a general case when dealing with hypotheses about the possible decisions which have been reached by an actor seems to have at least the five hypotheses generated by the tree above:

H_1 = A has decided to do x within time period y,

H_2 = A has decided (to not do x within time period y, but) to do x sometime after time period y,

H_3 = A has decided to not do x within time period y, but has not decided whether to do x sometime after time period y,

H_4 = A has made no final decision about doing or not doing x within time period y,

H_5 = A has decided to not do x at all, i.e., A has made a final decision to not do x.

One major difficulty with this list of hypotheses is that we really may not be able to take into account, when making our probability assessments, the length of time period y. A "y" of one month then may yield the same judgment as a "y" of six months. See the discussion below in problem 8iva and also in problems 6 and 7.

A second difficulty with this list of hypotheses is that things are beginning to get unwieldy. Five hypotheses are near the limit of what we want, and, if later we want (or need) to introduce further subhypotheses in our analysis we may find we have too many hypotheses for timely and comprehensible analyses.

In order to deal with this second difficulty, we can, at the price of some slight inaccuracy, collapse this set of five hypotheses into a set of four by combining hypotheses three and four into a new H_3 which reads "A has made no final decision about doing or not doing x." This yields the following, less accurate, set of four hypotheses:

H_1 = A has decided to do x within time period y,

H_2 = A has decided to do x sometime after time period y,

H_3 = A has made no final decision about doing or not doing x,

H_4 = A has made a final decision to not do x.

I frankly prefer the set of five hypotheses, because it really is more accurate, but I don't want to appear dogmatic on this point.

As was pointed out earlier, there is no "final" solution to the problem of "buried" subhypotheses when dealing with the possible decisions which have been reached by an actor because there always are further conditions whose presence or absence we could consider. One example is whether the actor is engaging in hedging (precautionary) actions which will enable the actor to do x if it eventually decides to do so. If we add on the possibility of hedging actions to our basic five hypotheses tree, we get the tree in Figure 4.7.

FIGURE 4.7
A General Tree for Possible Decisions by an Actor with Respect to Doing "x" within Time Period "y" when Hedging Actions are Possible

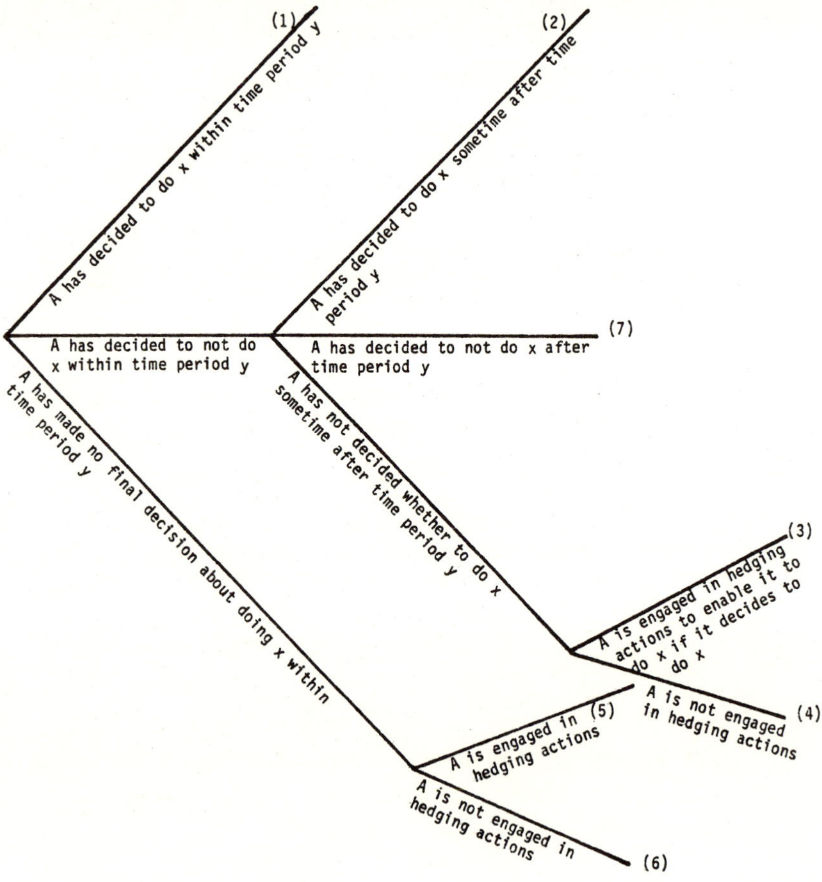

The tree, in turn, yields seven hypotheses:

H_1 = A has decided to do x within time period y,

H_2 = A has decided to do x sometime after time period y,

H_3 = A has decided to not do x within time period y, but has not decided whether to do x sometime after time period y, and A is engaged in hedging actions to enable it to do x if it decides to do x,

H_4 = A has decided to not do x within time period y, but has not decided whether to do x sometime after time period y, and A is <u>not</u> engaged in hedging actions to enable it to do x if it decides to do x,

H_5 = A has made no final decision about doing or not doing x within time period y, and A is engaged in hedging actions to enable it to do x if it decides to do x,

H_6 = A has made no final decision about doing or not doing x within time period y, and A is <u>not</u> engaged in hedging actions to enable it to do x if it decides to do x,

H_7 = A has made a final decision to not do x.

This is, indeed, a rather complicated set of hypotheses, with which it might be difficult to work.

If we take this same possibility of hedging actions, and start with the somewhat less accurate set of four hypotheses listed above, then we end up with only these five hypotheses

H_1 = A has decided to x within time period y,

H_2 = A has decided to do x sometime after time period y,

H_3 = A has made no final decision about doing or not doing x, but (and) is engaged in hedging actions to enable it to do x if it decides to do x,

H_4 = A has made no final decision about doing or not doing x, but (and) is <u>not</u> engaged in hedging actions to enable it to do x if it decides to do x,

H_5 = A has made a final decision to not do x.

Because of its shorter length, this is an easier set of hypotheses with which to work, but is less accurate than the set of seven hypotheses.

Suppose we are dealing with a situation where country A actually (but unknown to us) is worried about possibly having to do x, where x is an invasion of another country, but A has not made up its mind yet, and is engaged in hedging actions. Suppose we are receiving a stream of data about A, and that A eventually decides to invade.

Then, if we were using the last set of hypotheses developed above, in a Bayesian analysis, we probably would find the following to be the case. (1) In the early stages the revised probabilities of H_1 and H_2 would be close together, but the revised probability of H_1 would be larger (indeed, largest for all the hypotheses as the invasion grew near). (2) In the early stages the revised probability of H_3 would be close to the revised probabilities of H_1 and H_2, and in the later stages would be close to that of H_1, because we would see evidence as consistent with both hedging actions and deciding to invade by time y. It might be very close to the actual time of the invasion before the revised probability of H_1 became much larger than the revised probability of H_3.[3]

(3) Although both the revised probabilities of H_4 and H_5 would be much less than those of H_1, H_2 and H_3 at the beginning, the revised probability of H_4 would be larger than the revised probability of H_5 throughout and both would approach zero as time passed.

In summary, we would find at the beginning,

rev Prob $H_1 >$ rev Prob $H_2 >$ rev Prob $H_3 >$ rev Prob $H_4 >$ rev Prob H_5, and at the end

rev Prob $H_1 >$ rev Prob $H_3 >$ rev Prob $H_2 >$ rev Prob $H_4 >$ rev Prob H_5

both fairly big and close together (rev Prob H_1, rev Prob H_3); both close to zero (rev Prob H_4, rev Prob H_5)

Numbers respectively like

$0.50 > 0.43 > 0.04 > 0.02 > 0.01$ might be typical close to the end, and $0.90 > 0.09 > 0.005 > 0.004 > 0.001$ typical at the very end.

The question of whether to use the original five or four hypothesis sets <u>or</u> the seven or five hypothesis sets with hedging actions is a matter of our tastes for complexity and our expectations concerning the quality of data we will receive. The sophistication of the actor with which we are dealing may be important, too.

However, if we are dealing with hypotheses about the possible decisions which have been reached by an actor, we probably should at least use hypotheses of the kind listed in the four hypothesis set. While this will not solve the "buried" subhypotheses problem, it will alleviate the problem somewhat.

As noted before, if the hypotheses do not deal with the possible decisions which have been reached by an actor, but instead deal with what an actor, in the future, will decide to do or with non-decisional situations, we can use a tree to illuminate some of the possible buried subhypotheses. For example, if the original hypotheses are non-decisional of the form:

H_1 = A will do x within time period y,

H_2 = A will not do x within time period y,

then our tree would be the one in Figure 4.8.

FIGURE 4.8
General Tree for what an Actor will Do in the Future

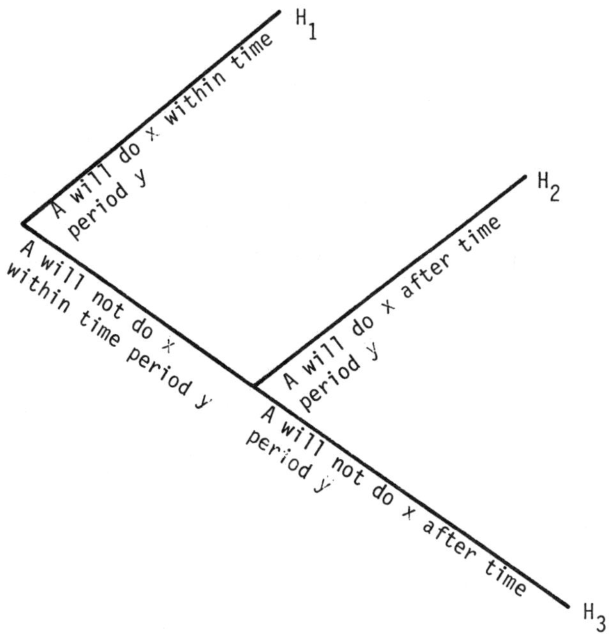

Our expanded list of hypotheses, which is more complete with respect to buried subhypotheses, is then:

H_1 = A will do x within time period y,

H_2 = A will do x after time period y,

H_3 = A will not do x at all.

In problem 8, where we deal more fully with the types of hypotheses which we can try to deal with in using Bayes, we will use trees repeatedly to discover "buried" subhypotheses. That section will add to the analysis of this section. See particularly problem 8iv and footnote eight there dealing with the use of the word "eventually."

PROBLEM 4. The optimal number of hypotheses.

The problem: The question is how many hypotheses are optimal to use in a given situation.

The solution: The nature of the hypotheses, the situation itself, and a number of idiosyncratic factors make it impossible, at this point anyway, to provide a definitive answer. However, three rules of thumb are that (1) if we are dealing with a situation involving possible decisions of an actor, then four or five hypotheses necessarily will be required; (2) more than seven hypotheses may be hard to handle, and (3) only two hypotheses may well lead to a meaningless residual hypothesis (which should be split up into meaningful hypotheses or subhypotheses).

Discussion: As noted in the last section, if we are dealing with decisions of an actor, then we necessarily are driven to using either the four or five subhypothesis sets (without hedging options) described if we wish to conduct a proper analysis. If we are not dealing with decisions of an actor, then there may be some flexibility regarding the number of hypotheses we can choose to use.

There are several reasons why it may be undesirable to have more than around seven hypotheses. For one thing, it may be somewhat difficult, because of the relatively fine distinctions which we are likely to have among the hypotheses, to assign conditional probabilities to events. It also is likely to be very time-consuming to make so many probability assignments for each event. (As we will find later, however, thinking in terms of odds rather than probabilities may speed up the process a bit.) As was the case in the poison gas example, use of a tree diagram may be helpful in reducing the number of hypotheses.

Reducing down to, or starting with, two hypotheses in a problem not involving decisions probably is a poor choice. The second, residual hypothesis, as we have seen, is likely to be meaningless as stated, and will require an analysis to discover the hypotheses (or subhypotheses) buried in it which should be made explicit.

Interestingly, an additional problem, pointed out in class by student Lloyd Feldman, may be that the primary, not residual hypothesis, also may be rather broad in scope and also may be in need of refinement. For example, in the "threat" and no threat dichotomy, "threat" itself may include a spectrum of possible hostile activities which are cause for concern and which should be specified.

PROBLEM 5. Multiple factors in hypotheses: the problem of how to develop hypotheses when we are dealing with multiple factors.

The problem: The problem is how to develop hypotheses when we are dealing with multiple factors which can be true or false or present or absent.

The solution: Use a tree diagram to develop the hypotheses.

Discussion: Developing hypotheses in a particular situation may not be at all easy. There may be multiple factors which can be true or false (or present or absent). Additionally, of course, we must end up with hypotheses that are mutually exclusive and exhaustive, which don't have meaningless hypotheses (residual or primary), and which are not too numerous. Use of a tree diagram, where we consider one factor at a time, is the best technique to develop hypotheses.

PROBLEM 6. Time and "sliding windows": the problem in phrasing hypotheses of whether to use a fixed time limit in the future or a "sliding window," or even to mention the factor of time.

The problem: Although not all hypotheses will require a time period, in phrasing hypotheses it seems clear that in many cases we will want to mention a time period. For example: "state A has decided to attack state B" seems to require some sort of time period within which the attack will occur. The choice is to use a fixed time limit in the future (e.g., "state A has decided to attack state B by 1200 hours on 30 June 19XY"), or to use a so-called "sliding window" (e.g., "state A has decided to attack state B within the next 30 days"). Use of a fixed time limit means that the analysis can "run out" if none of the action hypotheses occurs by the deadline, so we will have to start over if this happens and we want to keep using Bayes. A more serious problem is that use of a fixed time limit can result in distorted probability assessments as the time limit approaches.

The solution: Hypotheses which deal with matters of fact normally need not have any time frame attached. Hypotheses which involve a decision by some actor to take action in the future normally will need some sort of time period to be useful to the consumer. The use of the "sliding window" seems preferable to the use of a fixed time limit.

Discussion: Hypotheses which deal with matters of fact (e.g., "X is a spy") normally need not have any time frame attached. In order to be useful to the consumer, hypotheses which involve a decision by some actor to take action in the future normally will need some sort of time period specified, within which the action will occur. For example, "state A has decided to attack state B" is a hypothesis which seems to require some sort of time period within which the attack will occur. The time period will provide to the consumer of the analysis some feeling about how rapidly events are unfolding and about the time period in which (s)he has a chance to act. Also, such decisions by actors to take action normally really include time periods within which the action will take place.

The choices regarding the type of time period are to use a fixed time limit in the future (e.g., "state A has decided to attack state B by 1200 hours on 30 June 19XY") or to use a so-called "sliding window" (or "floating window") (e.g., "state A has decided to attack state B within the next 30 days").

Clearly, use of a fixed time limit implies that the analysis can "run out" if none of the action hypotheses occurs by the deadline (e.g., if 30 June 19XY arrives and state A has not attacked state B, then the analysis is over). We will have to start a <u>new</u> Bayesian analysis if we want to continue to use Bayes to consider the possibility of state A attacking state B.

Another problem associated with use of a fixed time limit is noted by Schweitzer: "...the passage of time and the reduction of the period remaining may itself be of significance, and a coordinator may choose to include an item to that effect for evaluation."[4]

I agree with the first part of his comment. Indeed, use of a fixed time limit may lead to serious distortions of probability estimates as the fixed time limit approaches because the time period for action to occur is short. For example, if we reach 28 June 19XY and seek to assess a value for

$$P\left(\begin{array}{l}\text{"Large scale military movements}\\ \text{in the neighborhood of the border}\\ \text{between A and B by A's soldiers}\\ \text{on 28 June 19XY"}\end{array}\middle/\begin{array}{l}\text{"State A had decided to}\\ \text{attack State B by 1200}\\ \text{hours on 30 June 19XY"}\\ \text{\& previous data}\end{array}\right)$$

then the value we arrive at may well be close to "1" simply because, if we assume that there is to be an attack by 30 June, time is running out for A. That is, A simply must do something if it is to attack by 30 June. So, the probability of military movements, assuming that the attack is coming (and given whatever sort of previous data we have) almost has to be assessed as close to "1." The effect of grinding in this probability of close to "1" in our Bayesian analysis may well be to increase significantly the revised probability of A attacking B, and lead the consumer of the Bayesian analysis to believe that war is near. But, all of this may well be an <u>artifact</u> of our self-imposed deadline of 1200 hours on 30 June. The deadline of our analysis may bear no relation to A's plans, but use of the deadline can lead to a misdiagnosis of the importance of evidence. Continued use of such a fixed deadline by an indications and warning (I&W) analyst might lead to increasingly high probabilities of A attacking B in the waning hours before 1200 hours on 30 June and a crisis situation could be proclaimed that was entirely unwarranted. Therefore, a fixed time limit solution has serious problems and should not be chosen if we have a better alternative.[5]

Such an alternative is the use of a "sliding window" or "floating window," e.g., "state A has decided to attack state B within the next 30 days." As time passes, the time frame of the "next 30 days" moves forward into the future, and the problem of artifactual probabilities associated with fixed deadlines does not occur.

A major problem with this "sliding window" idea is that we actually may not be able to take into account, when making our probability assessments, the actual length of the time period, e.g., "30 days." That is, a time period of "the next 30 days" may yield the same probability judgments as a time period of "the next 90 days" because we are focusing on what will or will not happen much more than on when it will happen. This issue has appeared before (in problem 3 after the set of five hypotheses) and will be raised again in problems 7 and 8iva.

Another issue, discussed later, is whether evidence that is older than the time period (e.g., over 30 days) should be dropped out of the analysis. This "life span of evidence" problem is treated at length below, but it should be noted now that there is no obvious logical reason why, if we choose to drop out old data, the age of the data has to match our "sliding window" time period. We could have, for example, a hypothesis that "state A has decided to attack state B within 30 days" and decide to drop out data older than 45 days, for example.

PROBLEM 7. Time periods in "sliding windows": the problem of the appropriate "sliding window" time period to use in hypotheses about decisions that have been made to do something in the future.

The problem: If we choose to use a "sliding window" in hypotheses about decisions that have been made to do something in the future, there is the problem of selecting the appropriate time period - 12 hours, 3 days, 3 months or whatever. Differences in the time period chosen might have a lot of impact on the conditional probabilities assigned to evidence. In general, the conditional probability that an event will happen, given a particular hypothesis, will increase as the time period increases (if the event is supportive of the action specified in the hypothesis).

Too short a sliding window period possibly may yield probabilities so small that the consumer may not see them as indicative of a hypothesized action. Too long an interval, in contrast, might yield probabilities that are higher--but at the price of not pinning down when the hypothesized action will occur.

The solution: The sliding window period should be chosen to be useful to the consumer. If there are different consumers with different interests, we may need to run parallel Bayesian analyses with different sliding window time periods specified in the hypotheses.

Discussion: There are three aspects to this problem. One, which was raised earlier in problems 3 and 6 and which will be discussed in problem 8iva below, is whether the assessor really can take into account different time periods when making conditional probability judgments. I think it is difficult to do so, but perhaps not impossible. This section will indicate the nature, i.e., the direction of the distinctions that the assessor will tend to make if any distinctions can be made.

The second aspect to this problem will be deferred to the end of this problem and to problem 9. It involves the question of how far into the future we can try to predict either future decisions or future events. (For example, does it make good sense to ask $P(E_i/H_i \& E_1 \& E_2 \& \cdots \& E_{i-1})$ where "H_i = A will invade B within 10 years"?)

The third aspect of the problem is stated in this section's topic sentence. The question here is what is the appropriate "sliding window" time period to use in hypotheses about decisions that have been made to do something in the future. A few examples may help to illustrate the problem. If we try to assess $P(E_i/H_1 \& E_1 \& E_2 \& \cdots \& E_{i-1})$, where H_1 = A has decided to attack B within 12 hours, and then assess $P(E_i/H_1^* \& E_1 \& E_2 \& \cdots \& E_{i-1})$ where H_1^* = A has decided to attack B within 3 days, the difference in the time interval of the two hypotheses might mean that the assessments of $P(E_i/H_1 \& E_1 \& E_2 \& \cdots \& E_{i-1})$ and $P(E_i/H_1^* \& E_1 \& E_2 \& \cdots \& E_{i-1})$ will <u>differ</u> greatly, and hence the revised probabilities of H_1 and H_1^* also might differ greatly. In general, if an attack actually has been decided on and if E_i is diagnostic of and supportive of the attack decision hypotheses, then the assessed value of $P(E_i/H_1 \& E_1 \& E_2 \& \cdots \& E_{i-1})$ will be <u>smaller</u> than that for $P(E_i/H_1^* \& E_1 \& E_2 \& \cdots \& E_{i-1})$.

Loosely, what we are saying is that the conditional probability that an event will happen, given a particular hypothesis about a decision that has been made to do something in the future, is likely to be smaller if the time interval specified in the hypothesis is shorter in length (and the event is supportive of the action specified in the hypothesis). At the extremes of time intervals, this loose argument seems valid. For example, if E_i = A's planes are being fueled, H_1 = A has decided to attack B within the next 2 hours, and H_1^* = A has decided to attack B within the next 24 hours, then $P(E_i/H_1 \& E_1 \& E_2 \& \cdots \& E_{i-1})$ will tend to be much smaller than $P(E_i/H_1^* \& E_1 \& E_2 \& \cdots \& E_{i-1})$. The latter may be assessed as nearly "1," while principally because of the short (2 hour) time interval, the former, i.e., $P(E_i/H_1 \& E_1 \& E_2 \& \cdots \& E_{i-1})$ might be assessed as considerably lower (e.g., "0.35"). The short time interval here tends to be more deterministic than the previous evidence.

Thus, the conclusion is that, if our hypotheses are about decisions that have been made to do something in the future, then the conditional probability that an event (which is supportive of the action specified in a particular hypothesis) will happen is likely to increase as the "sliding window" time interval specified in the hypotheses increases in length.

We still are faced with the question of what sliding window time period we should use. The ambiguous answer would appear to be that the interval should be chosen to be useful to the consumer. If there is more than one consumer, we may need to run parallel Bayesian analyses based on different time intervals. For example, a President interested in dissuading the North Koreans from attacking South Korea might want a Bayesian analysis with an invasion hypothesis on the order of two weeks. If it looked as if a decision had been made to attack within two weeks, (s)he would have time to warn the North Koreans not to invade and to mobilize resources to deter such an attack. If the time interval is extended to 3 months, 6 months, or even a year, the paradoxical result presumably is that the revised probabilities of attack will go up (if evidence favoring an attack decision is coming in), but the usefulness to the consumer may go down --(s)he is not nearly as certain <u>when</u> the attack will occur. That is, for a given event E_i, assessing

$$P\left(E_i \middle/ \begin{array}{l}\text{the North Koreans}\\ \text{have decided to}\\ \text{attack within a year}\end{array} \quad \& E_1 \& E_2 \& \cdots \& E_{i-1}\right) = 0.8$$

may give less useful information than assessing a lower value for a shorter time interval, say

$$P\left(E_i \middle/ \begin{array}{l}\text{the North Koreans}\\ \text{have decided to}\\ \text{attack within two weeks}\end{array} \quad \& E_1 \& E_2 \& \cdots \& E_{i-1}\right) = 0.3 \quad \text{or}$$

$$P\left(E_i \middle/ \begin{array}{l}\text{the North Koreans}\\ \text{have decided to}\\ \text{attack within three days}\end{array} \quad \& E_1 \& E_2 \& \cdots \& E_{i-1}\right) = 0.1$$

An indications and warning analyst needs to be most careful indeed about the selection of the time interval. If an attack really has been decided on, too short an interval might yield probabilities so small as to perhaps not be seen as indicating an attack, while too long an interval might yield probabilities that are higher but at the price of not pinning down <u>when</u> the attack actually will occur. It may well be that several Bayesian analyses with different time periods specified in the hypotheses will be required because there are different consumers; even a single given analyst (e.g., an I&W analyst) might want to run several such analyses to get a better feel for the emerging situation and as an internal check on the probabilities being assigned.

An additional aspect of the problem of deciding on the appropriate sliding window time period to use in hypotheses about decisions that have been made to do something in the future is that there is a question of whether we can assign probabilities with good justification if the sliding window is years in length. A sliding window of 3 years, 5 years, etc. may make it most difficult to assign probabilities to evidence. It appears, at this point, that much shorter periods - perhaps no more than a year - are appropriate.[6] This is, at best, however, an intuitive feeling. Much more analysis of this point is required.

PROBLEM 8. What types of questions can Bayes answer?

The problem: There is a very serious question as to what sorts of questions Bayes can answer and, in particular, as to whether it is reasonable to try to use Bayes to predict either future decisions by actors (such as states) or future events.

The solution: While no definitive answers can be given, the following tentative answers are offered:

(A) If there is a question of fact, e.g., whether oil is in the ground, whether x is a spy, then Bayes can be used.

(B) If there is a question of which decision has been reached, by a government or other actor i.e., what is the assumed present decisional situation, then Bayes can be used. The four or five hypothesis sets (without hedging options) developed in problem 3 can be quite useful models for formulating hypotheses. Evidence collected will be considered reflective of the decision that has been reached.

(C) The use of Bayes is much more questionable if we are trying to predict future states of affairs, rather than decide what is the present state of affairs. Other techniques, such as probability tree diagrams, may be much more applicable. In trying to use Bayes to predict future states of affairs, the best approach at this point seems to be to use systemic phrasing. If the question already involves possible future decisions by some actor, then the suggestion here is that the hypotheses should be phrased as "the systemic forces at work in the environment are such that the decision of the actor will be x (or not x)." If the original question does not involve decisions by some actor but simply is that some future event x will occur, then the suggestion here is that an attempt be made to rephrase the question as one involving future decisions, if that is possible; if that is not possible, then the hypotheses should be worded that "the systemic forces at work in the environment are such that x will occur." In all of these situations of (C) each item of evidence will be viewed as derivative of and reflective of the systemic forces at work, rather than possibly being an influence (or "causal" factor) on the possible future state of affairs. (This may do violence to our feeling about the so-called "evidence.") The discussion to follow must be viewed as only the beginning of the discussion of a complex question.

Discussion: This is a far more complex problem than it might appear to be at first, and cannot be definitively answered here.

There seems to be a feeling that Bayes can be used to answer such types of questions as these:

(A) Questions of fact:

(i) Is oil in the ground?

(ii) Is X a spy?

(B) Questions of which decision already has been reached:

(iii) What has state A already decided to do regarding a decision to attack state B within six months?

(C) Questions of future decisions or future events:

(iva) What will state A eventually decide to do regarding a decision to attack state B?

(ivb) Will Somoza decide to resign this year?

(ivc) Will King Hussein decide to sign a treaty with Israel?

(v) Will state A attack state B within six months?

(vi) Will war occur between states A and B within six months?

(viia) Will state A's leader die in an accidental plane crash this year?

(viib) Will a nuclear potential be developed by state A by 19XY?

(viii) Will government A be overthrown by a coup d'etat within six months?

(ix) Will party A win the upcoming election?

Each of these questions/issues will be discussed in turn.

(A) Questions of fact:

(i) "Is oil in the ground?"

Assuming that earthquakes or natural disasters which might affect the oil situation will not occur, and assuming that someone is not pumping oil out at a rapid rate, a geological exploration team might find Bayesian analysis a useful technique to use to answer the question of whether oil is in the ground. We might have something like

H_1 = oil is in the ground in economically exploitable quantities,

H_2 = oil is in the ground, but not in economically exploitable quantities,

H_3 = oil is not in the ground.

Evidence from air or satellite photos, as well as geological and seismological readings, all would be evidence about the state of nature and could easily and properly be put into Bayes. This is an example of a factual question about a relatively stable situation (the quantity of oil is not expected to change rapidly), and does not seem to raise any problems. Bayes, in short, seems quite applicable.

(ii) "Is X a spy?"

We have to be careful here. A tree diagram approach yields Figure 4.9.

FIGURE 4.9
Tree Diagram of Logical Possibilities Regarding X Being a Spy

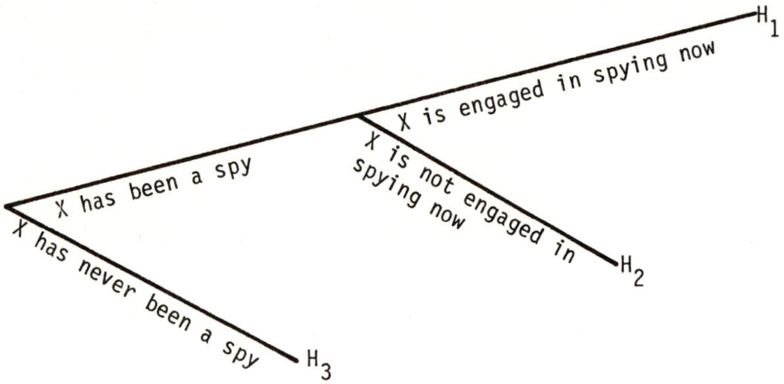

So, H_1 = X is engaged in spying now,

H_2 = X has been a spy, but is not spying now,

H_3 = X has never been a spy.

Evidence reflective of X's activities can be accommodated easily within this set of hypotheses. (See problem 19 below, however.) Bayes seems quite applicable to this factual question.

(B) Questions of which decision already <u>has been reached</u>:

(iii) "What has state A already decided to do regarding a decision to attack state B within six months?"

Problem 3 on unrecognized buried subhypotheses showed us that this sort of question can be dealt with by using either a set of four or a set of five hypotheses (ignoring hedging actions). The four hypothesis set will be as follows:

H_1 = A has decided to attack state B within six months,

H_2 = A has decided to attack state B sometime after six months,

H_3 = A has made no final decision about attacking or not attacking state B,

H_4 = A has made a final decision to not attack state B.

The five hypothesis set is somewhat more complicated. The first two hypotheses are the same as in the four hypotheses set. We have

H_1 = A has decided to attack state B within six months,

H_2 = A has decided to attack state B sometime after six months,

H_3 = A has decided to not attack state B within six months, but has not decided whether to attack state B sometime after six months,

H_4 = A has made no final decision about attacking or not attacking state B within six months,

H_5 = A has made a final decision to not attack state B at all.

In each set of hypotheses, we are referring to what can be termed as the possible "decisional situations" of country A. That is, each hypothesis specifies what state A already has decided to do about attacking state B. It makes sense to look for evidence that is reflective of the existing decisional situation. Evidence could be in the form of speeches, leaks to the press, etc. Bayes seems quite applicable to this type of situation, where we are trying to decide the existing decisional situation of a country, and where, in essence, we are assuming the present decisional situation matches what already has been decided.[7]

One more point needs to be made. Suppose that, based on our Bayesian analysis, we decide that H_1 is the most likely hypothesis. Note that H_1 simply states that a decision has been made to attack state B within six months. Bayes actually shows what has happened in terms of a decision to do something. We make one assumption when we assume that the present decisional state matches what has been decided, and we really are making a prediction - which goes beyond Bayes - when we say that A actually will invade B within six months. That prediction was not what was stated in hypothesis H_1 since the hypothesis referred only to a decision to do something. So it can be argued that the Bayesian analysis itself does not embody a prediction. The prediction occurs after the Bayesian analysis and in essence is a prediction that the decision that has been made will be implemented.

(C) Questions of future decisions or future events:

(iva) "What will state A eventually[8] decide to do regarding a decision to attack state B?

Now things get tricky. We argued in (iii) that it seemed reasonable to use Bayes to decide on the existing decisional situation (where we assumed that the existing decisional situation matched what had been decided). But, it can be argued that if the results of doing our analysis in (iii) are that no final decision has been made, then we still have no final answer to the question we really want answered - are they eventually going to decide to invade B or not?

The people in state A involved in the decision making process may well not know themselves what they eventually will decide to do, so it is reasonable to wonder if we actually can use Bayes to predict this. In other words, can we use Bayes to determine a future decisional situation in addition to determining what the present decisional situation is?

One way of trying to answer such a question is to imagine some of the items of evidence which might occur, and which we might grind into a Bayesian analysis. Suppose we take our hypotheses to be

H_1 = state A eventually will decide to attack state B,

H_2 = state A eventually will decide to <u>not</u> attack state B.

(It is reasonable to ask at this point why we have deleted mention of a time limit. Why isn't H_1 = "state A eventually will decide to attack state B within six months," for example? We will address this point at the end of this subsection.)

Suppose that

E_1 = the widespread publication of a poll (in state A) showing that sentiment in favor of war is building up among all sectors of society in state A.

Suppose we are using the odds-likelihood form of Bayes, which (we remember) is

$$\frac{P(H_1/E_1)}{P(H_2/E_1)} = \frac{P(H_1)}{P(H_2)} \cdot \frac{P(E_1/H_1)}{P(E_1/H_2)}.$$

Substituting our verbal descriptions, we have on the left side the revised odds:

$$\frac{P\left(\begin{array}{l}\text{state A eventually}\\ \text{will decide to}\\ \text{attack state B}\end{array}\middle/\begin{array}{l}\text{the widespread publication of}\\ \text{a poll showing that sentiment}\\ \text{in favor of war is building up in}\\ \text{all sectors of society in state A}\end{array}\right)}{P\left(\begin{array}{l}\text{state A eventually}\\ \text{will decide to}\\ \underline{\text{not}}\text{ attack state B}\end{array}\middle/\begin{array}{l}\text{the widespread publication of}\\ \text{a poll showing that sentiment}\\ \text{in favor war is building up in all}\\ \text{sectors of society in state A}\end{array}\right)}.$$

This is a perfectly reasonably sounding ratio. It seems to make sense, in other words, to ask the revised probabilities of the hypotheses in light of such "evidence." The prior odds of

$$\frac{P\left(\begin{array}{l}\text{state A eventually}\\ \text{will decide to}\\ \text{attack state B}\end{array}\right)}{P\left(\begin{array}{l}\text{state A eventually}\\ \text{will decide to }\underline{\text{not}}\\ \text{attack state B}\end{array}\right)}$$

also seem to make sense; the probabilities may be somewhat speculative, but they can be assigned on the basis of a subjective feeling about the direction of the systemic forces that will influence the eventual decision. When we get to the likelihood ratio, however, we run into problems. The likelihood ratio is

$$\frac{P\left(\begin{array}{l}\text{the widespread publication}\\ \text{of a poll showing that}\\ \text{sentiment in favor of war}\\ \text{is building up in all sectors}\\ \text{of society in state A}\end{array}\middle/\begin{array}{l}\text{state A eventually will}\\ \text{decide to attack state B}\end{array}\right)}{P\left(\begin{array}{l}\text{The widespread publication}\\ \text{of a poll showing that}\\ \text{sentiment in favor of war}\\ \text{is building up in all sectors}\\ \text{of society in state A}\end{array}\middle/\begin{array}{l}\text{state A eventually will decide}\\ \text{to }\underline{\text{not}}\text{ attack state B}\end{array}\right)}.$$

The question is this: do the conditional probabilities of the numerator and the denominator make any sense at all? I think they do not. Normally, in making such assessments there is a deductive logical connection between the event and the particular hypothesis so that we can ask how likely it is that the event will happen if the particular hypothesis is true. But that link is lacking here; indeed, the event can be seen as a possible "cause" of the particular hypothesis. That is, the poll itself may well have an influence on the decision making process and influence the eventual decision to attack or not attack. (The concept of "causal evidence" is treated later in detail in problem 15, and the "solution" suggested in the literature is in essence to forget about the likelihood ratio and just directly assess the revised odds! But, this means forgetting about Bayes! Such a solution seems rather questionable.)

Another approach, and one that offers hope of a solution, is to change the way we verbalize the conditional probabilities; we can phrase them in terms of the probability that the event will precede the hypotheses. That is, we can phrase the assessment of the numerator this way: "assuming that state A eventually will decide to attack state B, what is the probability that event E_1 (the widespread publication of a poll showing that sentiment in favor of war is building up in all sectors of society in state A) will precede such a decision?" The reasoning here is that some events are more likely (or less likely) to precede an attack decision than a non-attack decision.

It seems to me that it can be argued fairly that the assignment of a probability of this sort is extraordinarily difficult. If we knew that some event really was "necessary" for a decision to attack, then we could say, if we assumed an attack decision eventually was going to be made, that the probability that such an event would precede the attack decision (i.e., occur) would be "1." But the political theory necessary for identifying such "necessary" events seems to be lacking. Indeed, it seems reasonable to argue that the causal chains leading to an eventual decision to attack or not attack really are not known. On what basis can we justify saying, for example, that "if we assume that state A eventually will decide to attack state B, then there is a 0.7 probability that event E_1 (the widespread publication of a poll showing that sentiment in favor of war is building up in all sectors of society in state A) will precede the eventual attack decision"? There do not appear to be any grounds for making such a probability judgment.

Even more difficult, perhaps, is trying to assess the probability of event E_1 assuming that the eventual decision is to not attack. On what basis can we justify saying, for example, that "if we assume that state A eventually will decide to not attack state B, then there is a 0.4 probability that event E_1 (the widespread publication of a poll showing that sentiment in favor of war is building up in all sectors of society in state A) will precede the no attack decision"?

For the reasons just offered, the approach where we ask the probability that an event will precede what is eventually decided seems rather dubious. Another major approach can be attempted, however. We can try to modify the hypotheses. Suppose we rephrase them as follows:

H_1 = the systemic forces at work in the environment are such that state A eventually will decide to attack state B,

H_2 = the systemic forces at work in the environment are such that state A eventually will decide to not attack state B.

Looking at our E_1 event, we can now view it as an event coming from the environment and reflective (not causative) of the systemic forces at work, i.e., a "systemic" event. We have converted hypotheses strictly about a future decisional situation into hypotheses about the influence of systemic forces (including elements in the political system) on the future decision. (We will discuss the scope of the "environment" in problem 15.)

When we try to assign $P(E_1/H_1)$ this time, it seems to make more sense. The phrasing is this: "assuming that the systemic forces at work in the environment are such that state A eventually will decide to attack state B, the probability of the occurrence of event E_1 (the widespread publication of a poll showing that sentiment in favor of war is building up in all sectors of society in state A) is x."

This phrasing seems less awkward than the previous phrasing, and we might be comfortable assigning an x = 0.8, say. Using this phrasing, there does seem to be a deductive logical connection between the "systemic" event and the hypotheses. Nevertheless, the phrasing is rather complicated, and some events might still cause trouble when we try to assign probabilities.

A final variation is to change the hypotheses once more, this time in essence removing the decisional situation aspect. The hypotheses become entirely "systemic" in character. For example, we can let

H_1 = the systemic forces at work in the environment are such that state A eventually will attack state B,

H_2 = the systemic forces at work in the environment are such that state A eventually will _not_ attack state B.

Once again, while rather lengthy, the likelihood ratio conditional probability judgments make sense. The numerator will be as follows: "assuming that the systemic forces at work in the environment are such that state A eventually will attack state B, the probability that event E_1 (the widespread publication of a poll showing that sentiment in favor of war is building up in all sectors of society in state A) will occur is x." Assigning a value of 0.8 does not seem to be unreasonable.

Frankly, I think this last variation is neither necessary nor useful. The previous option where we talk about systemic forces influencing a future _decisional_ situation seems quite adequate. (See problem 17 below, _however_, for an exception to this conclusion.)

It was noted at the beginning of this subsection that the hypotheses did not have a time period included. For example, we did _not_ have H_1 = state A eventually will decide to attack state B _within six months_. The reason for _not_ including any time interval _is that it probably_ would be almost impossible when assessing $P(E_i/H_1\&E_1\&E_2\&\cdots\&E_{i-1})$ to make any distinction between an H_1 with a time period of six months and an H_1 with a different time period, say of one year. The judgments made will be focusing on the logical relation of the evidence to an "attack" or "no attack" eventual decision, and are hard enough to make without trying to include a time interval. In reality, I suspect that almost anyone trying to make such judgments actually would be unable to include a time interval (period).

The reader should ask him/herself when reading the example in the next subsection, where I have included a time interval (period) of "this year," if (s)he really could take the period into account in assessing the probabilities that are required. I have grave doubts, indeed, that the time period can be taken into account. (Compare our discussion here to the similar comments made in problem 3 with respect to the time period "y" included in the set of five hypotheses, and the comments made in problems 6 and 7.)

(ivb) Several more examples will now be presented to illustrate the use of systemic force concepts with decisional situations.

Back in the days before the late President Somoza resigned, and presumably before he had made up his mind to do so, an item of evidence might have been "Sandinista attack #X." If we had charged ahead and tried to use the odds-likelihood form of Bayes, we would have ended up as before with nonsense likelihood statements. Suppose, to make the example easier, this was our first item of evidence. Let

H_1 = Somoza will decide to resign this year,

H_2 = Somoza will not decide to resign this year,

and E_1 = Sandinista attack #X.

(It is clear that H_2 contains buried subhypotheses,[9] but we will go ahead with it since we are trying to keep the example simple.)

Using the odds-likelihood form of Bayes, we have

$$\frac{P(H_1/E_1)}{P(H_2/E_1)} = \frac{P(H_1)}{P(H_2)} \cdot \frac{P(E_1/H_1)}{P(E_1/H_2)}$$

or, in words,

$$\frac{P\left(\text{Somoza will decide to resign this year} \middle/ \text{Sandinista attack \#X}\right)}{P\left(\text{Somoza will not decide to resign this year} \middle/ \text{Sandinista attack \#X}\right)} = \frac{P\left(\text{Somoza will decide to resign this year}\right)}{P\left(\text{Somoza will not decide to resign this year}\right)} \cdot \frac{P\left(\text{Sandinista attack \#X} \middle/ \text{Somoza will decide to resign this year}\right)}{P\left(\text{Sandinista attack \#X} \middle/ \text{Somoza will not decide to resign this year}\right)}.$$

As before, the conditional probabilities (comprising the revised odds) to the left of the "=" sign make sense -- it makes sense to ask the probability of Somoza resigning (or staying on) this year after an attack. The prior probabilities to the right of the equals ("=") sign (comprising the prior odds) make sense too. But, the probabilities in the likelihood ratio once more seem to make no sense at all. What sense does it make to ask the probability of Sandinista attack #10 given that Somoza will decide to resign this year?

If we change our phrasing of the hypotheses to reflect systemic properties at work, the probabilities in the likelihood ratio will make more sense. Suppose that we let

H_1 = The systemic forces at work in the environment are such that Somoza will decide to resign during the coming year,

H_2 = The systemic forces at work in the environment are such that Somoza will <u>not</u> decide to resign during the coming year,

E_1 = Sandinsta attack #X has occurred.

Then the revised odds will be

$$\frac{P\left(\text{the systemic forces at work in the environment are such that Somoza will decide to resign during the coming year} \middle/ \text{Sandinista attack \#X has occurred}\right)}{P\left(\text{the systemic forces at work in the environment are such that Somoza will \underline{not} decide to resign during the coming year} \middle/ \text{Sandinista attack \#X has occurred}\right)}.$$

The revised probabilities make good sense -- we are calculating the probabilities of hypotheses about the impact of systemic forces on the basis of evidence considered to be reflective of those systemic forces. The prior odds make sense, too. The question is whether the probabilities of the likelihood ratio make sense. The likelihood ratio is

$$\frac{P\left(\text{Sandinista attack \#X has occurred} \middle/ \text{the systemic forces at work in the environment are such that Somoza will decide to resign during the coming year}\right)}{P\left(\text{Sandinista attack \#X has occurred} \middle/ \text{the systemic forces at work in the environment are such that Somoza will \underline{not} decide to resign during the coming year}\right)}.$$

The conditional probability in the numerator does seem to make sense. It asks the probability of a "systemic" event given the influence of systemic forces on a future decisional situation. We assign the probability on the basis of the logical consistency of the observed "systemic" event with the hypothesized direction and strength of the systemic forces. If Sandinista attack #X is of major proportions, then we can argue that the probability of its occurrence, given the assumption of strong enough systemic forces to compel Somoza to decide to resign, is high. On the other hand, looking at the denominator, if the attack is of major proportions, the probability that it would occur if the systemic forces are <u>not</u>

strong enough (eventually) to compel Samoza to decide to resign is low.

The point is that the "systemic" event can be viewed as actually reflective of and derivative from the actual strength and direction (trend) of the systemic forces. We thus have turned a problem involving just a future decisional situation into a systemic problem involving a future decisional situation in order to be able to incorporate systemic evidence and to be able to use Bayes to predict what decision eventually will be made. The future decisional situation is phrased as the outcome of a systemic process. Nevertheless, given the somewhat difficult phrasing, we might be better off using a decision making model or probability diagram rather than Bayesian analysis.[10]

(ivc) "Will King Hussein decide to sign a treaty with Israel?"

Another example where a problem involving a future decisional situation can be turned into a systemic problem involving a future decisional situation is the following. Suppose we are trying to predict whether King Hussein will decide to sign a proposed treaty with Israel. For a particular E_1, it may not make much sense to ask

P(E_1/King Hussein will decide to sign the treaty)

or

P(E_1/King Hussein will not decide to sign the treaty).

(Again, the second hypothesis contains buried subhypotheses.[11]) On the other hand, if we ask

P(E_1/The systemic forces at work in the environment are such that King Hussein will decide to sign the treaty)

and ask also

P(E_1/The systemic forces at work in the environment are such that King Hussein will not decide to sign the treaty)

then these probabilities may make sense.

Although this approach does seem to work, there still seems to be a question as to whether this sort of "tortured" phrasing does not indicate that we are disciples of the "Law of the Tool," one version of which says that if we give a child a hammer he will discover that everything in the house needs to be hammered.[12] We may, in short, be asking too much of Bayes, and trying to do inappropriate things with it. What we are trying to do could be done better, perhaps, with other tools. One such tool is the probability tree (i.e., tree diagram). Another tool would be an actual foreign policy decision making model. In the cases of both Somoza and King Hussein, for example, the models could be quite elaborate and include a host of variables that would be expected to influence their respective decisions.

With respect to the probability tree, it has its own set of

problems, but can be useful. The hypotheses are put at the end of the tree; the factors which we believe influence the hypotheses make up the rest of the tree and are put in the order we expect them to occur on in the causal order they impact on each other. (See Appendix A.)

(v) "Will state A attack state B within six months?"

This phrasing was dealt with somewhat in (iii). Note that the statement is entirely systemic. There is no mention of A eventually making some decision. We will examine the likelihood ratio with the existing phrasing, examine once more the option of phrasing the conditional probability by saying that an event will precede the hypothesized action, and look at effect of rephrasing the hypotheses with systemic language.

Using the same E_1 as before, i.e.,

E_1 = the widespread publication of a poll showing that sentiment in favor of war is building up among all sectors of society in state A,

the likelihood ratio will be

$$\frac{P\left(\begin{array}{l}\text{the widespread publication of}\\ \text{a poll showing that sentiment}\\ \text{in favor of war in building}\\ \text{up among all sectors of}\\ \text{society in state A}\end{array} \middle/ \begin{array}{l}\text{state A will}\\ \text{attack state B}\\ \text{within six}\\ \text{months}\end{array}\right)}{P\left(\begin{array}{l}\text{the widespread publication of}\\ \text{a poll showing that sentiment}\\ \text{in favor of war is building}\\ \text{up among all sectors of society}\\ \text{in state A}\end{array} \middle/ \begin{array}{l}\text{state A will}\\ \underline{\text{not}}\text{ attack}\\ \text{state B within}\\ \text{six months}\end{array}\right)}.$$

Once more, it is not clear that the probabilities can be assigned given the phrasing. The numerator, for example, is phrased as follows: "given that state A will attack state B within six months, the probability of event E_1(the widespread publication of a poll showing that sentiment in favor of war is building up among all sectors of society in state A) is x." Again, as was argued earlier, we appear to have no theory which would enable us to assign such a probability.

If we ask what the probability is that the event would <u>precede</u> the attack, the situation seems no better. The numerator would become as follows: "given that state A will attack state B within six months, the probability that event E_1 (the widespread publication of a poll showing that sentiment in favor of war is building up among all sectors of society in state A) will precede the attack is x." Again, we seem to have no grounds for assigning such a probability.

Use of systemic phrasing however, is once more helpful. If we change our hypotheses to

H_1 = the systemic forces at work in the environment are such that state A will attack state B within six months,

and H_2 = the systemic forces at work in the environment are such that state A will not attack state B within six months,

the numerator in the likelihood ratio becomes

$$P\left(\begin{array}{l}\text{the widespread publication}\\ \text{of a poll showing that}\\ \text{sentiment in favor of}\\ \text{war is building up}\\ \text{among all sectors of}\\ \text{society in state A}\end{array} \middle/ \begin{array}{l}\text{the systemic forces at work}\\ \text{in the environment are such}\\ \text{that state A will attack}\\ \text{state B within six months}\end{array}\right).$$

This probability does seem to make some sense. The numerator would be verbalized this way: "given that the systemic forces at work in the environment are such that state A will attack state B within six months, the probability of the occurrence of event E_1 (the widespread publication of a poll showing that sentiment in favor of war is building up among all sectors of society in state A) is x." There may be more grounds for assigning a probability this time than before we changed the wording of our hypotheses, but quite frankly I feel much less comfortable than in part (iii), where we were talking about systemic forces leading to a decision. The idea of a decision is lacking here; the hypotheses simply say that systemic forces at work in the environment are such that something will happen. Even though the event cited can be viewed as derivative from and indicative of the systemic forces at work, deducing and justifying the appropriate probability seems to me to be much more difficult than in part (iii), when we were talking about deducing and justifying the appropriate probability given that systemic forces would lead to a decision to attack. In short, switching from hypotheses involving systemic forces that lead to a decision to hypotheses involving systemic forces that lead to an outcome increases the abstractness of the situation (for me, at least) and makes it less clear that probabilities can be meaningfully assigned.

Quite frankly, my suggestion here is that, if we are faced with a question of the nature of (v), we try to turn it into a question involving future decisions such as we had in (iii). In other words, an attempt should be made to turn strictly systemic questions into systemic questions about future decisions. If this attempt fails, then it may be better to use tools such as probability diagrams (i.e., trees) rather than Bayes.

(vi) "Will war occur between states A and B within six months?"

The analysis here would be almost identical to (v). The hypotheses, if anything, are even more abstract than in (v), because they admit of B attacking A, of catalytic attacks by third parties which could trigger war, etc.

Moreover, there seem to be even more problems with trying to assess the probabilities of the likelihood ratio. The numerator would ask, for E_1,

$$P\left(E_1 \middle/ \begin{array}{l}\text{the systemic forces at work in the environment are such that} \\ \text{war will occur between states A and B within six months}\end{array}\right).$$

Such a probability might well be impossible to assign given today's underdeveloped international relations theory. Moreover, there are a number of such underdeveloped theories about the cause of war (e.g., equilibrium, balance of power, arms races), each focusing on particular sets of variables. In one theory, E_1 might be seen as important and reflective of and derivative from the systemic forces so that $P(E_1/H_1)$ might be assigned. But in a different systemic theory of war, E_1 might be of very little importance or even irrelevant. How, then, can any value be justified if there is no accepted body of theory which would specify which systemic conditions lead to war or no war? We have, in short, not enough theoretical knowledge to enable us to assign $P(E_1/H_1)$. It would be much better either to phrase questions in terms of decisional situations if we can, or to use a technique other than Bayes.

(viia) "Will state A's leader die in an accidental plane crash this year?"

The traditional, and seemingly valid, analysis of this sort of question is that we must be dealing with the type of developing situation which produces preliminary signs, and is not a chance or random event like a plane crash, in order to use Bayes. We simply must have a developing, non-chance, non-random situation where there are indicator items of evidence that can be used to revise our initial probabilities of hypotheses.

Somewhat more ambiguous is the question of whether Bayes could be used to predict the chances of accidental war between two states. In a period of sustained, high tension where data on performance errors on both sides were available, so that there was a stream of data, Bayes possibly might be useful.

(viib) "Will a nuclear potential will be developed by state A by 19XY?"

This appears to me to be a systemic outcome, i.e., the hypotheses could be

H_1 = the systemic forces at work in the environment are such that a nuclear potential will be developed by state A by 19XY,

H_2 = the systemic forces at work in the environment are such that a nuclear potential will <u>not</u> be developed by state A by 19XY.

Once again, I have grave doubts about the use of Bayes. A probability diagram or a system dynamics model might be much more useful.

(viii) "Will government A will be overthrown by a coup d' etat within six months?"

Once more, this appears to be amenable to being treated in systemic terms, but more usefully by a probability diagram rather than Bayes.

(ix) "Will party A win the upcoming election?"

We will spend a little time on what clearly is a completely systemic question in its original formulation. Let

H_1 = A will win the upcoming election,

H_2 = A will not win the upcoming election.

Suppose further that the first event, E_1, is riots by A's supporters:

E_1 = riots by A's supporters.

Using the odds-likelihood form of Bayes we have

$$\frac{P(H_1/E_1)}{P(H_2/E_1)} = \frac{P(H_1)}{P(H_2)} \cdot \frac{P(E_1/H_1)}{P(E_1/H_2)}, \text{ so}$$

$$\frac{P\left(\text{A will win the upcoming election} \Big/ \text{riots by A's supporters}\right)}{P\left(\text{A will not win the upcoming election} \Big/ \text{riots by A's supporters}\right)} = \frac{P\left(\text{A will win the upcoming election}\right)}{P\left(\text{A will not win the upcoming election}\right)} \cdot \frac{P\left(\text{riots by A's supporters} \Big/ \text{A will win the upcoming election}\right)}{P\left(\text{riots by A's supporters} \Big/ \text{A will not win the upcoming election}\right)}.$$

Once again, we find that the probabilities of the revised odds make good sense. It makes sense to ask whether A will win the upcoming election after riots have occurred. From a non-causal viewpoint, the riots can be seen as indicating the strength (or weakness) of the base of A's popular support and/or the level of popular frustration in the country, etc. (Of course, from the causal viewpoint the riots can be viewed as counterproductive, with a possibility that they could alienate voters, but as noted earlier we have no good way in Bayes of handling such causal evidence. If we really feel that the so-called evidence is causal in nature and should be so viewed, then Bayes probably is not the appropriate technique to use.)

The probabilities of the likelihood ratio are, however, once more nonsensical. What sense does it make to ask the probability that there will be riots by A's supporters given that A will win the upcoming election?

Once more, the solution to the problem can be found in systemic phrasing of the hypotheses.

Let H_1 = the systemic forces at work in the environment are such that the majority will support party A in the upcoming election,

H_2 = the systemic forces at work in the environment are such that the majority will <u>not</u> support party A in the upcoming election,

E_1 = riots by A's supporters.

The revised odds would be the expression

$$\frac{P\left(\text{the systemic forces at work in the environment are such that the majority will support party A in the upcoming election} \middle/ \text{riots by A's supporters}\right)}{P\left(\text{the systemic forces at work in the enrironment are such that the majority will } \underline{\text{not}} \text{ support party A in the upcoming election} \middle/ \text{riots by A's supporters}\right)}.$$

This expression does make some sense. We can ask, based on the evidence of the strength and nature of the riots by A's supporters, what the probability is that majority support for party A exists now or is developing

The probabilities in the prior odds make sense, too. Turning to the likelihood ratio, which is

$$\frac{P\left(\text{riots by A's supporters} \middle/ \text{the systemic forces at work in the environment are such that the majority will support party A in the upcoming election}\right)}{P\left(\text{riots by A's supporters} \middle/ \text{the systemic forces at work in the environment are such that the majority will } \underline{\text{not}} \text{ support party A in the upcoming election}\right)}$$

it does now seem that the probabilities make some sense. The event of the riots by A's supporters can be seen as reflecting the impact of the systemic forces at work in the society, i.e., an existing (or emerging) majority or minority. That is, the riots by A's supporters can be seen as reflecting an already existing majority, or an emerging majority (i.e., there will be a majority) or an existing minority, or an emerging minority (i.e., there will be a minority). Each of these possibilities is allowed for when we speak of the impact of the "systemic forces at work in the environment." The numerator, then, more exactly would be

$$P\left(\text{riots by A's supporters} \middle/ \text{the systemic forces at work in the environment are such that there is now a majority (or will be a majority) that will support party A in the upcoming election}\right).$$

This phrasing more readily allows for a dynamic systemic process at work in the environment. The riots, then, can be viewed as a sample of evidence from a dynamic systemic process and it does seem to make some sense to ask the probability that riots of their strength and nature would occur if we assume that the systemic forces at work are such that there is a majority now (or will be one) in favor of A. The denominator would be, more exactly,

$$p\left(\text{riots by A's supporters} \middle/ \begin{array}{l}\text{the systemic forces at work in the environ-}\\ \text{ment such that there is now a minority (or will}\\ \text{be a minority) that will support party A}\\ \text{in the upcoming election}\end{array}\right).$$

This also makes some sense. Nevertheless, the phrasing of both the numerator and denominator is quite tortured and the question can be raised again as to whether Bayes is the predictive tool most applicable to the situation. Political scientists have developed various election models and probability trees (diagrams) also may be more appropriate. (See Appendix A.)

A few more points need to be made regarding the question of what sorts of questions Bayes can answer. If we look at this section and the ways we have phrased our hypotheses in terms of systemic forces at work in the environment, we still will find that there will be items of evidence that cannot be considered to be reflective of or derivative from the systemic forces at work in the environment no matter how tortured our phrasing is, but instead necessarily are "causative." A good example of this is in problem 15 and involves a tactical situation where the event occurs of the weather turning bad. This is an event which certainly could influence an opponent's decision as to whether or not to attack, but it is <u>not evidence reflective of or derivative from the systemic forces at work in the environment</u> and no amount of tortured phrasing will make it so (as long as the opponent lacks weather making techniques). We simply cannot ask the probability of its occurrence given the systemic forces at work in the environment and have the question make any sense.[13] (As will be noted however, if the weather must be included, Zlotnick's solution for "causal" evidence can be used, even though the weather is not the exact sort of item Zlotnick meant by the term "causal".)

<u>PROBLEM 9</u>. How far in the future can Bayes predict?

<u>The problem</u>: The question is how far in advance we can use Bayes to predict either eventual future decisions or future states of affairs.

<u>The solution</u>: No definitive solution is offered here.

<u>Discussion</u>: I consider this to be an open question. The previous discussion in section 8 of what sort of questions Bayes can answer was not meant to be dogmatic on the issues raised. In that discussion, it was suggested that the prediction of a future state of affairs, whether a future decisional situation or a future event, can be handled by phrasing the hypotheses as involving

systemic forces at work in the environment that will lead to a decisional situation or will lead to a (systemic) event. If we can choose which of these formulations to use, it seems preferable to use the one involving future decisional situations. Some events inherently are non-decisional, however, and for those only a straight systemic formulation of the hypotheses is possible.

In using either formulation, the discussion largely ducked the issue of exactly when a decision would be made or an event would happen. Evidence was seen as reflective of and derivative from the systemic forces at work. We can ask if evidence reasonably could be expected to be reflective of and derivative from systemic forces which might eventuate in a decision or event, say, three years hence.

We first will discuss the question of time periods in the prediction of future events: If Bayes is viewed as a fever thermometer, showing the "movement" of the systemic forces, then we might phrase our hypotheses, for example, as

H_1 = the systemic forces at work in the environment are such that war will break out within three years,

H_2 = the systemic forces at work in the environment are such that war will break out after three years,

H_3 = the systemic forces at work in the environment are such that war will not break out.

We reasonably might be able to assign conditional probabilities to evidence (although the evidence might well be seen equally as likely for H_1 as for H_2 because of the lengthy period ahead). In terms of the time period chosen, it might be hard to distinguish between saying "...war will break out within three years" and "...war will break out within one year" when assigning conditional probabilities to events, however. Intuitively, I am bothered by trying to use Bayes to predict a systemic event of much more than six months to a year hence, simply because the conditional probabilities for the evidence are hard enough to assign without trying to justify the connection between an item of evidence about systemic forces and something that might occur in the distant ($>$ 1 year) future.[14]

We can repeat some of our earlier analysis about sliding windows for hypotheses about decisions that have been made to do something in the future, however, and say that in some cases it may make sense to talk of events which may happen within shorter specified periods. For example, we can have a hypothesis set where the first hypothesis is

H_1 = the systemic forces at work in the environment are such that A will attack B within the next 2 hours,

or, alternatively, a hypothesis set where the first hypothesis is

H_1^* = the systemic forces at work in the environment are such that A will attack B within the next 24 hours.

If E_i = A's planes are being fueled, then $P(E_i/H_1\&E_1\&E_2\&\cdots\&E_{i-1})$ will tend to be much smaller than $P(E_i/H_1^*\&E_1\&E_2\&\cdots\&E_{i-1})$. The latter may be assessed as nearly "1" while, principally because of the short time interval, $P(E_i/H_1\&E_1\&E_2\&\cdots\&E_{i-1})$ probably will be assessed considerably lower, say 0.35.

As in the previous analyses of sliding windows, the time interval should be chosen to be useful to the consumer.

If there is more than one consumer, we may need to run parallel Bayesian analysis based on different time intervals.

These last paragraphs dealt with sliding windows for hypotheses about future <u>events</u>, not future decisions. In trying to predict future decisional situations, several points need to be made. First, as was noted previously, when hypotheses are phrased in terms of decisions that have been made, the hypotheses can specify, for example, that a decision to attack within six months has been made. Then, the prediction, going beyond Bayes, is that there will be an attack within six months. The time period of the prediction is, then, six months. This is a "sliding window," of course, and the six months can advance.

When we try to predict a future decisional situation, e.g., that "the systemic forces at work in the environment are such that state A eventually will decide to attack state B," the word "eventually" intentionally ducks the issue of when the decision will be made. I have grave doubts about trying to build hypotheses that say, for example, "the systemic forces at work in the enrironment are such that state A will decide within x months to attack state B." I am not sure that in making such assessments we could differentiate between an x of 3 months and, say, an x of 9 months. That is, when we see evidence, it is hard enough to judge its diagnosticicity with respect to the direction and strength of the systemic forces at work without cranking in <u>when</u> the decision eventually will be made. It also should be noted again that it will be even more difficult to have the hypotheses state when the action will occur. For example, the hypothesis "the systemic forces at work in the environment are such that a decision will be made within x months to attack within y months" would be extremely difficult to work with. The tree for the logical set of hypotheses for this situation is the one shown as Figure 4.10.

FIGURE 4.10
Tree Diagram for Logical Possibilities of what State A will Decide to Do Regarding Attacking B, Including when the Decision will be Made and when the Action will Occur

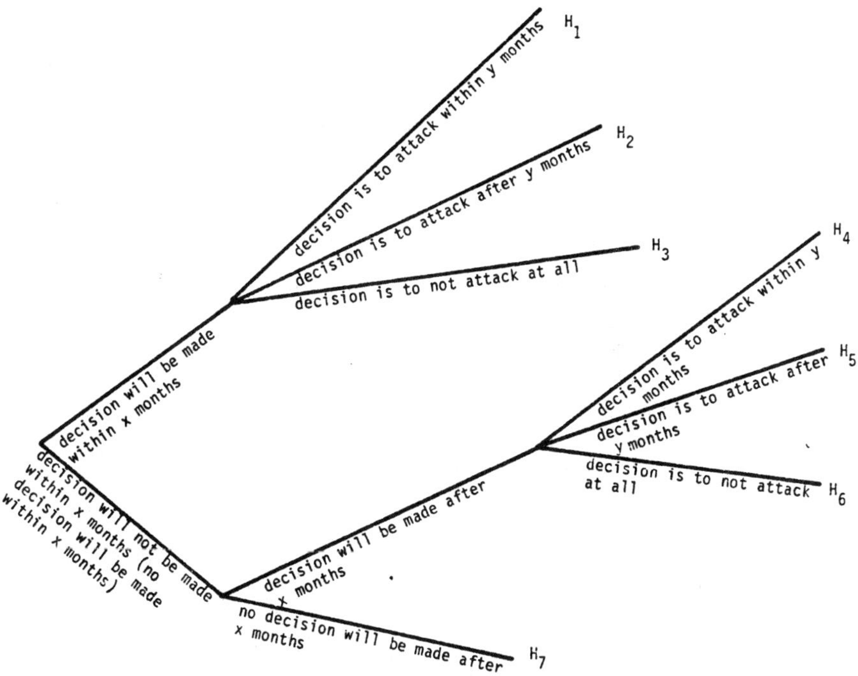

The set of seven hypotheses, then, is

H_1 = the systemic forces at work in the environment are such that state A will decide within x months that it will attack state B within y months,

H_2 = the systemic forces... state A will decide within x months that it will attack state B after y months,

H_3 = the systemic forces... state A will decide within x months that it will not attack state B at all,

H_4 = the systemic forces... state A will decide after x months that it will attack state B within y months,

H_5 = the systemic forces... state A will decide after x months that it will attack state B after y months,

H_6 = the systemic forces... state A will decide after x months that it will <u>not</u> attack state B,

H_7 = the systemic forces... state A will make no decision within x months or after x months about attacking state B.

This is much too complicated, I think, to be used. Note that it doesn't even include "hedging" hypotheses.

NOTES

Problem 1

1. Also, H_1, H_2 and H_3 could be true at the same time. Nor is the list exhaustive. Country A could be using another sort or poisonous gas, for example.

Problem 3

2. States really don't have "minds," of course. We are using an implicit "UAR" model here; i.e., we are assuming that the state is a unitary and rational actor, the most common assumption in political/military analyses.

3. Indeed, students in class have argued that we can't distinguish between H_1 and H_3, i.e., that the evidence will be as diagnostic of H_1 as of H_3. I think this will be true up until shortly before an attack. Some students very troubled by this have suggested that we consider abandoning Bayes and try to predict what will happen by instead using either a probability diagram (tree) (with an implicit unitary and rational actor model, or a systems model) or an explicit foreign policy decision making model. I think both of these alternatives may be highly useful, but feel that we should not give up so easily on trying to use Bayes.

Problem 6

4. Schweitzer (1978), p. 29.

5. I don't think Schweitzer's suggested item about the passage of time and the reduction of the time period remaining would get around the problem I have raised here, but I don't want to be dogmatic about this.

Problem 7

6. Dr. Hopple has pointed out that this line of argument is consistent with the literature on forecasting -- expert judgments are better for short term forecasts. Forecasts of five years or more in the future require time series data or similar methods, he notes. See problem 9, too.

Problem 8

7. If the state has changed its "mind," but no new evidence has emerged yet to show us this, our assumption may be wrong. See problem 51 below for a discussion of the non-stationarity of hypotheses problem.

8. Eventually (= finally) is a critical word here, because it helps to eliminate buried subhypotheses. Thus, we can write, as we do,

H_1 = state A eventually will decide to attack state B,

H_2 = state A eventually will decide to <u>not</u> attack state B

and not have the possibility of a hypothesis where state A has not yet made up its mind what to do.

9. The working tree for developing the set of hypotheses is as follows:

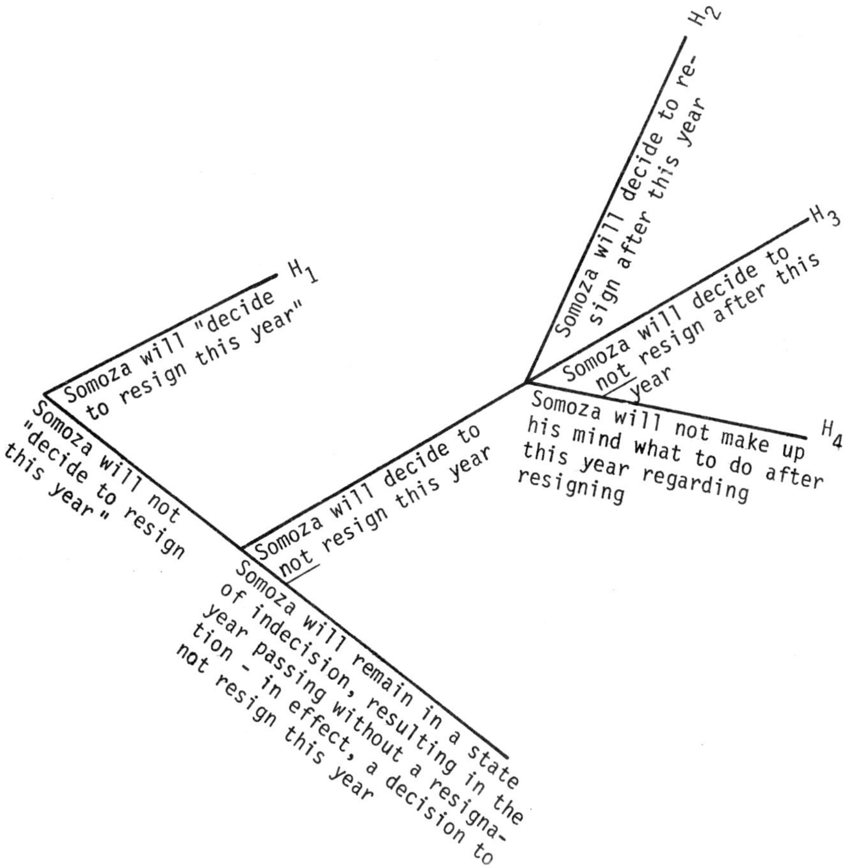

We thus end up with four hypotheses:

H_1 = Somoza will decide to resign this year,

H_2 = Somoza will decide to not resign this year but (and) to resign after this year,

H_3 = Somoza will decide to not resign this year and to not resign after this year,

H_4 = Somoza will decide to not resign this year but (and) will not make up his mind what to do after this year.

10. The present length of this manuscript prevents me from following through on Dr. Hopple's suggestion of another section (or even a chapter) on (i) Bayes vs other tools and (ii) Bayes with other tools. He notes he has often though about the potential utility of comparing Bayesian hierarchical inference results with causal modeling.

11. The buried subhypotheses can be discovered, as usual, by using a tree:

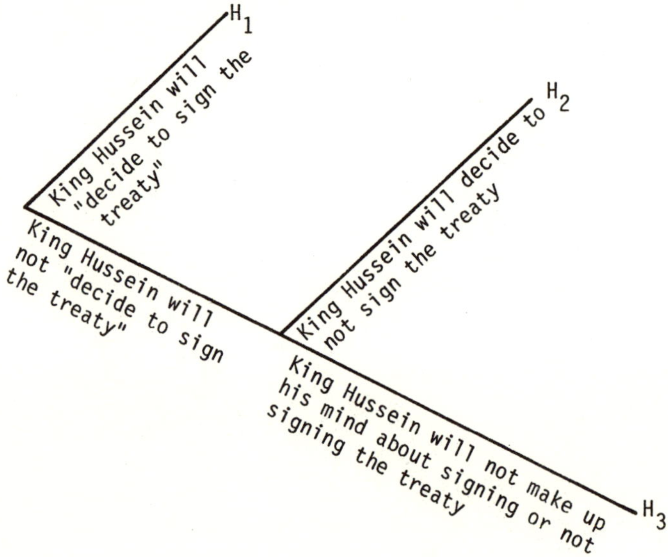

We thus end up with three hypotheses:

H_1 = King Hussein will decide to sign the treaty,

H_2 = King Hussein will decide to not sign the treaty,

H_3 = King Hussein will not make up his mind about signing or not signing the treaty.

Note that no time interval (period) is included in this set of hypotheses. Depending on the needs of the consumer of the analysis, it may or may not be necessary to include a time interval. However, as we argued at the end of subsection (8iva), it often is most difficult, indeed, really to take the time period into account.

12. I have heard Tom Murray mention this "Law of the Tool" several times in lectures, but I do not know with whom it originated. I also have seen it referred to as the "Law of the Hammer."

13. When I say it makes no sense to ask

$$P\left(\text{weather turns bad} \middle/ \begin{array}{l}\text{the systemic forces at work in the} \\ \text{environment are such that A will} \\ \text{attack}\end{array}\right)$$

I mean that it makes no sense if we ask this and require that we not be able to drop the hypotheses. In a Bayes problem we are asking $P(E_1/H_1)$, $P(E_1/H_2)$, etc., and expect to have at most conditional independence, i.e., at most to be able to drop previous evidence as in $P(E_2/H_1\&E_1) = P(E_2/H_1)$. We don't ever go to statistical independence where $P(E_1/H_1) = P(E_1)$. In this one instance, however, this is the only way it makes sense to ask something like

$$P\left(\text{weather turns bad} \middle/ \begin{array}{l}\text{the systemic forces at work in the envi-} \\ \text{ronment are such that A will attack}\end{array}\right).$$

We simply can say

$$P\left(\text{weather turns bad} \middle/ \begin{array}{l}\text{the systemic forces at work in the envi-} \\ \text{ronment are such that A will attack}\end{array}\right)$$

$$= P(\text{weather turns bad})$$

and $$P\left(\text{weather turns bad} \middle/ \begin{array}{l}\text{the systemic forces at work in the envi-} \\ \text{ronment are such that A will not attack}\end{array}\right)$$

$$= P(\text{weather turns bad})$$

because our definition of statistical independence, $P(B/A) = P(B)$ is that this is the case if the occurrence of A does not affect the probability of the occurrence of B. Clearly this is the case in terms of the effect of an "A" of "attack" (or an "A" of "no attack") on a "B" of "weather turns bad." Further since

$$P(\text{weather turns bad/attack}) = P(\text{weather turns bad})$$

and $$P(\text{weather turns bad/no attack}) = P(\text{weather turns bad}),$$

in a likelihood ratio the term

$$\frac{P(\text{weather turns bad/attack})}{P(\text{weather turns bad/no attack})}$$

will be

$$\frac{P(\text{weather turns bad})}{P(\text{weather turns bad})}$$

which equals $\frac{1}{1}$. This means of course, that the "evidence" (not evidence at all) of the weather turning bad has had no diagnostic impact, as we would like.

Problem 9

14. Again, Dr. Hopple points out that "this is also in line with the argument in the forecasting literature that expert data are better for short range forecasting."

5
Problems Involving Conditional Independence and Dependence

In this chapter we will deal with problems involved with conditional independence and dependence.

PROBLEM 10. Incorrect results from using conditional independence: the problem of mathematically invalid results because of the use of the conditionally independent form of Bayes rather than the conditionally dependent form.

The problem: In assessing the probabilities of items of evidence, E_i, we either can use the conditionally dependent form of Bayes, where previous items of evidence are considered and we assess $P(E_i/H_j\&E_1\&E_2\&\cdots\&E_{i-1})$ for each H_j, or we can use the conditionally independent form of Bayes, where previous items of evidence are ignored and we assess $P(E_i/H_j)$ for each H_j. The use of the conditionally independent form may well result in different assessments than if the conditionally dependent form is used. Mathematically speaking, the conditionally dependent assessments are the correct ones to use.

Because of a number of problems that are encountered if the conditionally dependent form of Bayes is used, if we could ignore the mathematical correctness of conditional dependence and were free to choose which form of Bayes to use, the choice well might be to use conditional independence.

The solution: In spite of the problems involved with using the conditionally dependent form of Bayes, the attempt should be made to use it.

Discussion: This is a very troublesome issue. Unfortunately, there appears to be a lack of realization, or at least a lack of discussion, of this issue in most of the recent internal writings on the use of Bayes in intelligence analysis. Indeed, there also has been an implicit assumption of conditional independence in most, if not all, of the intelligence applications in the last ten years of normal (i.e., non-hierarchical) Bayesian analysis. This is, it can be argued, a poor way of dealing with the problem.

If we were free to choose, on the basis of ease in using Bayes, whether to use the conditionally independent or conditionally dependent form of Bayes, the choice, for reasons to be discussed below, might well be to use the independent form. As will be argued below, however, a strong case also could be made for conditional dependence, if we were free to choose.

From the viewpoint of mathematics, however, we cannot choose. As has been seen in the probability problems presented, we must start out with the conditionally dependent form of Bayes. The conditionally independent form can be used only if it is justified mathematically. In simple probability problems of the coin, die, card, or urn type, there is little difficulty in determining if conditional independence is justified. Nothing is changed when coins and dies are tossed or thrown; hence, conditional independence is typical of Bayes problems dealing with them. When dealing with Bayes problems about drawing from urns, if there is no replacement, then there is conditional dependence, while if there is replacement, then there is conditional independence. (This is somewhat oversimplified, but does apply to the simple types of coin and die problems we have presented.)

When we move beyond these types of simple problems, however, matters become more complicated. Conditional independence and dependence actually can exist in the same problem. As an example, consider the (artifical?) situation where a doctor feels that a patient either has disease I or disease II (and the doctor has some justification for feeling that these alternatives are mutually exclusive and exhaustive). Suppose that two diagnostic tests have been ordered, and that each will show a result either of positive or negative. Figure 5.1 shows the tree for this.

FIGURE 5.1
Tree Diagram for Two Possible Diseases and Two Diagnostic Tests

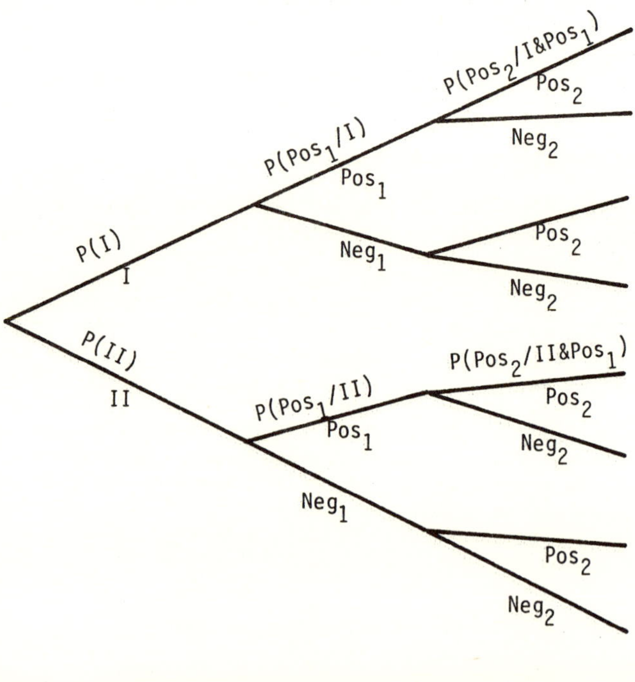

Now, it could be the case that data (based on frequency counts of past tests) show that, if the patient has disease I:

$P(Pos_2/I\&Pos_1) = P(Pos_2/I)$,

i.e., that we have conditional independence if the patient has disease I. On the other hand, data (again based on frequency counts of past tests) could show that, if the patient has disease II:

$P(Pos_2/II\&Pos_1) \neq P(Pos_2/II)$,

i.e., that we have conditional dependence if the patient has disease II. In this latter case, the correct value to use in our Bayes problem is the conditionally dependent value of $P(Pos_2/II\&Pos_1)$, rather than the conditionally independent value of $P(Pos_2/II)$. Note that if we <u>did</u> have conditional independence, i.e., <u>if</u> it were true that $P(Pos_2/II\&Pos_1) = P(Pos_2/II)$, then we could go ahead and use the value $P(Pos_2/II)$ in our calculations -- but then in actuality we really still would be using the conditionally dependent value of $P(Pos_2/II\&Pos_1)$ because it would equal the value $P(Pos_2/II)$ that we would be using. In this sense, then, if we are mathematically correct we always really will use the conditionally dependent <u>value</u>. It's just that, when we have conditional independence, we can simplify our <u>notation</u> and no longer <u>explicitly consider</u> prior evidence when we use the conditionally independent expression.

When we leave the objective or symmetric probability world of coins, dies, cards, and urns, and also pass beyond the world of frequency counts (which could be helpful in medical diagnosis work, etc.) and arrive in the almost completely subjective probability world of political/military problems, we still have the same mathematical rule. That is, if we assign subjective probabilities to $P(E_i/H_j\&E_1\&E_2\&\cdots\&E_{i-1})$ and to $P(E_i/H_j)$ and it turns out that $P(E_i/H_j\&E_1\&E_2\&\cdots\&E_{i-1}) \neq P(E_i/H_j)$, then we do <u>not</u> have conditional independence of E_i given H_j (or "under H_j"), and the value that <u>mathematics</u> tells us we <u>must</u> use is $P(E_i/H_j\&E_1\&E_2\&\cdots\&E_{i-1})$. If we do not use this value our calculations will be mathematically invalid. Only if our assigned subjective probabilities are such that $P(E_i/H_j\&E_1\&E_2\&\cdots\&E_{i-1}) = P(E_i/H_j)$, i.e., only if it turns out that conditional independence holds, can we go ahead and use the value that we have assigned for $P(E_i/H_j)$. (Note again that in this case we really are using the equal value of $P(E_i/H_j\&E_1\&E_2\&\cdots\&E_{i-1})$ -- it's just that the conditional independence notation is simpler and there is no explicit consideration of prior evidence.)

When all is said and done, then, <u>mathematically</u> speaking the <u>proper</u> thing always to do is to use the value $P(E_i/H_j\&E_1\&E_2\&\cdots\&E_{i-1})$. We never can go wrong this way. If it happens that $P(E_i/H_j\&E_1\&E_2\&\cdots\&E_{i-1}) = P(E_i/H_j)$, i.e., if it happens that we have conditional independence, then we still are using the value of $P(E_i/H_j\&E_1\&E_2\&\cdots\&E_{i-1})$ in our calculations. There actually is no point in even assessing $P(E_i/H_j)$. We only need to assess (and use) $P(E_i/H_j \& E_1 \& E_2 \& \cdots E_{i-1})$.

(The reader may object at this point by pointing out that although complete rigor would require the use of conditional dependence, the assumption of conditional independence may simplify analyses and calculations immensely and that for such reasons mathematicians routinely assume statistical independence, so why not assume conditional independence? While I accept this line of argument, my point is that complete rigor does indeed require conditional dependence, and that we can err significantly even if we openly state that after examining the issues we still will assume conditional independence. The example to follow is illustrative).

As another example, then, of how we can go wrong and get incorrect results if we use conditional independence, consider the following paraphrased from the writings of David A. Shum.[1] (His hypotheses, which, he notes, are only two in number for the sake of simplicity, are open to the criticism of not being properly formulated, but the example is his and should be presented as he expressed it.)

Consider two countries, A and B, which are in a state of "peace." Suppose that an analyst is running a Bayesian analysis where H_w is the hypothesis that country B is planning to wage war against country A in the very near future and H_p is the negative of H_w, i.e., H_p is "peace." Suppose that the analyst receives the following data: D_1 is the report of suspected acts of sabotage by agents of country B at a certain military installation in A, while D_2 is the report of the assassination by agents of country B of a member of country A's Joint Chiefs of Staff.

If we are using the conditionally independent form of Bayes, the judgments to be made are for $P(D_1/H_w)$, $P(D_1/H_p)$, $P(D_2/H_w)$, and $P(D_2/H_p)$. The judgments probably would be such that

$$P(D_1/H_w) > P(D_1/H_p) > 0$$

and $P(D_2/H_w) > P(D_2/H_p) > 0$

i.e., that D_1 and D_2 are more likely under the assumption of war than under the assumption of peace, and that no probabilities are 0.

From the above inequalities, we get:

$$P(D_1/H_w) \cdot P(D_2/H_w) > P(D_1/H_p) \cdot P(D_2/H_p) > 0$$

Dividing by $P(D_1/H_p) \cdot P(D_2/H_p)$, we get

$$\frac{P(D_1/H_w)}{P(D_1/H_p)} \cdot \frac{P(D_2/H_w)}{P(D_2/H_p)} > \frac{1}{1}.$$ That is,

the product of the likelihood ratios for D_1 and D_2 is greater than

$\frac{1}{I}$, which means that the revised probability for H_w in light of the evidence will be driven up. (This is intuitively what we would suspect, of course.)

But what about the assumption of conditional independence? This may be reasonable for the "peace" (i.e., no war) hypothesis. Schum notes that "even in what one refers to these days as 'peace' there are occasional acts of sabotage and attempted or actual assassinations of key personnel. If there is peace, one more often expects these events at unrelated times,"[2] and hence an assumption of conditional independence (which is an implicit judgment that $P(D_2/H_p\&D_1) = P(D_2/H_p)$, i.e., that knowing that D_1 has happened does not affect the assigned probability that D_2 would happen, under the peace hypothesis) is reasonable.

On the other hand, if it is assumed that B is planning to wage war in the near future, the [almost] "simultaneous occurrence of these events might be rendered more highly probable," and indeed the analyst, if asked, probably would say that

$$P(D_2/H_w\&D_1) > P(D_2/H_w).$$ [3]

In other words, the diagnostic value of D_2 has been altered by knowledge of D_1 when H_w is assumed; the analyst is saying that knowing that D_1 has happened increases the probability that D_2 would happen, when H_w is assumed.

What is the impact of this on the revised probability of war? We have

$$P(D_2/H_w\&D_1) > P(D_2/H_w).$$ Multiplying by $P(D_1/H_w)$ we get

$$P(D_1/H_w)\cdot P(D_2/H_w\&D_1) > P(D_1/H_w)\cdot P(D_2/H_w).$$

Dividing by $P(D_1/H_p)\cdot P(D_2/H_p)$ we get

$$\frac{P(D_1/H_w)}{P(D_1/H_p)} \cdot \frac{P(D_2/H_w\&D_1)}{P(D_2/H_p)} > \frac{P(D_1/H_w)}{P(D_1/H_p)} \cdot \frac{P(D_2/H_w)}{P(D_2/H_p)}.$$

Thus, the product of the likelihood ratios is larger when we consider conditional dependence between D_1 and D_2 under H_w. This will drive up the revised probability of H_w more than before. We thus may be led more quickly towards acceptance of H_w by considering conditional dependence between D_1 and D_2. Mathematically, of course, this is the proper thing to do, as we have argued.

An implication of this example is that indications and warning (I and W) analysts may do a disservice to decision makers if the conditionally independent form of Bayes is used. This is because, as the example indicated, the conditionally independent form may understate the proper probability revisions of the "war" or "attack" hypothesis.

A further implication is that in many Bayes problems when we go ahead, as suggested, and assess all our probabilities using the conditionally dependent form of Bayes, we will find that the "peace" or "no attack" sort of hypothesis will be such that we will discover that we have conditional independence, so that although we do consider previous items of evidence, they will have no impact.

Although conditional dependence is the mathematically correct choice in using Bayes, there are a number of practical and psychological considerations that may make it difficult to do the proper thing, or even advantageous from the view of ease (rather than from the view of correctness) to just go ahead and use the conditionally independent form of Bayes. That is, for a number of reasons, we may want to assign a subjective value to and use $P(E_i/H_j)$ without ever even considering whether it is justified, i.e., without even assigning subjective values to $P(E_i/H_j\&E_1\&E_2\&\cdots\&E_{i-1})$ and $P(E_i/H_j)$ and checking to see whether or not $P(E_i/H_j\&E_1\&E_2\&\cdots\&E_{i-1}) = P(E_i/H_j)$.

It probably is worth a little time to discuss the pros and cons of conditional independence and dependence from the viewpoint of practical and psychological considerations. So, if we forget, for a moment, what is correct mathematically and suppose that we simply could <u>choose</u> either to use the conditionally independent or dependent form of Bayes, what might our arguments be.

Considering first the <u>advantages of conditional independence</u>, the <u>first advantage</u> is that, if we are using the conditionally independent form of Bayes and find out subsequent to a probability judgment for an item of evidence, E_{21}, say, that the event E_{21} never happened (or was quite different in nature from what we thought it was), we need only eliminate or reassess $P(E_{21}/H_j)$ for all H_j. If we were using the conditionally dependent form of Bayes, however, we would have to reassess all the subsequent items of evidence (events) that we had considered before finding out that E_{21} was improperly reported, i.e., we would have to reassess $P(E_{22}/H_j\&E_1\&E_2\&\cdots\&E_{21})$ for all H_j and reassess $P(E_{23}/H_j\&E_1\&E_2\&\cdots\&E_{22})$ for each H_j etc. up to our latest event in Bayes.

A <u>second advantage</u> often claimed for the conditionally independent form of Bayes is that we should make the same probability assessments, and hence get the same final products, regardless of the order in which we consider the items of evidence. People assert then that "it really makes no difference in what order we consider the evidence."

This looks fairly reasonable at first glance, but we might ask if we really can assess $P(E_5/H_j)$, say, without subconsciously remembering E_1, E_2, E_3, and E_4 and somehow letting these events color our assessment of $P(E_5/H_j)$. We are not computers, but fallible humans, and may subconsciously reintroduce conditional dependence into the calculations we make. If we do so, order is quite likely to make a difference. So the assertion that, when we assess $P(E_1/H_j)$, $P(E_2/H_j), \cdots$, $P(E_n/H_j)$ the order in which we consider the evidence makes no difference is true mathematically, but probably is not really true. (We should not end our discussion of this point without noting that it can be shown (and will be later in problem 12 after Figure 5.3) that the order of the evidence (events), strictly speaking, should make no "difference" even if we have conditional dependence. The order will affect the individual probabilities assigned but their product should be the same. Again, because of the fallibility of humans, in practice order will make a difference.)

A third advantage cited for conditional independence is that it is simply much easier to assess $P(E_7/H_j)$, say, than to assess $P(E_7/H_j\&E_1\&E_2\&\cdots\&E_6)$. That is, there are limits, according to psychologists, as to the number of variables humans can consider simultaneously. The reasonable range is 7 ± 2, (i.e., between 5 and 9), according to some research.[4] Trying to process $P(E_{50}/H_i\&E_1\&E_2\&\cdots\&E_{49})$ is very difficult, at best, and probably impossible in the sense that we really cannot keep in mind H_j and all 49 previous items of evidence when we assess $P(E_{50}/H_j\&E_1\&E_2\&\cdots\&E_{49})$. What we almost certainly do in making such an assessment is to ask the probability of E_{50} given the particular H_j, and pick out or remember a small number (5 to 9?) of what appear to be the most pertinent (with respect to assigning the probability of E_{50}) of the previous items of evidence. If we are systematic, as we should be, in assigning the probability of E_{50}, we should read through the previous 49 items of evidence, methodically pick out those items which are most relevant to assigning the probability of E_{50}, and provide a written (or taped) justification of the probability we assign, citing the roles of the selected items of evidence.

While the above constitute some of the arguments in favor of independence, there are a number of arguments against it, or disadvantages. First, we can argue that it probably is very unrealistic to use conditional independence, to ask, say $P(E_4/H_j)$, because we should consider the previous items of evidence. That is, we really should want to and should indeed calculate $P(E_4/H_j\&E_1\&E_2\&E_3)$. (Which is, of course, correct mathematically, but remember that we are forgetting the mathematical imperatives and simply considering conditional independence and dependence on their practical and psychological merits.) Political/military analysis, it can be argued, is a process of fitting the bits and pieces of evidence together, and a process which tells us to consider evidence in isolation may miss the mark by a wide margin. (See the example added in footnote 1.)

One author, perhaps unwittingly, undermined a mindless choice of conditional independence when he noted the implication that its usage means that the probability of the occurrence of a unit of evidence (given a hypothesis) is no different if the evidence appears in November than if it appears in October, or September, say. Our immediate reaction might well be -- "but what about all the things that happened between September and November?"

Moreover, the possible patterns of development linking all those things that have happened will not be taken into account if conditional independence is used. This pattern could be of significant diagnostic value. If we treat the issues of conditional independence vs. conditional dependence as a choice to be made on practical and psychological grounds, then this sort of problem where previous evidence is ignored makes conditional independence look like a questionable choice indeed.

Second, in addition to arguing that we really should consider previous items of evidence when assessing the conditional probability of our latest item of evidence, we can argue that we really can't ignore the previous items of evidence. Even if subconsciously -- because we know the previous items of evidence-- they will enter into our calculations. We probably would be better off simply to recognize this and explicitly consider the previous items of evidence by using the conditionally dependent form of Bayes. Then, we would get the rigor of an explicit consideration instead of the ambiguity and potential bias of a subconscious consideration.

In an attempt to gain the purported advantages of conditional independence, (i.e., the ease in modifying the calculations if an item or processed data is later found to be bad, the supposed irrelevance of the order in which the evidence is considered, and the claimed ease in making the probability assessments when one ignores previous items of evidence), while at the same time facing up to the unreality of ignoring previous data, some analysts try to have the best of both conditional independence and dependence by saying, in essence, "Well, although we formally use the conditionally independent form of Bayes, we really do take into account previous items of evidence when we assign our conditional probabilities to events." We can counterargue that if we are going to consider previous items of evidence, then we should do so explicitly by using the conditionally dependent form of Bayes and face up to the possible problems of using conditional dependence.

Let us now turn to an examination of the advantages of and problems with the conditionally dependent form of Bayes. The first and very obvious advantage is the reality of the assumption of conditional dependence. Political/military analysis involves the fitting together and simultaneous consideration of a number of items of evidence, and assessing and using $P(E_6/H_j)$, say, may be very unrealistic, whereas assessing and using $P(E_6/H_j\&E_1\&E_2\&\cdots\&E_5)$ may be far more realistic. The use of the conditionally dependent formulation allows us to look at the previous items of evidence and to incorporate into $P(E_i/H_j\&E_1\&E_2\&\cdots\&E_{i-1})$ patterns of development of the evidence which may be of diagnostic value. In short, as student Earl Blake noted, "...prior evidence may be in fact essential to making a valid probability assessment. Therefore, using conditional dependence would seem to be the most accurate and realistic approach."[5] (Again, see the example added in footnote 1.)

The _first major disadvantage_ of conditional dependence is that, if we discover, some time after we have assessed $P(E_i/H_j\&E_1\&E_2\&\cdots\&E_{i-1})$ for all H_j, that E_i is "bad" in the sense of (1) not having occurred after all, or (2) having happened differently than we thought so that E_i really is very different from our previous conception of it, then we must _initially_ either (1) drop out the assessments $P(E_i/H_j\&E_1\&E_2\&\cdots\&E_{i-1})$ for all H_j in the case that E_i didn't happen or (2) reassess the values $P(E_i/H_j\&E_1\&E_2\&\cdots\&E_{i-1})$ for all H_j in the case that E_i is very different from our previous conception of it.

Unfortunately, this is only the start of our reassessments. We _next_ must go on and reassess $P(E_{i+1}/H_j\&E_1\&E_2\&\cdots)$, $P(E_{i+2}/H_j\&E_1\&E_2\&\cdots)$, _etc._ for all H_j. That is, we have to redo _all_ our assessments for _all_ events subsequent to E_i that we had made before we found out that E_i was "bad." This can amount to a considerable degree of work.

Moreover, there is a serious question concerning the reliability of such reassessments. That is, our reassessments may well be affected by the _subsequent_ events that we know have occurred. For example, if we have made assessments for 10 events and later find out that event #4 never happened, how can we reassess our conditional probabilities for events 5, 6, 7, etc. and not have our judgments colored by events subsequent (rather than just prior) to them that we know have occurred? It may, for example, be difficult indeed to reassess for event #5 given events 1, 2, and 3, without at the same time at least subconsciously remembering that events 6-10 also have occurred. Ideally, we should make assessments only once, and it may well be that any reassessments are inherently flawed by knowledge of subsequent events.

Occasionally, someone suggests that a solution to the problem of an E_i that turns out to be bad because it never happened simply is to delete the assessments of $P(E_i/H_j\&E_1\&E_2\&\cdots\&E_{i-1})$ for all H_j and not bother to do any reassessing for $P(E_{i+1}/H_j\&E_1\&E_2\&\cdots)$, $P(E_{i+2}/H_j\&E_1\&E_2\&\cdots)$, etc. The judgments of subsequent events E_{i+1}, E_{i+2}, etc. probably would be little affected if one item, E_i, were removed, it is argued. This argument seems much more reasonable if "i" is fairly large, say 7 or greater. For example, if we are considering E_7, then, when we assess $P(E_8/H_j\&E_1\&E_2\&\cdots\&E_7)$, there are 7 items of data being considered, so if one (e.g., E_7) is deleted, the net effect may be little --it may well be that $P(E_8/H_j\&E_1\&E_2\&\cdots\&E_7)$ is very close to or identical to $P(E_8/H_j\&E_1\&E_2\&\cdots\&E_6)$. There are three obvious basic flaws in this simplistic line of argument, however.

The first is that unless we actually do go ahead and assess, say, $P(E_8/H_j\&E_1\&E_2\&\cdots\&E_6)$, we won't know for sure if it really is essentially the same as $P(E_8/H_j\&E_1\&E_2\&\cdots\&E_7)$. In other words, we just can't be sure. E_7 might be critical in assessing

$P(E_8/H_j\&E_1\&E_2\&\cdots\&E_6)$. We must reassess to find out.

The second flaw is that even if it were the case that E_7 might not be critical in assessing, say, $P(E_8/H_j\&E_1\&E_2\&\cdots\&E_6)$, there is no way in advance of predicting if it might not be critical in assessing, say, $P(E_{21}/H_j\&E_1\&E_2\&\cdots\&E_{20})$. Finding out that E_7 didn't happen might be critical for the reassessment of $P(E_{21}/H_j\&E_1\&E_2\&\cdots\&E_6\&E_8\&\cdots\&E_{20})$; we never will know unless we do the reassessing.

A third flaw is that it also is conceivable that under reassessments it might turn out that for some H_j it would make no difference if E_7 were included or left out, while for other H_j it would make a difference. Many patterns are possible, and probably not predictable.

The rebuttal line of argument offered to these three objections is that, if we have given detailed explanations for the probabilities assigned, then we only need to reassess for those probabilities where the bad E_i influenced the judgments. For example, if E_7 is found not to have happened, then after deleting $P(E_7/H_j\&E_1\&E_2\&\cdots\&E_6)$ for all H_j the suggestion then is to reassess only where E_7 is mentioned in the verbal justifications for the conditional probabilities assigned. This seems to be a much more defensible line of argument and at first glance seems to be a fairly reasonable suggestion. The fatal flaw, however, is that even if for some $E_i (i>7)$, it was true that $P(E_i/H_j\&E_1\&E_2\&\cdots\&E_{i-1})$ depended on a linkage of say, E_3 to E_4 to E_5 to E_6 to E_7 to E_8 to E_9, etc., and that E_7 was not mentioned explicitly, if subsequently E_7 had to be deleted and the chain broken, it might be that $P(E_i/H_j\&E_1\&E_2\&\cdots\&E_6\&E_8\&E_9\&\cdots\&E_{i-1})$ would be rather different. That is, when E_7 was included, it was not one of the major factors in assigning most probabilities, and would not be mentioned in the list of items of evidence most salient for assigning probabilities to subsequent events. But, the deletion of E_7 might be critical in breaking up patterns and reprioritizing the influence of previous items of evidence in assigning probabilities for subsequent events. The moral, when all is said and done, still is to redo all assessments.

A second major disadvantage of conditional dependence has been mentioned before. It is that the human mind really can only consider a limited number of variables at a time, perhaps on the order of five to nine, and there is a serious question if it is meaningful to ask, say, $P(E_{50}/H_j\&E_1\&E_2\&\cdots\&E_{49})$. The human mind simply cannot consider 49 previous items of evidence all at once. As noted before, what probably happens is that a relatively few of the most salient (to the assessor) items in the list of 49 are considered (by being remembered, or, better, consciously selected) and greatly influence the assessment. So, all the evidence really is not considered when the probability judgments are made.

In doing the analysis, it normally will be the case that the most salient items of previous evidence will vary as probability assessments are made. For example, in assessing, say, the probability $P(E_{50}/H_j\&E_1\&E_1\&\cdots\&E_{49})$ for a given H_j, it might be that H_j and E_6, E_7, E_{21}, E_{29}, E_{47} were the most important elements, while in assessing $P(E_{51}/H_j\&E_1\&E_2\&\cdots\&E_{50})$ for the same H_j, it might be that H_j and E_{12}, E_{15}, E_{23}, E_{37}, and E_{41} were the important elements.

This selection of the most "salient" items of previous evidence means, as noted, that not all the previous evidence will be considered when the conditional probability judgments $P(E_i/H_j\&E_1\&E_2\&\cdots\&E_{i-1})$ are made. This raises significant questions about the reliability of such judgments. Different individuals assessing the same probability almost certainly will have different sets of salient events in mind when they assess the probabilities. So, the old problem of analysts disagreeing over the evidence that has been considered will re-emerge. Moreover, even the same analyst may, from day to day, change his or her mind as to what the most salient items of evidence are, so that, if there is debate over a particular assignment of a probability, the analyst should come prepared with documentation for the justification of the original choice of the "salient" items of evidence. This will help the analyst retrace his or her original line of thought.

Another possible difficulty involved with choosing only a small number of salient events is that if, for reasons of carelessness, fatigue, or whatever, there is no complete look at all the evidence to select the most salient items each time an assessment is made, then we can have a problem of continuing to use the <u>same</u> set of salient events. In this situation, this set of salient events may bias subsequent probability assessments, not letting the other items of evidence carry the weight they should (and would if the human mind <u>could</u> consider all the evidence).

A <u>third major disadvantage</u> of conditional dependence is that, if we consider items of evidence in different orders we will end up with different results. As we noted earlier, however, even if the independent form of Bayes is used, the <u>subconscious</u> recall of items of evidence previous to the one being assessed will introduce, in effect, conditional dependence, with the net result that order is quite likely to make a difference. It might turn out, however, that an <u>explicit</u> consideration of evidence under the assumption of conditional dependence might be more sensitive to the order of the evidence than would be the case if the same evidence were considered under the explicit assumption of conditional independence. That is, we might have greater differences from the different orders if we used conditional dependence. (A research design for such an experiment would be challenging.)

The point should be made again that it can be shown (and will be, later in problem 12 after Figure 5.3) that order <u>should</u>, <u>mathematically</u> speaking, make no difference whether we have conditional independence or dependence. In the case of conditional independence, the values assigned for any given item of evidence _should be_ identical for all orders of the evidence. In the case of conditional dependence, the order would affect the individual probabilities assigned, but the products _should be_ the same for a given hypothesis. The products will not be the same because of the fallibility of the human mind. We thus will end up with inconsistent results if we consider the data in different orders.

A summary of our arguments about the practical and psychological advantages and disadvantages of conditional independence and dependence can be presented in matrix form as follows (where "A" means an advantage and "D" a disadvantage):

	Conditional Independence	Conditional Dependence
A:	If we find out an E_i either didn't happen or is different from what we thought, reassessments are easy.	It is realistic to consider previous items of evidence.
	We will make the same probability assessments regardless of the order in which we consider the items of evidence. (Probably not completely true, but _perhaps_ more so than if we have conditional dependence.)	We really can't ignore previous items of evidence - they will enter our calculations subconsciously even if we try to ignore them. We probably are better off explicitly considering them.
	It is much easier to assess $P(E_i/H_j)$ than to assess $P(E_i/H_j \& E_1 \& E_2 \& \cdots \& E_{i-1})$	
D:	It's not realistic - political/military analysis is a process of fitting bits and pieces of evidence together. Ignoring previous items of evidence leads to inadequate analysis.	It is very time-consuming to make the necessary reassessments if an item of evidence E_i is later found not to have happened or to have been different from what we thought. Such reassessments may be unreliable in that they may be affected by the subsequent events that are known to have occurred. Reassessments must be made for _all_ subsequent events.

We really can't ignore previous items of evidence - they will be remembered subconsciously. We are better off considering them consciously where we get the rigor of an explicit consideration instead of the ambiguity of a subconscious consideration.

The human mind can consider only a limited number of items of evidence at a time, so when we assess $P(E_i/H_j\&E_1\&E_2\&\cdots\&E_{i-1})$ where $i > 7$, we really will not consider all the items of evidence prior to E_i, but instead only a subset (of perhaps 5-9 items) of the most salient items of evidence.

If we consider items of evidence in different orders, we essentially are certain of obtaining different probability judgments (which will be inconsistent). The inconsistency may be greater than the inconsistency when conditional independence is used.

We should close our discussion of conditional independence and dependence by pointing out again that, from a mathematical perspective we are <u>not</u> free to <u>choose</u> whether or not to use conditional dependence or independence. In essence, the correct values always are the conditionally dependent ones.

Because of the purported advantages of conditional independence, or because of lack of knowledge of the distinction between conditional dependence and conditional independence, or because of confusion between the concepts of distinct items of evidence and conditional independence (the next problem to be discussed), we find that the conditionally independent form of Bayes is the form almost universally used in intelligence analyses. This is most unfortunate. Despite the greater complexity inherent in the conditionally dependent form of Bayes, its advantages warrant attempting to use it.

<u>PROBLEM 11</u>. Confusing "conditional independence" and "distinct": the problem of incorrectly equating the concepts of "conditional independence" and "distinct" items of evidence.

<u>The problem</u>: Because of an awareness of some of the difficulties involved in using the conditionally dependent form of Bayes, some years ago it was suggested that it might be useful to try to "segment" items of "conditionally dependent evidence" into clumps or units of evidence that would be "conditionally independent with

respect to one another." Unfortunately, this approach appeared to ignore the fact (as the Schum example in the previous section demonstrated) that there could be conditional independence of evidence under one hypothesis, but conditional dependence of evidence under another hypothesis. Even more unfortunately, this approach has been misinterpreted, with some later practitioners incorrectly believing that, if they could segment evidence verbally into "distinct" clumps or units of evidence, those clumps or units then would be "conditionally independent." This is definitely wrong!

The solution: While it often is useful to combine reports on the same subject matter into one unit or item of "evidence" so that there will not be multiple reports on the same basic event, this will not necessarily mean that "conditional independence" has been achieved. "Conditional independence" and "dependence" are properties that also involve hypotheses, not just the evidence. "Distinct" items of evidence may be conditionally independent under some hypotheses and conditionally dependent under other hypotheses, or even conditionally dependent under all hypotheses (or conditionally independent under all hypotheses). Do not equate "distinct" and "conditionally dependent."

Discussion: This problem is closely related to the last problem. It appears that this problem has arisen because of Jack Zlotnick's efforts to "segment" items of "conditionally dependent evidence" into clumps or units of evidence that would be "conditionally independent with respect to one another." (Since the Schumm example showed that there could be conditional independence of evidence under one hypothesis but dependence of evidence under another, this effort, which ignored the role of the hypotheses, was doomed to failure.)

The basic problem is that Zlotnick gave a rule to use to see if we have distinct units of evidence, where distinct does not mean mutually exclusive but rather completely different with respect to content. This conception of "distinct" is absolutely critical. The items of evidence, whether distinct or not, definitely are not mutually exclusive -- each has indeed occurred. If we apply the rule and the results are "positive," then we have conditionally independent units of evidence, which are also distinct. However, if Zlotnick's rule is applied and the results are "negative," we have conditionally dependent units of evidence which may or may not be distinct. When all is said and done, "distinct" and "conditionally independent" are overlapping, but not identical concepts. Some analysts, however, seem to have been misled into thinking that if they can "segment" their evidence into "distinct" units (when they do so they generally do not use any Zlotnick type test, but just verbally segment the evidence) then it will follow that the units of evidence will be conditionally independent. This simply is not the case.

PROBLEM 12. Results affected by the order of the evidence: the problem that the order in which we consider the items of evidence is itself an important factor in determining the revised probabilities of hypotheses when we use the conditionally dependent form of Bayes.

The first sub-problem: Items of evidence often will be received where the time of occurrence of each is specified, but where the items are received out of sequence and assessed in this order of receipt rather than in the order of their occurrence. Because the items of evidence do include the time when they occurred, it turns out that every ordering of the evidence represents the same composite[6] event. It then follows that, mathematically, it should make no difference in which order the evidence is considered -- each ordering should lead to the same revised probabilities for the hypotheses.[7]

In practice, however, because we are humans and not computers, order will make a difference - we will assign inconsistent probabilities for different orders of evidence. The nub of the problem is that, if we assess items of evidence out of the time order in which they occurred, the revised probabilities for our hypotheses may be considerably different from what they would have been if we had assessed the items of evidence in the time order in which they actually occurred. (This issue was raised before in our problem 10 discussion of conditional independence and dependence.)

The solution to the first sub-problem: Attempts should be made to consider the items of evidence in the order in which they occurred, but this is not always possible. As student James McGowan pointed out, evidence will not necessarily be collected in the same sequence in which it occurred, nor will it necessarily arrive at the analyst's desk in the sequence in which it occurred. Particularly in crisis situations, evidence may be received out of order. If at all possible, once the correct time order is known, we should put the evidence in order and go back and reassess. (Use of the conditionally independent form of Bayes is a possible solution, but it is not advised.)

The second sub-problem: Items of evidence sometimes will be received where the time of occurrence of each is not specified. Because the items of evidence do not include the time when they occurred, it turns out, because of the way we reason when assigning conditional probabilities, that every ordering of the items of evidence represents a different composite event. In this case the problem of "inconsistency" of results for different orders of the evidence disappears -- we can say honestly that since different orders of evidence actually represent different composite events, there is no reason to expect consistency. Unfortunately, however, a new problem arises. If different orders really do represent different composite events, the problem is which order (with its resultant revised probabilities of hypotheses) we should choose for the items of evidence in the absence of information about when they occurred.

The solution to the second sub-problem: Unfortunately, there is no obvious solution to this problem. It should be pointed out again that our discussion is based on the assumption that the conditionally dependent form of Bayes is being used. If the conditionally independent form of Bayes is used, problems concerning the order of evidence almost completely disappear. I still do not suggest abandoning conditional dependence and going to conditional independence, however. This is, of course, a matter of judgment, and I do not mean to be dogmatic on this point.

Discussion of the first sub-problem: Consider items of evidence E_1, E_2, and E_3 where the time of occurrence of each is specified and, for the sake of simplicity there are, only two hypotheses, H_1 and H_2. (The illustration to follow can be easily expanded to multiple hypotheses.)

If we use our traditional tree format, we can illustrate the situation in Figure 5.2.

FIGURE 5.2
Probability Tree Diagram for Two Hypotheses and Events E_1, E_2, E_3 Considered in that Order

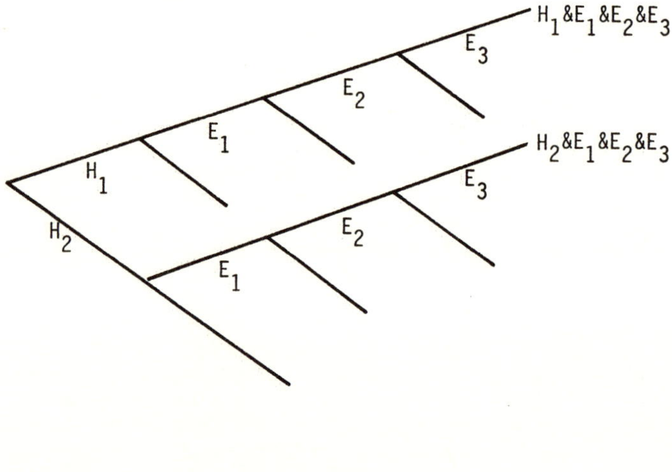

Suppose, on the other hand, we considered the data in this order: E_3, E_2, E_1. Then, the tree to illustrate this situation would be the one in Figure 5.3.

FIGURE 5.3
Probability Tree Diagram for Two Hypotheses and Events E_1, E_2, E_3 Considered in the Order E_3, E_2, E_1

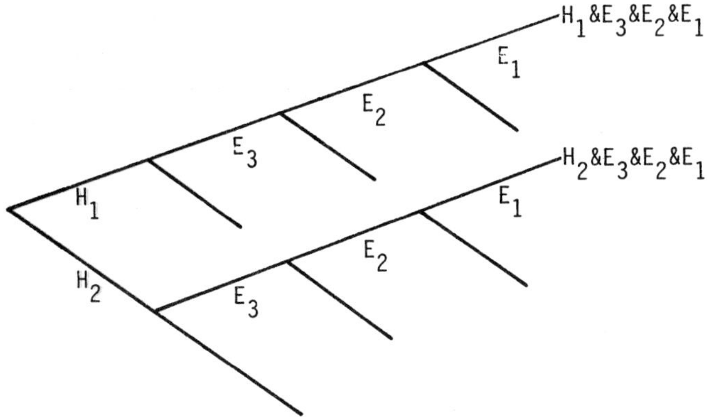

Notice next that the top paths in each tree are the same composite event, i.e., that the first tree's H_1 & E_1 & E_2 & E_3 equals the second tree's H_1 & E_3 & E_2 & E_1. That is, because each top path has exactly the same items of evidence on it, each top path is the same composite event.

But if each path is the same composite event, the probabilities of the paths should be the same. So $P(H_1\&E_1\&E_2\&E_3)$ on the first tree should equal $P(H_1\&E_3\&E_2\&E_1)$ on the second tree. Since the probability of a path equals the product of the probabilities of its branches, we should have

$$P(H_1\&E_1\&E_2\&E_3) = P(H_1\&E_3\&E_2\&E_1), \text{ or}$$

$$P(H_1) \cdot P(E_1/H_1) \cdot P(E_2/H_1\&E_1) \cdot P(E_3/H_1\&E_1\&E_2)$$

$$= P(H_1) \cdot P(E_3/H_1) \cdot P(E_2/H_1\&E_3) \cdot P(E_1/H_1\&E_3\&E_2).$$

Similarly, the bottom paths on both trees are the same composite event, so the probabilities should be the same. i.e., we should have

$$P(H_2\&E_1\&E_2\&E_3) = P(H_2\&E_3\&E_2\&E_1), \text{ or}$$

$$P(H_2) \cdot P(E_1/H_2) \cdot P(E_2/H_2\&E_1) \cdot P(E_3/H_2\&E_1\&E_2)$$

$$= P(H_2) \cdot P(E_3/H_2) \cdot P(E_2/H_2\&E_3) \cdot P(E_1/H_2\&E_3\&E_2).$$

In using the Bayes formulas then, where the entries come off the described trees, it should make no difference which tree (and thus which data order) is used. The revised odds using the first tree are

$$\frac{P(H_1/E_1\&E_2\&E_3)}{P(H_2/E_1\&E_2\&E_3)} = \frac{P(H_1)}{P(H_2)} \cdot \frac{P(E_1/H_1)}{P(E_1/H_2)} \cdot \frac{P(E_2/H_1\&E_1)}{P(E_2/H_2\&E_1)} \cdot \frac{P(E_3/H_1\&E_1\&E_2)}{P(E_3/H_2\&E_1\&E_2)}.$$

The revised odds using the second tree are

$$\frac{P(H_1/E_3\&E_2\&E_1)}{P(H_2/E_3\&E_2\&E_1)} = \frac{P(H_1)}{P(H_2)} \cdot \frac{P(E_3/H_1)}{P(E_3/H_2)} \cdot \frac{P(E_2/H_1\&E_3)}{P(E_2/H_2\&E_3)} \cdot \frac{P(E_1/H_1\&E_3\&E_2)}{P(E_1/H_2\&E_3\&E_2)}.$$

Clearly, the numerators of the revised odds of both equations are the same, i.e., $P(H_1/E_1\&E_2\&E_3) = P(H_1/E_3\&E_2\&E_1)$ since $E_1\&E_2\&E_3 = E_3\&E_2\&E_1$. For the same reason, the denominators of the revised odds of both of the equations are also the same, i.e., $P(H_2/E_1\&E_2\&E_3) = P(H_2/E_3\&E_2\&E_1)$. Further, the numerators on the right of both equations are the same because they are the probabilities of the top paths of the two trees (probabilities which were shown above to be equal). Likewise, the denominators on the right of both equations are the same.

All this explains why we claim that the order in which we consider exactly the same items of evidence should make no difference. While the individual conditionally dependent probabilities of the items of evidence would be different, the final products should be the same. Thus, the revised probabilities of the hypotheses should be the same.

In practice, of course, since humans are not computers, for different orders of the items of evidence not only will the individual conditional probabilities of the evidence be different, but the final products also will be different. So, we will get different revised probabilities for our hypotheses. Note that even if items of evidence are reported to have occurred simultaneously, different orders in considering them probably will yield inconsistent results. For example, if E_1 and E_2 are reported to have occurred simultaneously, then assessing $P(E_1/H_j)$and $P(E_2/H_j\&E_1)$ will yield a product of $P(E_1/H_j)\cdot P(E_2/H_j\&E_1)$. This product is most unlikely to equal the product of $P(E_2/H_j)\cdot P(E_1/H_j\&E_2)$ resulting from assessing $P(E_2/H_j)$ and $P(E_1/H_j\&E_2)$.

(One interesting point was made by student James McGowan. Alluding to Schum's example in the section on conditional independence and dependence, McGowan noted that the effects of considering evidence in different orders may be much greater for a "threat" hypothesis than for a "peace" hypothesis because, as in the Schum example, the evidence actually may be conditionally independent under the "peace" hypothesis and different orders therefore may be largely irrelevant.)

There seems to be no really good solution to this first sub-problem. Ideally, we would want to consider the items of evidence exactly in the order in which they occurred. Unfortunately, items of evidence do not always arrive in the order in which they occurred, and if we process each item as it arrives, we often will be considering evidence out of order. This is particularly likely to be true in crises.

A logical question to ask at this point is whether it would be advisable, in order to solve this problem, to abandon conditional dependence and to use the conditionally independent form of Bayes. This seems a very high price to pay. We argued before, in problem 10, that conditional dependence had the great advantage of being realistic. Conditional independence is a strong and unrealistic assumption, so I do not advise its usage as a solution to this problem.

The better possible solution is to go back, once we know the correct order, put the items of evidence in order, and then reassess our conditional probabilities. There are several difficulties with this possible solution. One is that there simply may not be enough time to do so; this may be particularly true in a crisis situation. Second, it may be very laborious to go back and do all the necessary reassessing. For example, if the 50th item of evidence is determined to belong just before E_{20}, then a good deal of reassessing will be required (student Keith Oatman pointed out). Third, there is the problem of contamination of results. That is, when we reassess for a given E_i that has been "moved up" in the rearranged list of events, although we would ask, for each H_j, $P(E_i/H_j \& E_1 \& E_2 \& \cdots \& E_{i-1})$, we now would know some events that had occurred <u>after</u> E_i and this could contaminate our assessment. Nevertheless, since the sequence in which the items actually occur is part of the jigsaw puzzle of fitting pieces of evidence together, it seems advisable to go back once the correct order is known, rearrange the items of evidence in this correct (temporal) sequence, and reassess.

An even stronger argument for going back and reassessing is that it is conceivable that, if the items of evidence have been received out of order, the calculated revised probabilities of hypotheses may be <u>widely</u> at variance from what they would have been for the correct order, and decision makers acting on these revised probabilities may have taken actions they would not otherwise have pursued. This line of reasoning implies the need for reassessment once the correct order for the items of evidence is known. There must, however, be judgment here. We could spend all our time reassessing if we decided to reassess every time we decided some items of evidence had been considered out of sequence. No definitive advice on <u>when</u> to reassess can be given: it is another murky, judgmental situation. At some point, however, reassessment does seem highly desirable.

<u>Discussion of the second sub-problem</u>:

In the discussion of the first sub-problem it was asserted

that if humans consider exactly the same items of evidence in different orders and use the conditionally dependent form of Bayes, then, while the individual probabilities of the items of evidence will vary, the products of the probabilities should be the same, although they will not be in practice since humans are not computers and therefore lack internal consistency.

We showed why the products of the probabilities should be the same by using several trees with items of evidence E_1, E_2, E_3 on each tree. Part of the discussion depended on composite event $E_1 \& E_2 \& E_3$ being equal to composite event $E_3 \& E_2 \& E_1$, i.e., it depended on the items of evidence E_1, E_2, and E_3 really being exactly the same items in each ordering so that both orders really represented the same composite event.

We were sure the items of evidence were exactly the same items in each ordering because we assumed the time of occurrence was specified for each. We will show now that, if the time of occurrence is not attached to the items of evidence, then different orders of the items (evidence) really will represent different composite events. That is, we will show that different orders will represent the same composite event if each of the items of evidence making up the composite events includes the time when the item happened. If the times are missing, however, we will show that, because of the way we reason when assigning conditional probabilities, different orders really represent different composite events. In this case, then, the problem of "inconsistency" of results for different orders disappears -- we can say honestly that, since different orders actually represent different composite events, there is no reason to expect consistency. Unfortunately, however, a new problem arises - if different orders really do represent different composite events, the question is which order we should choose for the items of evidence in the absence of information about when they occurred.

It should be pointed out again that our discussion is based on the assumption that the conditionally dependent form of Bayes is used. If the conditionally independent form of Bayes is used, discussions about the order of the evidence almost completely disappear. I still do not suggest going to independence, however.

The following concrete example will serve as an informal proof of the above assertions.

Let H_1 = Egypt has decided to attack Israel within the next seven days,

H_2 = Egypt has decided to not attack Israel at all,

H_3 = Egypt has decided to attack Israel after seven days,

H_4 = Egypt has decided to not attack within the next seven days, but has not made up its mind what to do after seven days,

H_5 = Egypt has not made up its mind about what to do.

Suppose that

E_1 = Phone call between Mubarak and Hussein where they agree to attack Israel within a week,

E_2 = Egyptian air force goes on alert,

E_3 = An Egyptian army corps is sent to Suez.

If the items of evidence E_1, E_2, and E_3 are considered in that order these assessments might be made (where, for convenience, we have ignored the assessments in each case for H_3, H_4, and H_5):

$P(E_1/H_1)$ = 0.2 (assessed as a low probability because of probable fear of a leak),

$P(E_1/H_2)$ = 0.01 (there is a slight chance that Mubarak actually might expect a leak and might want it in order to put pressure on Israel, even though he really doesn't plan on attacking and might fear an Israeli overreaction if they believe he plans on attacking),

$P(E_2/H_1\&E_1)$ = 0.95 (high because of the consistency with both H_1 and the phone call which is <u>assumed to have occurred</u>),

$P(E_2/H_2\&E_1)$ = 0.05 (still a low probability, but the alert also might be used by Mubarak to put more pressure on Israel),

$P(E_3/H_1\&E_1\&E_2)$ = 0.95 (because of consistency of E_3 with H_1 and the phone call and the air alert which are <u>assumed to have occurred</u>),

$P(E_3/H_2\&E_1\&E_2)$ = 0.10 (because of the consistency with a deception campaign involving the previously occurring E_1 and E_2, to put pressure on Israel, but still not very likely because of possible Israeli overaction if it is believed).

Suppose now that the data are considered in this order: E_3, E_2, E_1. If we consider the data in this order, we might make these assessments

$P(E_3/H_1)$ = 0.9 (Egypt needs to send up troops if it is going to attack soon),

$P(E_3/H_2)$ = 0.05 (Egypt might want to put pressure on Israel; probability is fairly low because of possible Israeli overreaction),

$P(E_2/H_1\&E_3)$ = 0.95 (high, because Egypt probably also needs to go on alert at some point if it is going to attack soon and because of the consistency with E_3 which is assumed to have occurred),

$P(E_2/H_2\&E_3)$ = 0.05 (again, Egypt might want to put pressure on Israel; probability is fairly low because of possible Israeli overreaction),

$P(E_1/H_1\&E_3\&E_2)$ = 0.3 (the event makes sense in light of H_1 and events E_3 and E_2 which are assumed to have occurred, but since a leak might be feared it still is not a very likely event),

$P(E_1/H_2\&E_3\&E_2)$ = 0.1 (this could be seen as one more event in a possible pattern of description involving the previously occurring E_3 and E_2, but it still is not likely because of possible Israeli overreaction if it is believed).

Now,

$$P(E_1/H_1)\cdot P(E_2/H_1\&E_1)\cdot P(E_3/H_1\&E_1\&E_2) = (0.2)\ (0.95)\ (0.95) = 0.1805$$

while

$$P(E_3/H_1)\cdot P(E_2/H_1\&E_3)\cdot P(E_1/H_1\&E_3\&E_2) = (0.9)\ (0.95)\ (0.3) = 0.2565$$

which are different, as we suggested they would be. Also,

$$P(E_1/H_2)\cdot P(E_2/H_2\&E_1)\cdot P(E_3/H_2\&E_1\&E_2) = (0.01)\ (0.05)\ (0.10) = 0.00005$$

while

$$P(E_3/H_2)\cdot P(E_2/H_2\&E_3)\cdot P(E_1/H_2\&E_3\&E_2) = (0.05)\ (0.05)\ (0.1) = 0.00025$$

which again are different.

BUT ARE THE COMPOSITE EVENTS THE SAME? That is, is composite event $E_1\&E_2\&E_3$ the same composite event as $E_3\&E_2\&E_1$? Nationally, the answer is "yes." But clearly, the answer is "yes" only if the items of evidence E_1, E_2, and E_3 of the first composite event really are exactly the same items as the E_3, E_2, and E_1 of the second composite event. A convincing case can be made that they will be exactly the same only if we can provide the time or time sequence in which each of the items of evidence occurred.

An example of this for items of evidence E_1, E_2, and E_3 making up the composite event $E_1\&E_2\&E_3$ could be this:

E_1 = On July 7 there is a phone call between Mubarak and Hussein from 3 to 3:50 p.m. where they agree to attack Israel within a week,

E_2 = On July 8 at 11:30 a.m. the Egyptian AF goes on alert,

E_3 = On July 9 at 2 p.m. an Egyptian army corps begins to move to Suez.

Consider now the composite event $E_3\&E_2\&E_1$. If the items of evidence, in sequence, are

E_3 = On July 9 at 2 p.m. an Egyptian army corps begins to move to Suez,

E_2 = On July 8 at 11:30 a.m. the Egyptian AF goes on alert,

E_1 = On July 7 there is a phone call between Mubarak and Hussein from 3 to 3:50 p.m. where they agree to attack Israel within a week,

then the items of evidence really are completely identical to the E_1, E_2, and E_3 of the first composite event. The E_1's, E_2's, and E_3's of the two sequences are exactly the same and the composite event "$E_1\&E_2\&E_3$" of the first sequence really is the same as the composite event "$E_3\&E_2\&E_1$" of the second sequence.

If we look at the judgments that have to be made for the second composite event, they are rather strange. That is, after assessing $P(E_3/H_1) = 0.9$ (they need to send up troops) when we try to later assess $P(E_1/H_1\&E_3\&E_2)$ this time, where E_1, E_3, and E_2 are specified in terms of when they happened, the assessment can't be reasoned as it was before where, when we asked $P(E_1/H_1\&E_3\&E_2)$, we explicitly looked for the logical consistency of E_1 with H_1 and an E_3 and E_2, both of which were assumed already to have occurred! This time, we have to ask for the logical consistency of E_1 with H_1 and an E_3 and E_2 which happened after E_1.

We must underline the absolutely critical implication of the discussion of the last paragraph. The implication is that, given the original wording of E_1, E_2, and E_3, the composite event $E_1\&E_2\&E_3$ turned out to be really not the same composite event as $E_3\&E_2\&E_1$. This is because, when we made our probability judgments involving composite event $E_1\&E_2\&E_3$, we were assuming that E_2 followed E_1 and E_3 followed E_2, while, when we made our probability judgments involving composite event $E_3\&E_2\&E_1$ we were assuming that E_2 followed E_3 and E_1 followed E_2. For example, when we assessed $P(E_3/H_1\&E_1\&E_2)$ (for composite event $E_1\&E_2\&E_3$), we justified our assessment in terms of the consistency of E_3 with a deception campaign involving the previously occurring E_1 and E_2. But when we assessed $P(E_1/H_1\&E_3\&E_2)$ for composite event $E_3\&E_2\&E_1$, we justified our assessment in terms of the consistency of E_1 with a deception campaign involving the previously occurring E_3 and E_2.

Student Duane Miller suggested in a class discussion that the original two composite events could be written better by describing the first as "'E_1'&'E_2(after E_1)' & 'E_3(after $E_1\&E_2$)'" and the second as "'E_3'&'E_2(after E_3)' & 'E_1(after $E_3\&E_2$)'" in order to make the point clearer that the two sequences of items of evidence do not represent the same composite event.[8] Some wording which will make clear the actual sequence of the constituent items of evidence (if it is known - see below) should be included in the description of the events.

The moral of our discussion seems to be this: if no time order is explicitly stated for the items of evidence E_1, E_2, E_3..., different orderings of these items actually will turn out to be different composite events because the justifications for the probabilities we assign will implicitly or explicitly assume that the items of evidence happened in that particular order! The justifications for the probabilities we assign will depend on the patterns revealed by the particular assumed or explicit sequence of the items. Moreover, since the different sequences then really will represent different composite events, the mathematical requirement that the products of the probabilities of the individual judgments be the same for different orders no longer applies. If different orders lead to different probability products we no longer really care -- because the different orders really will represent different composite events.

The immediate question then becomes, of course, which of all the possible sequences of items of evidence represents the composite event in which we are interested. The answer is that if we can't -- even after further attempts to determine the time order -- tell which ordering represents the actual time sequence of the items of evidence, then we can't really say which ordering is the one in which we are interested. (See below for more discussion on this question.)

[If we do have the information on when each of the items of evidence E_1, E_2,..., occurred, then the thing to do is to put them in order and assess their probabilities in the order in which they occurred.[9]

If we are specific in describing the items of evidence in terms of time order so that different sequences really will be the same composite event, then it now will be true that the products of the individual probability judgments <u>should</u> be the same. However, as noted earlier, because we are <u>not computers</u>, if we assign probabilities for different orders we actually will be inconsistent in the probabilities we assign and we can expect that the products of the individual probability judgments will <u>not</u> be the same.

We should note that a composite event involving items of evidence not in sequential order will require us to make some rather strange probability assessments if we ever have to make them. As an example, consider the judgment of $P(E_1/H_1\&E_3\&E_2)$ that we were discussing earlier. It was noted that this was really asking for the probability of E_1 in terms of its logical consistency with H_1 and an E_3 and E_2 which happened <u>after</u> E_1. Such a strangely worded question actually can be asked. We actually can ask "What is the probability that on July 7 there is a phone call between Mubarak and Hussein from 3 to 3:50 p.m. where they agree to attack Israel within a week, given the assumption that Egypt is planning on attacking Israel within the next seven days and given the data that on July 9 at 2 p.m. an Egyptian army corps begins to move to Suez and on July 8 at 11:30 a.m. the Egyptian AF goes on alert." This is very awkwardly phrased, but does make sense mathematically[10] and hence can be asked, even though it is so awkward that it might be very difficult to assess.

In practice, there normally is no reason to make such an awkward judgment, but it is possible that in a crisis we could get a number of items of evidence known to have occurred at certain times, and later receive an item of evidence which had occurred earlier than some of the previous items of evidence. In this case, we either would have to make the "strange" sort of probability judgment talked about here, or go back, put the new item of evidence in its correct order, assess it and then <u>reassess</u> for the previously assessed but now time-subsequent items of evidence. This latter solution is the better one, I think. There is a tradeoff between having to make a strange probability judgment or to run the contamination risks of reassessment. I suspect the latter risks are preferable.]

It should be pointed out again that our discussion in this section assumed that we were using the conditionally dependent form of Bayes. If we use the conditionally independent form of Bayes, almost all the problems involved with order disappear. The word "almost" is included because, as was argued in the section on conditional independence and dependence, even if the conditionally independent form of Bayes is used, there will be some conscious or subconscious recollection of items of evidence previously considered and some possible subconscious or informal inclusion of elements of previous evidence when assigning the conditionally independent probabilities. However, the impact of elements of previous evidence is likely to be far less than in the case when conditional dependence is used. Nevertheless, I still would not suggest abandoning use of conditional dependence.

The question remains of how to deal with the different probability judgments that will result when no specific time order can be specified for the items of evidence under consideration. As we have seen, if no specific time order can be specified for the items under consideration, then different orders really will represent different composite events and we can expect to get different final products for the different probability judgments. The question is how to deal with and reconcile these different products.

This assumes, of course, that, once we realize that different orders really represent different events, we will go to all the trouble of (1) determining all the different orders and (2) bothering to make the probability judgments for each order.

Now it is relatively easy, in principle, to do (1), i.e., to determine all the different orders. A simple tree will do the job for us. If, for example, we have received a batch of 4 items of evidence which we arbitrarily label as E_1, E_2, E_3 and E_4, then we simply can make a tree like Figure 5.4 to get all 4! = 24 orders.

With respect to doing (2), however, it is to be seriously doubted that anyone is going to assess $P(E_i/H_1)$, $P(E_j/H_1\&E_i)$, $P(E_k/H_1\&E_i\&E_j)$, $P(E_l/H_1\&E_i\&E_j\&E_k)$ for all 24 orderings of i, j, k, l, and for each hypothesis, too; i.e., to make 24 judgments for H_1, and then 24 judgments for H_2, etc. If there are only 3 hypotheses, this would entail $24 \cdot 3 = 72$ judgments. We might as well forget this sort of brute force approach. If we ask what we should do, however, the answer is hardly clear.

One answer simply would be to abandon the assumption of conditional dependence and avoid the whole issue. We simply would assess $P(E_1/H_1)$, $P(E_2/H_1)$, $P(E_3/H_1)$, $P(E_4/H_1)$, $P(E_1/H_2)$, $P(E_2/H_2)$, $P(E_3/H_2)$, etc, and would expect roughly similar results no matter which order we picked for considering the items of evidence. ("Roughly," because, as we noted in our discussion of conditional independence, it is difficult, even if we assume conditional independence as an intellectual choice, to suppress totally our knowledge of previously considered evidence when we are making subsequent probability judgments.)

Another approach would be to sample, for each hypothesis, a few judgments based on different orderings of the evidence. If the results exhibit fairly low standard deviations (e.g., 0.1 or less), we may feel that we can use the average values. If not, we have a serious problem. I invite reader suggestions.

FIGURE 5.4
Tree Diagram of All the Possible Orders of the Four Items of Evidence E_1, E_2, E_3, E_4

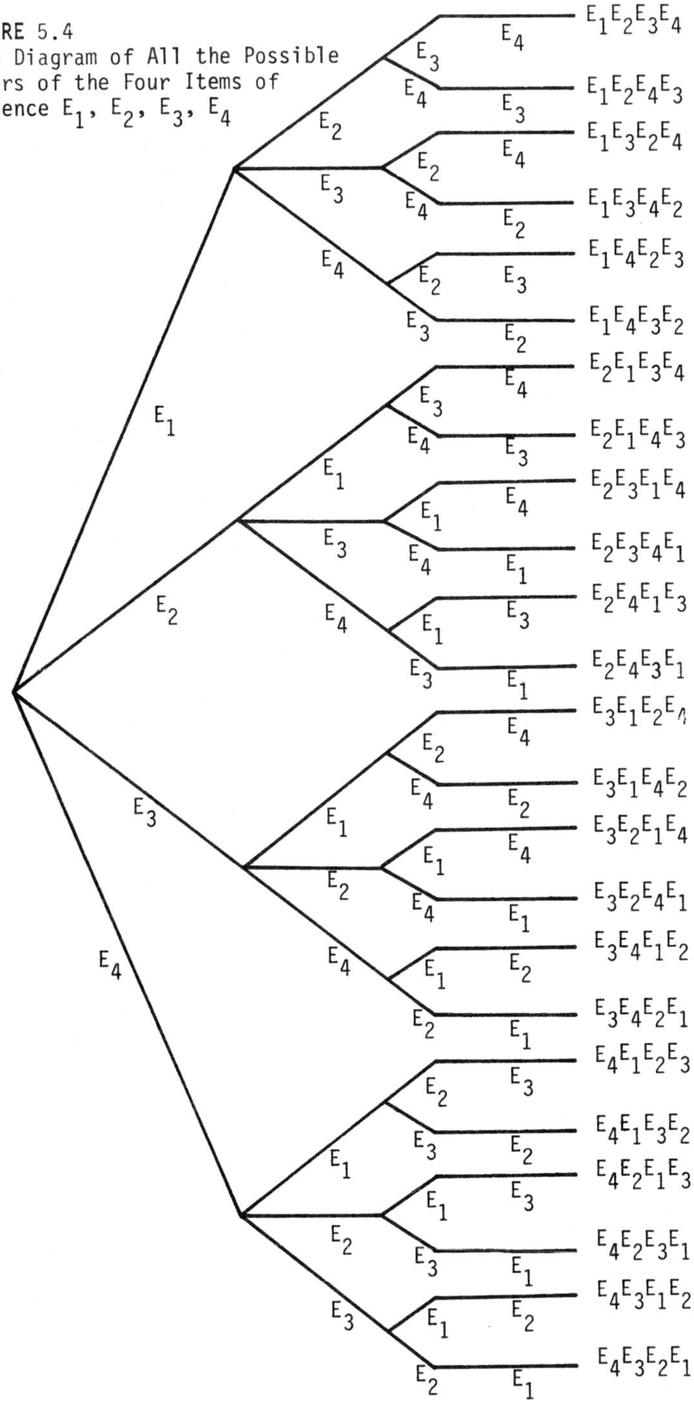

NOTES

Problem 10

1. SCHUM, DAVID A. (1966a) "Inferences on the Basis of Conditionally Nonindependent Data." Aerospace Medical Research Laboratories Report AMRL-TR-66-215, July. Ohio. AD645588. This also appears (with the same pagination as the government document and with the same title) as SCHUM, DAVID A. (1966b) "Inferences on the Basis of Conditionally Nonindependent Data." Journal of Experimental Psychology 72(3): 401-409.

The following dramatic example of how the use of conditional independence can lead to a terribly erroneous analysis came to my attention too late to add to the main text. (In the example, I have added probabilities and made some minor modifications.)

Suppose that an armed robbery, which required a 15 minute block of time to commit, occurred some time between 3:00 a.m. and 3:30 a.m. on a particular day. Suppose that the police have a suspect, Jones, and let

H_1 = Jones is guilty of the crime, and

H_2 = Jones is not guilty of the crime.

Suppose, to be fair, we make these initial assessments:

$P(H_1) = 0.01$

$P(H_2) = 0.99.$

Suppose further that the first item of evidence is this:

E_1 = at 3:10 a.m. Jones was in a car near the scene of the crime. (Based on a report by a very reliable witness.)

Given a reasonable knowledge of Jones's nocturnal habits, including the geographical areas in which he usually spent his early morning hours, we might make these assessments:

$P(E_1/H_1) = 0.8$, $P(E_1/H_2) = 0.1.$

Suppose that the next item of evidence is this:

E_2 = at 3:20 a.m. Jones was in a bar located near the scene of the crime. (Reported by another very reliable witness.)

Suppose that we ignore E_1 and make the required conditionally independent judgments for E_2. The following judgments seem consistent with the judgments for E_1:

$P(E_2/H_1) = 0.8$, $P(E_2/H_2) = 0.1.$

Our tabular array and revised probabilities are as follows:

j	$P(H_j)$	$P(E_1/H_j)$	$P(E_2/H_j)$	Row Product	Revised Probability of H_j
1	0.01	0.8	0.8	0.0064	0.39
2	0.99	0.1	0.1	0.0099	0.61
	1.00			0.0163	1.00

So, by using conditional independence, the probability that Jones is guilty has been revised upward to almost 0.4.

The results are dramatically different if we use conditional dependence, however. In this case, since we know that the crime took place between 3:00 a.m. and 3:30 a.m. and took an uninterrupted 15 minutes to perform, we must assess

$$P(E_2/H_1\&E_1) = 0$$

because there is no way E_2 also could occur if Jones is guilty (H_1) and was in a car at 3:10 (E_1) -- there would be no open 15 minute period in the 30 minutes in which to commit the crime. Regardless of our assessment of $P(E_2/H_2\&E_1)$ (call it "x"), the revised probability of H_1 must go to 0, and the revised probability of H_2 must go to 1, as the following tabular array shows.

j	$P(H_j)$	$P(E_1/H_j)$	$P(E_2/H_j\&E_1)$	Row Product	Revised Probability of H_j
1	0.01	0.8	0	0	0
2	0.99	0.1	x	0.099x	1
	1.00			0.099x	1

So the probability that Jones is <u>innocent</u> has gone to 1, i.e., certainty! (This assumes perfect reporting by the witnesses. See chapter 8.) This dramatically different result from what we obtained using conditional independence clearly demonstrates the type of disaster that can occur if we don't use conditional dependence. Clearly and unequivocally, the conditionally dependent assessments are the valid ones to use.

A more subtle point that emerges from this example is that if we use conditional dependence, two (or more) items of evidence -- which taken individually (in using conditional independence) might seem to support a particular hypothesis -- when considered together may lessen support for the hypothesis. (See the very interesting article from which this example was drawn: FALK, RUMA and BAR-HILLEL (1983) "Probabilistic Dependence Between Events." The Two-Year College Mathematics Journal 14(3) June: 240-247. My thanks to COL Dennis K. Hodsdon for bringing this article to my attention. See also FAIRLEY, WILLIAM B. and MOSTELLER, FREDERICK (eds.) (1977) Statistics and Public Policy. Reading, Massachusetts: Addison-Wesley Publishing Company. The section on statistical models in legal settings is quite interesting, and issues of independence and dependence are very central to the discussions there.)

2. *Ibid*, p. 402.

3. *Ibid*.

4. MILLER, GEORGE A. (1956) "The Magic Number Seven, Plus or Minus Two: Some Limits on Our Capacity For Processing Information." Psychological Review 63 (2): 81-97.

5. This was said by him in class at some point. Hereafter, since I generally did not document *when* a student said something, I will not put down a footnote. "In class" will mean in a class conducted by me at the Defense Intelligence College sometime between 1978 and 1982.

Problem 12

6. As was noted in footnote 4 of Chapter 1, only in the discussion of this problem do I find that clarity demands the use of the term "composite event" (instead of just "event").

7. This is because, although the individual conditional probabilities for items of evidence will vary when the conditionally dependent form of Bayes is used, their final products should, for mathematical reasons stated below after Figure 5.3, be the same, and thus yield the same revised probabilities for the hypotheses.

8. Indeed, this convincing argument of his that sequences $E_1\&E_2\&E_3$ and $E_3\&E_2\&E_1$ really were different led to core of the argument presented here. Students had been arguing for years that the sequences were different, but his was the final, convincing argument (to me).

9. As a technical note, it should be pointed out again that, if we are specific in describing the items of evidence in terms of when they happened, then different sequences will be the same composite event. Using the *explicit* revised descriptions of items of evidence E_1, E_2, and E_3 given above, for example, we noted above that it is clear that in this case $E_1\&E_2\&E_3$ is the same composite event as $E_3\&E_2\&E_1$. (We can see this easily by simply writing out the two sequences.)

10. An analogous urn question would be this: What is the probability of drawing a red ball on the second draw, given that a white ball was drawn on the first draw and a green ball was drawn on the third draw in drawing three balls without replacement from an urn having three red, four white, and seven green balls?

6 Problems Involving the Evidence to Use

In this chapter we will discuss problems concerning what kinds of information to use as items of evidence in a Bayesian analysis.

PROBLEM 13. The types of evidence to use: the question of whether to use "all" relevant evidence or "just" indicator relevant evidence.

The problem: This really is more a statement of a choice than a problem. The choice is whether we will be willing to use all evidence we decide is relevant, after determining its applicability as it comes to our attention, or whether we will use only evidence that appears to relate to a pre-selected, structured set of "indicators."

The solution: There is no "school solution" to this question. The type of indicators used in indications and warning analyses may be very well structured and easily lend themselves to Bayesian analysis. We would use only "indicator"-relevant evidence in this case. For most other types of analysis, an "all" relevant evidence approach probably will be more helpful. The steps to be followed, if we choose to use indicators, are found in the following discussion.

Discussion: This is not so much a problem as it is a cryptic reference to a choice we can make. We can choose to make Bayes quite unstructured with respect to the evidence we will consider in the sense that we will examine all the evidence that comes to our attention and then process all the evidence that appears to us to be relevant to the hypotheses. We have no preconceived notions about what evidence we will use, and place no limitations or specifications on what evidence we will consider. We determine the applicability of each piece of evidence as it comes to our attention.

Alternatively, we can decide to collect and process only evidence relating to a pre-selected, specific set of events, which are called "indicators."

The steps to follow if indicators are used is as follows:

First, determine the hypotheses and their prior probabilities.

Second, decide on a structured, comprehensive set of indicators.

Third, set up a separate scale for each indicator of the qualitative or quantitative values (which can be called "settings") it can take on, including the setting for what we consider to be the "normal" condition. For example, if an indicator is "hostility toward us in public statements by country B's leaders," the scale might be "highly hostile, moderately hostile, slightly hostile, neutral, slightly friendly (which in this case may be the normal setting), moderately friendly, highly friendly."

Fourth, when the latest evidence comes in, decide on the indicator to which it applies and then decide what the setting of the indicator should be, based on the evidence. Using the above example, after getting evidence involving public statements about us by country B's leaders, we would decide it applies to the hostility indicator; then, we would analyze the evidence and determine what the new hostility setting should be, e.g., "moderately hostile."

Fifth, for each indicator, assign conditional probabilities to the setting we have chosen. Note, at this point, that we might want to interchange steps four and five, but that we can do this only if we decide not to consider previous data settings. That is, if we are sure we in no way want to consider previous data settings, then we can assign all the possible probabilities in advance by assigning for each hypothesis the probability values of all the possible settings for each indicator. (This is what is done sometimes in hierarchical Bayesian analysis, by the way.) If we have many indicators and need to use them in crises, where we will have little time, this pre-assignment of probabilities may be the thing to do. If we do this, then as soon as we select each indicator setting, we know the (pre-determined) probability to use.

The sixth and final step is to calculate the revised probability of each of our hypotheses by using Bayes's formula or its embodiment in the tabular array.

If we choose to use indicators, see the next problem for a discussion of the type of previous data and the prior probabilities to use.

PROBLEM 14. Previous data: the problem of the type of "previous data" with respect to previous indicator settings to use and which prior probabilities of hypotheses to use if we choose to use indicators.

The problem: If indicators are used, we may choose either to use the conditionally independent or dependent form of Bayes. Each choice raises issues of what sort of "previous data" about previous indicator settings, if any, to use and what our prior probabilities of the hypotheses will be for each round of "setting the indicators," as we continue to update the settings of the indicators. (Each new determination of all the settings can be called a "round" of "setting the indicators.")

The solution: If we choose to use conditional independence, there is no "previous data" about indicator settings from previous rounds to consider. For each round, we simply use the same original set of prior probabilities for the hypotheses and set the indicators on the appropriate values. (If, in spite of using conditional independence, we still want to consider the data of indicator settings from previous rounds, as a last "indicator" for each round of "setting the indicators" we can use the patterns of previous data settings.)

If we choose to use the conditionally dependent form of Bayes, the complexities involved probably will require us to use only a partial form of dependence where we consider for each indicator the previous data settings of just that particular indicator (but do not consider anything about other indicators). In this case, for each round of "setting the indicators," we should use as our prior probabilities of the hypotheses the revised probabilities of the hypotheses from the previous round.

Discussion: The notion of possibly considering "previous data," that is, data on indicator settings in previous rounds of setting the indicators and data on the settings in the same round of other indicators is, figuratively, a landmine. If conditional independence is used, then there is no "previous data" of any sort to consider. We simply choose our setting for each indicator and assign the probability of that setting given each of the hypotheses, one by one. (See below, however.)

If we do choose to use the conditional independence format, we can go through this procedure as often as we wish -- once a day, once a week, or whatever -- simply by choosing the appropriate setting for each indicator, making our probability assessments for the particular settings, and then calculating the effect on the fixed prior probabilities with which we started the whole analysis. That is, we should not take the revised probabilities of the previous round and use them as the prior probabilities in the calculations for a subsequent round of setting the indicators. The reason is that if we start off with our priors, and all the indicator settings are normal, then if we take the resulting revised probabilities and put them in as the prior probabilities for the next round of setting the indicators and find that all the indicator settings are still normal, the net effect will be to process the original priors twice when "nothing" really has changed -- the indicator settings will be unchanged, staying at normal for the second round, but the revised probabilities will not be the same as after the first round.

But this is counter to what we want to happen. If the indicator settings are unchanged from those of the previous round, we want the revised probabilities also to remain the same as after the previous round,[1] so we must use the original set of prior probabilities in each round, simply changing our indicator settings.

Here is an example. Suppose we have three hypotheses (H_1, H_2, and H_3) and two indicators (A and B) and suppose further that the prior probabilities of the hypotheses are 0.1, 0.4, and 0. 5. Suppose that, on our first setting of the indicators, each is at normal. Suppose that we make these assessments

$P(A\ normal/H_1) = 0.2$ $P(B\ normal/H_1) = 0.1$

$P(A\ normal/H_2) = 0.6$ $P(B\ normal/H_2) = 0.4$

$P(A\ normal/H_3) = 0.9$ $P(B\ normal/H_3) = 0.7.$

Then, using a tabular array, the revised probabilities are shown in Table 6.1.

TABLE 6.1
Tabular Array for Revised Probabilities of the Hypotheses after First Setting of the Indicators

j	$P(H_j)$	$P(A_{normal}/H_j)$ on 1st setting	$P(B_{normal}/H_j)$ on 1st setting	Row Product	Revised Prob of H_j
1	0.1	0.2	0.1	0.002	0.00484
2	0.4	0.6	0.4	0.096	0.23245
3	0.5	0.9	0.7	0.315	0.76272
				0.413	1.00001

Now, suppose that after receiving some new evidence, we feel the proper new settings for A and B are still the normal settings. The question at this point, then, is whether we should take the revised probabilities and use them as the prior probabilities again, as in Table 6.2 (where the conditional probabilities for A normal and B normal are the same as before because nothing has changed):

TABLE 6.2
Tabular Array for Revised Probabilities of the Hypotheses after Second Setting of the Indicators (and Using Latest Probabilities of H_j's)

j	Latest probs of H_j	$P(A_{normal}/H_j)$ on 2nd setting	$P(B_{normal}/H_j)$ on 2nd setting	Row Product	Rev Prob of H_j
1	0.00484	0.2	0.1	0.0000968	0.00018
2	0.23245	0.6	0.4	0.055788	0.10400
3	0.76272	0.9	0.7	0.4805136	0.89581
				0.5363984	0.99999

The results here are <u>not</u> what we really want. That is, although on the basis of new evidence we did not change the data settings, our use of the previous revised probabilities has led to a new set of revised probabilities, driving up the probability of H_3 even more while driving down the probabilities of H_1 and H_2. Thus, although we can argue that "nothing" has changed, the revised probabilities have changed again.

The solution to this is to use always the original priors (which, in this example are 0.1, 0.4, and 0.5). If we had done this, then Table 6.2 would have been Table 6.3.

TABLE 6.3
Tabular Array for Revised Probabilities of the Hypotheses after Second Setting of the Indicators (and Using Original Probabilities of H_j's)

j	Original probs of H_j	$P(A_{normal}/H_j)$ on 2nd setting	$P(B_{normal}/H_j)$ on 2nd setting	Row Product	Rev prob of H_j
1	0.1	0.2	0.1	0.002	0.00484
2	0.4	0.6	0.4	0.096	0.23245
3	0.5	0.9	0.7	0.315	0.76272
				0.413	1.00001

In this case, if we leave the indicator settings at normal, feeling that nothing has changed, then the revised probabilities of the hypotheses stay the same, as we would like. Again, the rule is that if we are using an indicator list and are assuming conditional independence, then we always use the original set of priors for the calculations of each round of setting the indicators.

Philosophically, what we are doing here is equivalent to saying that we have two categories of evidence, A and B, and are continually reexamining our view as to what the evidence is in each category. If the evidence is seen as unchanged, then our assessment of the revised probabilities should not change (and will not if we use the suggested procedure). When and if our view as to the evidence changes, we will change the indicator settings and the revised probabilities will then change. But, then, if subsequently our view of what the evidence is changes back to a previous "view," then our revised probabilities will be identical to those earlier ones. All this comes about by keeping the same set of prior probabilities for each round.

There is, however, one technical problem with saying that since "nothing has changed" the values of $P(A_{normal}/H_j)$ for the second setting are the same as the values of $P(A_{normal}/H_j)$ for the first setting. The technical problem is that there may be diagnostic impact in the event of "no change." If the settings of all the indicators stay the same, at normal, <u>this may be diagnostic in and of itself</u>. In addition, of course, any fixed pattern, even if not at normal, can be diagnostic. (Worse, a change in the pattern can be diagnostic.) To take into account the nature of the pattern, we might want to sort of introduce a type of conditional dependence via the back door and introduce as a last "indicator" a description of the particular pattern of the settings of all the indicators and assess the conditional probability of this pattern given the various hypotheses and given the previous pattern settings. Thus, we would examine the latest pattern of settings itself and also look at how it had been changing from past patterns and assess the diagnostic impact of all this.

An alternative approach would be to input the latest setting for an indicator as "no change from the previous time period" or "an increase of x from the previous time period" or "a decrease of y from the previous time period" so that we could directly take into account the pattern of the evidence and address the diagnosticity of that pattern. This, however, seems awkward and inconsistent with the idea of the settings of an indicator being fixed quantities and I don't recommend it.

If we decide to use conditional dependence, we have several logical choices:

(1) We can go dependent "part way" by considering only the previous settings for each indicator in the previous rounds or

(2) we can go further by considering both the previous data settings for each indicator as we are dealing with it and all the present data settings for those indicators in the indicator list preceding the indicator we are dealing with or

(3) We can go all the way by considering the previous data settings for each indicator as we are dealing with it and both the present data settings for those indicators in the indicator list preceding the indicator we are dealing with and the past data settings of all the other indicators. This third choice, unfortunately, will require much more work than is possible if there are many indicators preceding the one with which we are dealing and/or if there are many rounds of prior data settings to consider. Even the second choice may be too much if there are many rounds of previous data settings of the indicator we are dealing with and/or many preceding indicators.

Choice (1) appears to be reasonable: if possible, we should at least consider the previous data settings for the indicator with which we are dealing. In practice, however, our choice may well be to use the conditional independence format; it simply may be too troublesome to explicitly consider any sort of "previous" data settings.

If we do feel we have the time and try to use this first alternative, there is the question of what our prior probabilities should be for each round of setting the indicators. We will argue that this time we want to use the latest revised probabilities of hypotheses as our priors because, if the setting of an indicator is, say, unchanged from the previous setting then we will take this pattern of lack of change into account by changing the conditional probability of the setting for each hypothesis. Furthermore, if these latter probabilities are changed, then the previous revised probabilities of the hypotheses - reflective of the first data setting - should be inputted as the priors for the new diagnostic probabilities to work on; otherwise, if we were to return to the original priors, we would ignore the evidence of the first data setting.

As a concrete example, consider a situation where

H_1 = there will be a large scale war within four weeks,

H_2 = there will be limited scale war within four weeks,

H_3 = there will be no war within four weeks,

and there is an indicator of "degree of mobilization." Suppose that the "normal" setting is "no mobilization this week," and that the first mobilization setting is "normal." Because we feel that mobilization generally is needed to prepare for a large scale war within four weeks, we may assess P(mobilization/H_1) = 0.8. We then can calculate P(no mobiliz this week/H_1) indirectly as P(no mobiliz this week/H_1) = 1-P(mobiliz/H_1) = 1-0.8 = 0.2.

Similarly, we indirectly might assess $P\left(\begin{matrix}\text{no mobilization}\\\text{this week}\end{matrix}\Big/ H_2\right)$ = 0.6. We can argue that mobilization is somewhat needed to prepare for a limited scale war within four weeks (but not needed as much as it is for large scale war), and therefore assess

P(mobilization/H_2) = 0.4. Then

$$\begin{aligned}P(\text{no mobilization this week}/H_2) &= 1\text{-}P\ (\text{mobilization}/H_2)\\ &= 1\text{-}0.4 = 0.6.\end{aligned}$$

Finally, we indirectly might assess P(no mobilization this week/H_3) = 0.9. We can argue that mobilization largely is unnecessary if there is to be no war within four weeks, and therefore assess P(mobiliz/H_3) = 0.1. Then

$$P\left(\begin{matrix}\text{no mobilization}\\\text{this week}\end{matrix}\Big/ H_3\right) = 1\text{-}P(\text{mobilization}/H_3) = 1\text{-}0.1 = 0.9.$$

Next, suppose that after, say, a week we find that the mobilization indicator setting is still at normal. Then, our assessment of probabilities might look like this:

$$P\left(\begin{matrix}\text{no mobilization}\\\text{this week}\end{matrix}\Big/ \begin{matrix}\text{large scale war}\\\text{within 4 weeks}\end{matrix} \ \& \ \begin{matrix}\text{setting of no mobiliz}\\\text{last week}\end{matrix}\right) = 0.1$$

because if there is to be a large scale war within four weeks and there was no mobilization last week, then the conditional probability of "mobilization" this week should be higher because of the need to prepare in time for war, and hence the conditional probability of "no mobilization" (= 1-Prob (mobilization)) should be lower than its previous value of 0.2.

Similarly, we indirectly might assess

$$P\left(\text{no mobilization this week} \middle/ \text{limited scale war \& no mobilization last week}\right) = 0.5.$$

Finally, in reasoning through the assessment of

$$P\left(\begin{matrix}\text{no mobilization}\\ \text{this week}\end{matrix} \middle/ \begin{matrix}\text{no war within}\\ \text{4 weeks}\end{matrix} \;\&\; \begin{matrix}\text{no mobilization}\\ \text{last week}\end{matrix}\right) = 0.95,$$

we can argue that if we assume that there is to be no war and we have in addition an event of "no mobilization last week" which is consistent with the hypothesis of no war, then the conditional probability of "no mobilization this week" should be even higher than previously.

Suppose that the prior probabilities of H_1, H_2, and H_3 were assessed by us as 0.3, 0.6, and 0.1, respectively. Then, for the first normal setting of "no mobilization" the tabular array will be as in Table 6.4.

TABLE 6.4
Tabular Array for Revised Probabilities of the Hypotheses after First Setting of the Indicator of Degree of Mobilization

j	$P(H_j)$	$P\left(\begin{matrix}\text{"no mobilization"}\\ \text{1st setting}\end{matrix} \middle/ H_j\right)$	Row Product	Revised Prob of H_j
1	0.3	0.2	0.06	0.118
2	0.6	0.6	0.36	0.706
3	0.1	0.9	0.09	0.176
	1.0		0.51	1.000

For the second calculation, if we use the revised probabilities from Table 6.4 as the priors, our entries will be as in Table 6.5.

TABLE 6.5
Tabular Array for Revised Probabilities of the Hypotheses after Second Setting of the Indicator of Degree of Mobilization (and Using Revised Probabilities of H_j's as the Priors)

j	Probability of H_j's from previous calculation	$P\left(\begin{matrix}\text{"no mobilization"}\\ \text{2nd setting}\end{matrix} \middle/ H_j \;\&\; \begin{matrix}\text{"no mobilization"}\\ \text{1st setting}\end{matrix}\right)$	Row Product	Revised Prob of H_j
1	0.118	0.1	0.0118	0.022
2	0.706	0.5	0.3530	0.664
3	0.176	0.95	0.1672	0.314
			0.5320	1.000

Philosophically, what we are doing here is equivalent to saying that our first setting of the mobilization indicator is a first "item of evidence" and our subsequent second setting of the indicator is a "second item of evidence" in a Bayesian analysis. If we had waited to run the analysis until we had assigned values to both items of evidence, our tabular array would have been Table 6.6.

TABLE 6.6
Tabular Array for the Revised Probabilities of the Hypotheses after First and Second Settings of the Degree of Mobilization

j	$P(H_j)$	$P\left(\text{"no mobilization" 1st setting} / H_j\right)$	$P\left(\text{"no mobilization" 2nd setting} / H_j \text{ \& "no mobilization" 1st setting}\right)$	Row Product	Revised Prob of H_j
1	0.3	0.2	0.1	0.0060	0.022
2	0.6	0.6	0.5	0.1800	0.663
3	0.1	0.9	0.95	0.0855	0.315
	1.0			0.2715	1.000

Thus, we have the argument that we do _not_ return to our original set of prior probabilities when we use conditional dependence this way. In each round, as we continue to add on the evidence of additional settings of the indicator, we take the latest set of revised probabilities from the previous round and put them in our tabular array as our prior probabilities.

PROBLEM 15. "Causal" evidence: the problem of how to deal with actions which are intended to alter behavior.

The problem: Suppose we have developed some hypotheses about the possible future behavior of an actor, A. The literature on Bayes contains the notion of so-called "causal" evidence - meaning an action by another actor, B, specifically intended to alter A's behavior in a particular direction. The problem, according to the literature, is that, if we treat the "causal" evidence as reflective of the trends in the systemic forces at work in the environment (with respect to the possible behavior of A), then the (purportedly paradoxical) effect is to change the probability of the hypothesis that B wanted to affect in a direction opposite to that desired by B. The solution offered for this (purportedly paradoxical) result is to directly assess the revised probability of the hypotheses in the direction sought by B.

Although this is the way the problem is presented in the literature, I feel that the analysis is off-target. I think that it would be quite proper to include such "causal" evidence _if it really were_ reflective of and derivative from the systemic forces at work in the environment (even if it were intended to alter the direction of the systemic forces). If, later, the event _did_ lead to a change in A's policy, we would expect this to be shown by events which would occur whose probabilities, when assessed, would show this latent effect. (Note, by the way, that the actual later impact of the "causal" action could be contrary to its desired direction.)

My real question is whether such "causal" evidence properly can be viewed as reflective of and derivative from the systemic forces at work in the "environment," even if this latter concept is defined broadly. The "environment" seems to me to involve the government and society of the state within which actor A is based. To include actions (the "causal" evidence) by actor B in our analysis seems to me to stretch the umbrella of "the environment" too far.

The solution: I think that we perhaps can broaden our scope of the definition of so-called "causal" evidence to include both consciously directed efforts to influence the course of events, and non-consciously directed events or actions which might influence the course of events. We also should divide these categories into evidence (events) external and internal to the environment.

Given this categorization of evidence (events) into four categories, we probably should deal with evidence (events) as follows. "Causal" events (evidence) that are external to the environment should be completely left out of our Bayesian analysis. "Causal" events (evidence) internal to the environment, whether they are consciously directed efforts to influence the course of events or non-consciously directed events or actions which might influence the course of events, should be treated, if possible (through appropriate wording of the hypotheses), as reflective of and derivative from the systemic forces at work in the environment, even though they may (eventually) have a (causal) impact. If and when "causal" events do have an impact, other events reflective of that impact can be expected to occur and to be assessed appropriately.

(If we are highly interested in the possible impact of "causal" evidence, either external or internal to the environment, there may be a problem in that we may have to wait some time for valid diagnostic date to emerge that is reflective of the systemic forces, student Duane Miller pointed out. A probability diagram (tree) is an alternative probability tool which can be used to predict the possible impact, if we need to do so.)

Discussion: This problem can be viewed as really being part of the overall problem of what sort of evidence or events can be included in a Bayesian analysis, which in turn is part of the problem of what sort of questions Bayesian analysis can answer. We first will present the normal explanation of the problem of "casual" evidence offered in the literature of Bayes. As an example, then, of what is meant by the phrase "'causal' evidence," consider two countries, call them A and B, and a speech by a leader of country B warning country A's leaders not to do a certain thing, i.e., not to take action C. Then, some writers consider the warning in the speech an item of evidence, E_1. But E_1 is a curious sort of item of evidence. It does not directly provide any indication as to whether A's leaders will decide to take action C (call this H_1) or will decide not to take action C (call this H_2). It really is indicative that B's leaders are concerned that state A may take action C and it is intended to cause the leaders of state A not to take action C. It is "causal" evidence in the sense that it is intended to cause the opponent to act in a particular way.

How can we include such strange "evidence" in our Bayesian analysis? Suppose we let

H_1 = the systemic forces at work in the environment are such that state A's leaders eventually will decide to take action C,

H_2 = the systemic forces at work in the environment are such that state A's leaders eventually will decide to not take action C,

and E_1 = Leader "L" of state B has warned the leaders of country A not to take action C.

Then, if we try to use the odds-likelihood form of Bayes in purely mechanical fashion we would have

$$\frac{P(H_1/E_1)}{P(H_2/E_1)} = \frac{P(H_1)}{P(H_2)} \cdot \frac{P(E_1/H_1)}{P(E_1/H_2)}.$$

Something very curious will happen, however, if we put in the numbers on the right hand side that seem warranted if we assume E_1 to be reflective of and derivative from the systemic forces at work in the environment. Note that the whole point of the warning by the leader of country B was to <u>reduce</u> the probability of H_1; that is, in our terms, the <u>desired</u> effect of the speech was to make $P(H_1/E_1) < P(H_1)$. But, if we quite reasonably assign values to the probabilities in the likelihood ratio,

$$\frac{P(E_1/H_1)}{P(E_1/H_2)}$$

we certainly would have to say that $P(E_1/H_1) > P(E_1/H_2)$; that is, the probability that E_1 will happen certainly is higher under the assumption that A's leaders will decide to do C than under the assumption that A's leaders will decide to not do C. That is, statement E_1 is more likely to be made, the argument goes, if the systemic forces at work in the environment are such that A is going to decide to do C than if A is going to decide to not do C. But, if we put the numbers into our likelihood ratio so that $P(E_1/H_1) > P(E_1/H_2)$, then the inexorable <u>mathematical</u> result is to drive <u>up</u> the revised probability of H_1 -- a mathematical result exactly opposite to what B wants and opposite to what B expected the result of the warning to be!

This result is considered in the literature to be paradoxical and undesired. Jack Zlotnick suggested a way around this inability to mathematically input such "causal" evidence. On the surface, the idea seems rather clever. Zlotnick argued that, since we cannot input a mathematically meaningful likelihood ratio, the thing to do is to directly assess the result of the "causal" evidence on the target. In terms of the equation

$$\frac{P(H_1/E_1)}{P(H_2/E_1)} = \frac{P(H_1)}{P(H_2)} \cdot \frac{P(E_1/H_1)}{P(E_1/H_2)}$$

this means directly assessing the left hand side of the equation. That is, it means <u>directly</u> assessing the revised odds of

$$\frac{P(H_1/E_1)}{P(H_2/E_1)}$$

<u>instead</u> of calculating them based on the data entered into the right hand side of the equation. Once we have directly assessed the revised odds, if we assume for the sake of argument that the prior odds of

$$\frac{P(H_1)}{P(H_2)}$$

equals $\frac{1}{1}$, then we would have

$$\underbrace{\frac{P(H_1/E_1)}{P(H_2/E_1)}}_{\substack{\text{directly}\\ \text{assessed}}} = \underbrace{\frac{P(E_1/H_1)}{P(E_1/H_2)}}_{\substack{\text{can't be}\\ \text{assigned directly}}}.$$

Now, we simply take the value of the impossible-to-assign

$$\frac{P(E_1/H_1)}{P(E_1/H_2)}$$

to be equal to the directly assessed

$$\frac{P(H_1/E_1)}{P(H_2/E_1)}.$$

In this way, we can get a value for the likelihood ratio to plug into a stream of evidence. (In shorthand notation, Zlotnick was arguing that, if revised odds R equal prior odds P times likelihood ratio L, i.e., R = PL, and we take P = 1 so that R = L, then if evidence is "causal" we can directly assess the result R and take this value to be equal to L.)

Zlotnick's procedure, however, as he noted himself, seems to do violence to the spirit of Bayes. In using Bayes we are looking at evidence reflective of and derivative from the systemic forces at work in the environment, and assigning probabilities based on this evidence. The probabilities thus assigned lead to a determination of the revised probabilities of the hypotheses. To not wait for evidence to emerge from the environment, but instead to directly assess the revised probabilities seems counter to the whole Bayesian procedure (and may reintroduce the analyst's ego involvement and bias the results). Moreover, while speculation on the possible effects of attempts to influence other states is necessary, the uncertainty about such reactions is quite great. To introduce such grave uncertainties into the Bayes equation -- rather than waiting for evidence to emerge and then evaluating that evidence -- seems a highly dubious procedure.

We can go beyond Zlotnick's own criticism of this approach, however, to argue that, if the "causal" evidence really were from the environment, then the original results where the revised probability of H_1 increased would not be paradoxical at all but would be proper (even if not desired by state B)! The reason is that, if the "causal" evidence were from the environment, then it properly could be viewed as reflective of and derivative from the systemic forces at work in the environment; the warning itself, although intended to alter the direction of the systemic forces at work in the environment, should be viewed as just one (more?) item of evidence indicative of the trends in the systemic forces. That is, we could argue that even though the intention of the speech was to cause state A's leaders to not take action C, the fact that the speech was made (i.e., the occurrence of E_1) could be seen as reflective of the systemic forces at work in the environment. In this sense, then, nothing would be wrong at all. If we were to input the conditional probabilities for this item of evidence reflective of the systemic forces at work in the environment, the probability that state A's leaders eventually would decide to do C would go up, and properly so. The answer to the objection that the speech by B's leaders was intended to reduce the probability of A doing "C" is that this might be the eventual result of the speech, and, if it were, then there eventually would be items of evidence that, once assessed, would lead to a decrease in the probability of H_1.

However, I feel that something fundamentally is wrong with the "causal" evidence of the example. I really don't think that it properly can be viewed as emanating from the "environment," even if this latter concept is defined broadly. The "environment" with respect to our hypotheses about what state A's leaders will decide to do seems to me to involve the government and society of state A. To include actions (the "causal" evidence) by state B in our analysis seems to me to stretch the umbrella of the "environment" (in which the systemic forces are at work) too far.

I think that it may be possible and useful to broaden our scope of the concept of so-called "causal" evidence to include not only its present meaning of consciously directed efforts by some party to influence the course of events, but also to include non-consciously directed events or actions which might influence the course of events. (An example of the latter is the event of riots by party A's supporters in the problem 8 discussion of what sort of questions Bayes can answer.) We also should divide these categories into events external and internal to the environment (i.e., the government and society of the state within which actor A is based). Given this categorization of events into four categories, we probably should exclude "causal" events not in the environment from our Bayesian analysis.

Zlotnick's self-criticism of his proposed solution seems valid; the approach of directly assessing the revised probabilities does indeed bypass the Bayesian approach. If and when such "causal" evidence does have some impact, we can expect events (evidence) reflective of that impact to emerge and be assessed in a way properly reflective of that impact. It should be noted, by the way, that the actual impact might be in a direction not intended, in the case of consciously directed efforts. (If, for one reason or another, it is deemed important to assess immediately the possible impact of the "causal" evidence, a better technique than Zlotnick's solution is the use of a probability diagram (tree) with the possible reactions to the "causal" evidence at the start of the tree and the possible actions of the hypotheses at the end of the tree. (See problem 16 and Appendix A.) A decision making model may be appropriate, too.)

"Causal" evidence or events internal to the environment, on the other hand, whether consciously directed efforts to influence the course of events or non-consciously directed happenings or actions (deemed by us as "causal" because they might influence the course of events) should be treated, if possible (through appropriate wording of our hypotheses), as reflective of and derivative from the the systemic forces at work, even though they (eventually) may have a (causal) impact. If and when such internal "causal" evidence does have an impact, events reflective of that impact can be expected to occur and to be assessed appropriately.

Before leaving this discussion of the analytical scope of the environment and the conclusion that casual evidence external to the environment should be excluded, I want to present a very reasonable counter-argument to the analysis; interestingly, it reaches the same conclusion (but for different reasons) about excluding the "external" causal evidence.

The counter-argument is this. With respect to future decisions (or even future events), the trouble with what we have termed causal evidence external to the environment is not so much that it is not reflective of and derivative from the trends in the actual systemic forces at work, but instead that it is so difficult to assign reasonable probabilities to this evidence.

The concept of systemic forces at work, the argument goes, should refer to all of the factors which will influence the ultimate decision by actor A; if this is so, then the statement by actor B's leaders really is one element reflective of those systemic forces at work which will impact on the ultimate decision by actor A. (International relations theorists will note that all this can be related to discussions of the concept of "power" and its scope.) Therefore, any attempt to restrict the scope of the environment from which evidence is drawn to the government and society of the state within which actor A is based is invalid. The environment, properly defined, will be the international political system and any factors that can influence the decision making process of actor A. Items of evidence will be considered as reflective of and derivative from the systemic forces at work in this (broadened) environment. The problem is that when we get "evidence," we want to be able to say reasonably how consistent the evidence is with the alternative directions of the systemic forces, i.e., we want to assess reasonably P(evidence/H_j). However, when dealing with what we have termed above "casual evidence external to the environment," making such assessments seems to be very difficult, if possible at all. For example, it does not seem possible to assess reasonably

$$P\left(\begin{array}{l}\text{state B's leaders will}\\\text{make a speech warning}\\\text{state A not to take}\\\text{action}\end{array}\middle/\begin{array}{l}\text{the systemic forces at work in the}\\\text{broadened conception of the environ-}\\\text{ment are such that state A eventu-}\\\text{ally will take action C}\end{array}\right).$$

That is, looking at the statement of B's leaders as an item of evidence from the stream of systemic forces at work, it seems to be very hard or impossible to make and justify any probability assessment. It is all too vague, it seems. The conclusion would seem to be that we should leave out items of casual evidence external to our original conception of the environment, but because they are so hard to assess and not because they are invalid items. When and if these deleted items of evidence have an impact on the systemic forces at work in the (originally defined) environment of actor A, evidence reflective of and derivative from the systemic forces at work can be expected to emerge and to modify the probabilities of the hypotheses.

(The reader may judge for him(her)self which argument concerning the scope of environment is correct. Let me note that actors such as terrorist organizations and international businesses make it difficult to adopt a conception of the environment restricted to the government and society of just one state.)

Before leaving this discussion of "causal" evidence, we should note once more that there may be cases where an action or event can be "causal" in nature, and yet no phrasing of our hypotheses, no matter how tortured, will enable us to view it as being derivative from and reflective of the systemic forces at work in the environment. A good example is a tactical military situation where the weather suddenly turns bad. We may feel that this will have a profound impact on the enemy's plans to attack, but unless we use Zlotnick's trick we can't include the evidence in a Bayesian analysis.

To see why, suppose the hypotheses about the enemy's courses of action are H_1 = the enemy plans on attacking within 12 hours and H_2 = the enemy does not plan on attacking within 12 hours and E_1 = the weather turns bad.[2] In an odds-likelihood equation where

$$\frac{P(H_1/E_1)}{P(H_2/E_1)} = \frac{P(H_1)}{P(H_2)} \cdot \frac{P(E_1/H_1)}{P(E_1/H_2)}$$

the probabilities in the revised odds make complete sense. That is, it certainly makes sense to ask the (numerator) revised probability of attack assuming the weather has turned bad. But, when we look at the required probabilities in the likelihood ratio, we find that they make no sense. What sense does it make to ask the (numerator) probability that the weather will turn bad assuming there will be an attack, or the (denominator) probability that the weather will turn bad assuming no attack? The problem is that the evidence is not reflective of or derivative from the systemic forces at work in the environment of the enemy. (The question would only make sense if the enemy had and was using weather modification equipment. Then the evidence could be linked to the possible decisional situations of the enemy and the probabilities required in the likelihood ratio would make good sense.)

Again, one way of handling this problem of "causal" evidence would be to use Zlotnick's solution. In this case, where there might be a good deal of insight into the enemy's tactical use of bad weather, Zlotnick's solution might be warranted. The real question to ask is how much uncertainty we have about possible enemy reactions. If there is a fair amount of uncertainty, then we probably are much better off not to use Zlotnick's solution but instead to wait for evidence to emerge from the environment. Alternatively, we could try to use a probability diagram where one of the variables determining possible enemy courses of action would be the weather condition. (Appendix A would apply.)

Another point should be made. Student Jerome Rapin pointed out that there is a possible link between "causal" evidence external to the environment of a target and what will be termed, in problem 55, a "paradox of warning" situation. He noted that, if Bayes is being used in an indications and warning situation, the decision maker may act on the basis of the Bayes data by taking an action (e.g., a threat) toward the target country the Bayesian analysis deals with. This "causal" evidence, external to the environment of the target, if it is effective, may lead to the decision makers in the target country changing their minds. If the decision maker asks the I&W analyst before taking the action how the action will affect the Bayesian analysis, the I&W analyst may feel pressed to use Zlotnick's solution to answer the question. This is not a good solution at all, as we have argued. The analyst probably should try to say that for such questions a probability diagram (tree) is a better solution. Possible actions by the target in response to the proposed course of action can be put at the beginning of the tree, and hypotheses about possible final actions of the target country put at the end of the tree, and appropriate other variables put in-between. (See problem 16.)

PROBLEM 16. The one-sided view: the problem of the one-sided view of a complex interaction process that results if we do not utilize "causal" evidence external to the environment in our Bayesian analyses.

The problem: In the previous problem we argued that we should not utilize "causal" evidence external to the environment of an actor in our Bayesian analyses. If we are looking at a complex interaction process, this means that in a sense we have a one-sided view of the process - our hypotheses almost always will involve the possible actions of only one of the two sides, and all our evidence must be reflective of and derivative from the systemic forces at work in the environment of that same side.

The solution: Instead of utilizing Zlotnick's approach to include "causal" evidence external to the environment of an actor, we are much better off if we want to utilize the items of external "causal" evidence (or feel we must use them for one reason or another) if we abandon Bayesian analysis and utilize (as suggested earlier) a probability diagram (tree) approach *or* utilize a foreign policy decision making model *or* some sort of action-reaction model.

Discussion: An example of the sort of situation that can arise is as follows. Suppose on day I of a crisis, target country A, which the Bayesian analysis hypotheses concern, attacks some of country B's property, but that B does not respond. Suppose then on day II that country A again attacks some of country B's property, and this time country B responds in kind. If our hypotheses are strictly about country A, then only country A's actions would be considered reflective of the systemic forces at work in the environment. Country B's responses would be external "causal" evidence and, we have argued, should not be included in the Bayesian analysis.

If we feel that the "causal" evidence somehow should be taken into account, one possible response, of course, would be to utilize Zlotnick's solution. This is not a very good solution, we have argued before, and we probably would be better off simply abandoning Bayesian analysis and utilizing one of the alternatives suggested in the solution section above. Probability diagrams (trees) can be a very useful probability tool in modeling interaction processes. In fact, there is a special term for trees used in tracing an interaction process: "Events sequence network analysis." The tree is shown in Figure 6.1.

This is a very powerful, flexible instrument. One contractor, CACI, Inc.-Federal, has done a number of studies for the government using this technique.[3]

FIGURE 6.1
Events Sequence Network Analysis Tree Diagram

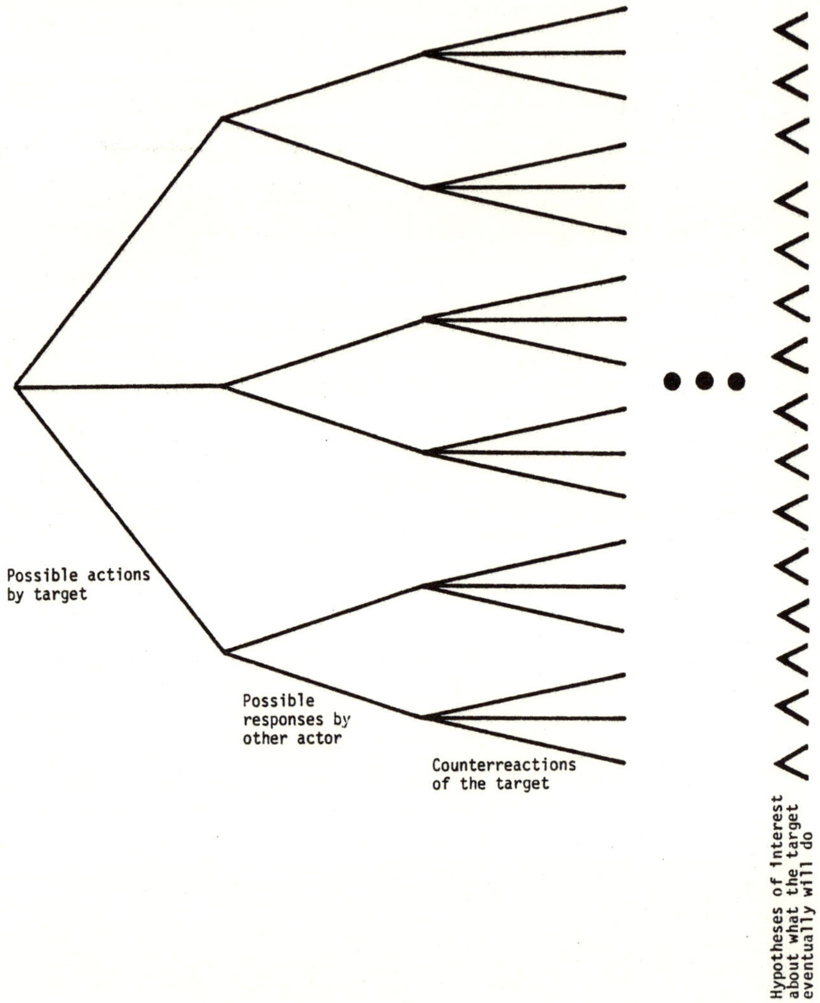

PROBLEM 17. Predicting with "mixed" items of evidence: the problem of trying to predict a future event when evidence is a mixture of (1) items of evidence which indicate the state of policy at present and (2) items of evidence which may be the "causes" of a possible policy change in the future.

The problem: Stated above.

<u>The solution</u>: Use of a straight systemic approach will tend to accommodate both types of evidence. In particular, the "causal" evidence can be reinterpreted as derivative of and indicative of the systemic forces at work in the environment.

<u>Discussion</u>: An example (suggested in class by Dr. John Kringen) of this sort of mixture of items of evidence is the following:

Suppose we are interested in predicting whether state A will attack state B. Using a systemic decisional situation approach we can phrase the hypotheses as

H_1 = the systemic forces at work in the environment are such that state A's leaders eventually will decide to attack state B,

H_2 = the systemic forces at work in the environment are such that state A's leaders eventually will decide to <u>not</u> attack state B.

(Once again, it might be better to put in a time period.)

Suppose our first items of evidence are

E_1 = (a report that) nation A's leaders have decided to launch an attack within the next month if weather permits,

and E_2 = (a report that) opposition against the war is continuing to build up among all sectors of society in state A, even within the government.

Then E_1, in a sense, says what the "present" policy is and seems to answer the question posed to say that "eventually" is now and the decision has been made. E_2, on the other hand, suggests that such a decision might be reversed in the future, i.e., to use the language of problem 51 below, that a future non-stationarity of hypotheses situation might occur.

Use of a straight systemic formulation can accommodate both of these items of evidence.[4] If the wording is

H_1 = the systemic forces at work in the state of nature are such that state A eventually will attack state B,

H_2 = the systemic forces at work in the state of nature are such that state A eventually will <u>not</u> attack state B,

then E_1 can be seen as derivative of and reflective of the systemic forces at work and will lead to a major upward revision of the probability of H_1. But this time, it does not imply that the question posed by H_1 is totally answered. Other items of evidence may occur. E_2 can be interpreted also as an item of evidence derivative of and reflective of the systemic forces at work and will tend to lead to a downward revision of H_1.

PROBLEM 18. Recognizing invalid "evidence": the problem of identifying "evidence" not really from the environment.

The problem: The concept of "evidence" not really from the environment was raised in the section on "causal" evidence. Two major types of such "evidence" are (1) external "causal" evidence and (2) what we can term physical occurrences/events not subject to anyone's control, such as the weather in the example provided. The reason for saying there is a problem is that, if we do not recognize such evidence as invalid, we may go ahead and try to include it in a Bayesian analysis, either by using Zlotnick's solution or by considering the evidence to be derivative from and reflective of the systemic forces at work in the environment and trying to assess it as such. Invalid or inaccurate revised probabilities of our hypotheses then may result. This problem turns out to be far more common than the reader might guess.

The solution: If we ask ourselves, for each item of evidence, "can this be seen as reflective of and derivative from the systemic forces at work in the environment (government and society of the actor) or is it instead external 'causal' evidence or a physical occurrence or event not subject to anyone's control?" we probably can recognize such invalid evidence. Occasionally, there will be borderline cases where it will not be completely clear. With practice, recognizing invalid evidence will become easier.

Discussion: The basic points have been made already, but it may be useful to give an example of what some might consider a borderline situation. Consider these hypotheses[5]

H_1 = Country A will intervene militarily in country B to prevent the ouster of the current pro-A government,

H_2 = Country A will not intervene militarily in country B to prevent the ouster of the current pro-A government.

Note first that the hypotheses are not phrased in decision making terms. That is, they are not written "country A has decided to (or to not) intervene military." So, the implicit model is not one where we will be picking up evidence which we will interpret as reflective of a decision already made. The tone is more systemic, and we can rephrase the hypotheses to be systemic, e.g.,

H_1^* = the systemic forces at work in the environment are such that country A's leaders eventually will decide to intervene militarily in country B to prevent the ouster of the current pro-A government,

H_2^* = the systemic forces at work in the environment are such that country A's leaders eventually will decide to not intervene militarily in country B to prevent the ouster of the current pro-A government.

We then can interpret evidence in terms of its reflection of the direction of the systemic forces. The first item of evidence presented is

E_1 = An announcement in the government newspaper in country A saying that country A considers country B to be within the group of nations led by country A.

E_1 is a legitimate item of evidence from the environment concerning country A, although E_1 appears to be so innocuous that we might well argue that

$$P(E_1/H_1) = P(E_1/H_2) \text{ or } P(E_1/H_1^*) = P(E_1/H_2^*).$$

The next event is

E_2 = Report in a third country's press that internal opposition to the present government of country B is increasing.

The problem with E_2 is that it does not seem to be evidence from the environment, i.e., it does not seem to be reflective of and derivative from the systemic forces at work in country A. It is not about country A's decision making or society or anything to do with country A. It is about country B, and can be viewed as an event which could influence country A. It seems to be what we have termed external "causal" evidence. However, if we go ahead and ask $P(E_2/H_1)$, we are asking, in words, "the probability that there would be a third country report that internal opposition to the present government of country B is increasing given that country A will intervene militarily in country B to prevent the ouster of the current pro-A government." As stated, this seems to be a nonsense question.

Using our systemically written hypothesis H_1^* does not seem to help much, because $P(E_2/H_1^*)$ is read "the probability that there would be a third country report that internal opposition to the present government of country B is increasing given that the systemic forces at work in the environment are such that country A's leaders eventually will decide to intervene militarily in country B to prevent the ouster of the current pro-A government." The event still seems to me to bear no relation to the first hypothesis, even with the revised wording of the first hypothesis, and it appears impossible to assess $P(E_2/H_1^*)$.

This is, however, as close to a borderline case as I have seen, and some might argue for a broader definition of the environment, where this item of evidence might be seen not as external "causal" evidence but instead as really reflective of and derivative from the systemic forces at work in that more broadly conceived environment. I would be interested in reading arguments as to how to assign the probabilities required, however.

Another example of "evidence" which seems to me to be defective because it seems to be external "causal" evidence rather than evidence from the environment is this:

> E_3 = A report that rebels in country B have surrounded a particular provincial capital and that sympathetic army units are prepared to turn over the city and join forces with the rebels (reported by a usually reliable source).

Asking $P(E_3/H_1)$ or $P(E_3/H_1*)$ seems to result again in nonsense questions: again, it seems impossible reasonably to make the probability assessments requird.

(An attempt to rescue this sort of "evidence" by asking the probability of the "evidence" <u>preceding</u> intervention or non-intervention has been treated earlier under the topic of what sort of questions Bayes can answer. This sort of attempt really seems to provide no solution either.)

Invalid evidence, in the sense used above, where it is equated to evidence not really from the environment and not reflective of nor derivative from the systemic forces at work in the environment, can be spotted when we try to phrase our probability judgments but find that they make little or no sense and seem difficult or impossible to make -- when we ask $P(E_1/H_1)$ and find out, no matter how we word the event and the hypotheses, that H_1 has no relevance for assigning a probability to E_1. We should try to reword the hypotheses and the evidence to see if wording can be found that does show the event as "coming from" the environment in the sense that it is evidence somehow reflective of and derivative from the systemic forces at work in the environment. (After a while, when the user has experience doing Bayes, invalid evidence will "leap out" at him or her.)

<u>PROBLEM 19</u>. "Statistical" evidence: the problem of evidence not really about an event or person, but actually about the class to which the event or person belongs.

<u>The problem</u>: There sometimes is an attempt to use evidence which really is not about the particular event or person under investigation but instead actually is about the class to which the event or person belongs. This sort of "actuarial" or "statistical" evidence may be good (from the insurance companies' viewpoint) for selling insurance policies, but is a biased, prejudiced, bigoted way of dealing with specific events or people in a Bayesian analysis.

<u>The solution</u>: Statistical or actuarial evidence should <u>not</u> be used in a Bayesian analysis. Evidence should be things <u>that</u> are

about a specific event or person, not the class to which they belong. In terms of a person, evidence should be about things which that specific person has said or done.

Discussion: While the examples in the previous section on evidence not really from the environment seemed to me to be rather clear-cut, other examples of deficient evidence may not be at all obvious. A very good example of this is the following. Suppose, with respect to a U.S. male citizen, "X,"

H_1 = X is a Soviet spy,

H_2 = X is not a Soviet spy,

and E_1 = X's parents were born in the Soviet Union.

(These hypotheses are slightly simplified from those in problem 8Aii to keep our discussion here a little easier.)

We should ask, what are the "states of nature?" The answer is that, since the hypotheses specify the mutually exclusive and exhaustive possibilities with respect to the states of nature, the states of nature simply are that X is a spy or is not a spy. The next question, then, is what sort of evidence is reflective of the true state of nature? Information about X's behavior, presumably, since it could be indicative of whether he is a spy or not. But what about biographical data about his parents? Should such data be considered evidence or events reflective of the true state of nature concerning X? Should we distinguish among contemporary data about his parents, data about his parents prior to his birth, and data after his birth but prior to contemporary data? These are not easy questions to answer.

One way of looking at E_1 is to formally attempt to use it and to see if we can find any phrasing of either the evidence (or the hypotheses) where it may lead to a reasonable probability question.

If we think of our discussion above about the phrasing of probabilities of events in percentage terms, we might say

$P(E_1/H_1)$ = percent of time a person's parents were born in the Soviet Union given that the person is a spy,

$P(E_1/H_2)$ = percent of time a person's parents were born in the Soviet Union given that the person is not a spy.

The questions themselves do appear to make some sense. We would expect, however, that this item E_1 would not be diagnostic. That is, we would expect that $P(E_1/H_1) = P(E_1/H_2)$, which is simply saying that spies and non-spies are just as likely to have parents from the Soviet Union. But we could imagine situations where in great numbers spies had parents from particular countries. Even clearer, consider Israel, and not spies, but saboteurs.

If H_1 = X is a saboteur,

H_2 = X is not a saboteur,

and E_1 = X's parents were Palestinians

then we would expect that

$P(E_1/H_1)$ = percent of time a person's parents were Palestinians given that the person is a saboteur,

would be greater than

$P(E_1/H_2)$ = percent of time a person's parents were Palestinian given that the person is not a saboteur.

We still may be uncomfortable about such "evidence" in that it is clearly statistical and actuarial in nature and really reflects a good deal of prejudice. It is about the class of people to which "X" belongs, but not about the individual "X." Nevertheless, it does seem to be diagnostic. Moreover, a slight change in the phrasing of E_1 will make it more about X than about X's parents and hence more relevant. We can phrase the E_1 of the first example as

E_1 = X is of Soviet ancestry,

and in the second example,

E_1 = X is of Palestinian ancestry.

In the first example, then, $P(E_1/H_1)$ would now read as "the percent of time that a person is of Soviet ancestry given that the person is a spy" and $P(E_1/H_2)$ would now be read as "the percent of time that a person is of Soviet ancestry given that the person is not a spy." These two probability judgments still are statistical and actuarial in nature and still reflect substantial bias and prejudgment; they still are about the class of people to which "X" belongs rather than about individual "X." Nevertheless, they do seem to make sense and to be potentially diagnostic. The same is true for the saboteur example where we could have a reworded

E_1 = X is of Palestinian ancestry.

Our probabilities for E_1 then would read

$P(E_1/H_1)$ = percent of time that a person is of Palestinian origin given that the person is a saboteur,

$P(E_1/H_2)$ = percent of time that a person is of Palestinian origin given that the person is not a saboteur.

Continuing on with the main example where

H_1 = X is a Soviet spy,

and H_2 = X is <u>not</u> a Soviet spy

suppose that

E_2 = X is born in country Z (which is an ally of the Soviet Union).

Again, strictly from a statistical or actuarial viewpoint, it does make sense to ask the conditionally independent questions of

$P(E_2/H_1)$ = percent of time that a person is born in country Z (which is an ally of the Soviet Union) given that the person is a spy,

and $P(E_2/H_2)$ = percent of time that a person is born in country Z (which is an ally of the Soviet Union) given that the person is <u>not</u> a spy.

Such questions are, once again, purely statistical and actuarial in nature, but do seem to make sense. We would expect that $P(E_2/H_1) = P(E_2/H_2)$, i.e., we would expect E_2 not to be diagnostic.

If we asked the questions more properly in conditionally dependent form as

$P(E_2/H_1\&E_1)$ = percent of time a person is born in country Z (which is an ally of the Soviet Union) given that the person is a spy and the previous datum that the person is of Soviet ancestry.

and $P(E_2/H_2\&E_1)$ = etc.,

we again have questions that are quite statistical and actuarial in nature and do seem to make some sense, but again they are likely to not be diagnostic because we would expect that $P(E_2/H_1\&E_1) = P(E_2/H_2\&E_1)$. Suppose next that we have

E_3 = X's father was a Soviet agent.

When we phrase the conditionally independent probability judgments, we have

$P(E_3/H_1)$ = percent of time a person's father was a Soviet agent given that the person is a Soviet spy,

and $P(E_3/H_2)$ = percent of time a person's father was a Soviet agent given that the person is <u>not</u> a Soviet spy.

Here again we have something which could be diagnostic, in a strictly statistical and actuarial sense. At a deeper level, we might feel that overtly or subtly a father who is a spy might influence the son. (British spies in Egypt in the 20th century reportedly sometimes were descendents of generations of previous British agents.) So, diagnostic value may be there even though the evidence is not about something X himself has done or said. (We probably should ask instead $P(E_3/H_1\&E_1\&E_2)$, $P(E_3/H_2\&E_1\&E_2)$, which would be more complicated statistical questions, but would make sense.)

When all is said and done, however, we really should be rather dubious about using such "statistical," actuarial" evidence. We really are looking for evidence about the things a person himself has said or done. The sort of "statistical" or "actuarial" evidence used above is the sort misused by biased people to prejudge others. It is about the class to which "X" belongs, not about "X" as an individual. Moreover, given the great sensitivity of Bayes to even mildly diagnostic evidence, a series of such items of "statistical" evidence might lead to "conclusions" highly unwarranted in the particular case under scrutiny. I hesitate for fear of misinterpretation to present the following example, but perhaps it will show the terrible misuse that could result from such "statistical" evidence. Suppose

H_1 = X is a thief,

H_2 = X is not a thief,

E_1 = X is a member of minority Y.

Then if we ask $P(E_1/H_1)$ it can be read

$P(E_1/H_1)$ = percent of time a person is a member of minority Y given that the person is a thief

while

$P(E_1/H_2)$ = percent of time a person is a member of minority Y given that the person is not a thief.

These questions do make sense statistically, and the probabilities may be diagnostic in the sense that $P(E_1/H_1)$ may not equal $P(E_1/H_2)$. But, if we go ahead and use them to modify our prior probabilities, perhaps using the odds-likelihood form of Bayes

$$\frac{P(H_1/E_1)}{P(H_2/E_1)} = \frac{P(H_1)}{P(H_2)} \cdot \frac{P(E_1/H_1)}{P(E_1/H_2)}$$

then we would modify our prior probability that someone is a thief on the basis of his membership in a minority group. While such a modification might make some "statistical" sense, any attempt to use it for a specific person would exhibit the worst sort of bias, prejudice, and bigotry. I present this example of the clearcut misuse of "statistical," "actuarial" evidence to argue against its usage. I think that such "evidence" should be labeled as "statistical" and "actuarial" in nature and not included in a Bayesian analysis.[6]

"Statistical" evidence may be somewhat more difficult to spot than what we have termed invalid "evidence," i.e., evidence not really from the environment. (In a sense, we might say that "statistical" evidence is not really from the "environment" either, since it can be viewed as being not about the subject to which the hypotheses refer, but about the class to which the subject belongs.) Perhaps the best way to spot such "evidence" is to see if casting the probability judgments in "percent" terms makes sense, but still leaves us with a situation where the "evidence" is not really something about an event or about what a person himself actually has said or done. If this is the case, the evidence is statistical and should not be used. But if the evidence is about an event itself or about something a person has done or said, then it most likely is not "statistical" in nature.

PROBLEM 20. "Negative" evidence: the problem of the need to consider items (events) of non-occurrences of events in particular time periods.

The problem: This has been referred to by Schweitzer as the problem of negative evidence.[7] He points out that there may be great significance attached to the non-occurrence of "positive evidence" and the continuance of the status quo.

The solution: Negative evidence should be included in Bayesian analyses.

Discussion: The fact that something does not occur (i.e., the status quo is maintained) in a particular time period can be viewed as a legitimate event in its own right. The non-occurrence of an event which would be expected to occur if a particular hypothesis is true can be very important in revising our judgments about the probabilities of hypotheses, e.g., as to whether A has decided to attack B or not. Student Martin Joyce commented as follows:

> Most intelligence analysts routinely consider events as they occur as bits of evidence leading to possible conclusions. This is true of normal Bayesian analysis also. However, all too often, the non-occurrence of events, in itself an important piece of evidence, may go unnoticed. Normal Bayes suffers from this afflication also.
>
> When an analyst develops his hypotheses for Bayesian analysis, he probably already has a good feel for the types of data he

will watch for to crank into the model. If, as time passes, he realizes that he has not received a piece of data that he might have expected to see, he is faced with the problem of how to input that fact (i.e., the non-occurrence) into his model.

An example might be the following: Hypothesis I is that Country A is going to war against Country B in the next two weeks. Incoming SIGINT [signals intelligence] data and HUMINT data are used to come up with revised probabilities. We might expect that PHOTINT [photo intelligence] would reveal troop relocations, logistical activity, etc. For some reason, in this situation, PHOTINT reveals nothing out of the ordinary. This is significant and must be counted in the problem.

Note that, in all of this, there is the assumption that we know for sure that the event (i.e., item of evidence) has not occurred in a particular time period. We do not consider the possibility that it may have occurred, but not been observed, or, if observed, not yet reported. So, perfect observation and reporting are assumed. (See Chapters 7 and 8 where these issues are discussed.)

PROBLEM 21. Developing lists of negative evidence: the problem of how to alert analysts to the need to consider negative evidence and how to develop lists of items of negative evidence for a particular set of hypotheses.

The problem: As stated.

The solution: The problem of how to alert analysts probably can be solved by providing them with examples which will convince them of the need to consider negative evidence. Developing lists of negative evidence is a more difficult matter, but can be approached by asking the analysts to consider, for each hypothesis, if there are any items of evidence they think would have to occur if the particular hypothesis was true. A consolidated list of these "necessary" events would then comprise the list of items to be looked at. If we get a substantial list, we might simply decide to use an indicator list approach, using the necessary items as a subset of the indicator list.

PROBLEM 22. Deciding the relevance of data: the problem of deciding which items of data clearly are "relevant," or "peripheral," or "irrelevant" with respect to the hypotheses.

The problem: There is a real question as to what constitutes evidence, that is, which items of data are relevant with respect to the hypotheses. Items may be viewed as clearly relevant, "peripheral," or irrelevant.

The solution: There is no schoolbook solution. The underlying, almost certainly implicit, cognitive model that each analyst is using, and which will vary from analyst to analyst, means that the judgments made as to the degree to which items of data are relevant necessarily are subjective and selective. There probably will be agreement on the status of the majority of the items of data considered, but for a minority of data there will be disagreement as to whether it is clearly relevant, or of "peripheral" relevance, or irrelevant.

There is some suggestion that perhaps the "peripheral" items of evidence should be given less weight,[8] but the procedures to yield less "weight" are not clearly specified, nor is it likely that agreement on what is "peripheral" can be obtained.

Discussion: There well may be a good amount of agreement on the majority of the data received as to whether the items are clearly relevant or "peripheral" or irrelevant with respect to the hypotheses. But there usually will be a minority of data about which there will be disagreements. The underlying, almost certainly implicit, cognitive model that each analyst is using (and which will vary from analyst to analyst) means that the judgments made regarding relevance will be (possibly highly) subjective and selective.

When the CIA has run centrally controlled Bayesian analyses, Schweitzer reports that the control element, when consolidating a list of items of evidence submitted by various analysts, has imposed "very little editorial judgment," and that "any item which appears to be even marginally related is included for evaluation."[9] The analysts, however, when supplied the consolidated lists, were allowed to ignore any items considered irrelevant. This, of course, reintroduces the problem of different analysts not considering the same evidence. A better solution when working with such master lists is to ask each analyst to consider all the evidence, and point out that if the analyst considers a particular item of evidence E_i to be "irrelevant," the implication is that $P(E_i/H_j\&E_1\&E_2\&\cdots\&E_{i-1})$ is the same for each H_j, and therefore has no impact. So the analyst simply needs to choose (any) value for $P(E_i/H_1\&E_1\&E_2\&\cdots\&E_{i-1})$ and make this the value for every H_j, in order for this item of evidence to have no diagnostic value.

Schweitzer raises the question as to whether, if certain "peripheral" items are included, they then should carry less weight than the other items.[10] We can answer that "peripheral" is a subjective judgment, so that weighting may well be inappropriate. Also, it is not clear how the weighting will be done. Finally, we can add that the probabilities themselves provide some weighting; items that everyone agrees are not "peripheral" may still lack diagnosticity, but an item viewed by an analyst as "peripheral" is very likely to be rather undiagnostic. We would only need to worry about weighting if we were considering a fair number of what we felt to be "peripheral" items and these were being assessed in such a way as to have some real diagnostic impact. We then well might decide that the items were not so "peripheral" after all. (This problem is closely related to Problem 44 which discusses events (i.e., items of evidence) logically distant from the hypotheses.)

NOTES

Problem 14

1. However, see below after the example to follow.

Problem 15

2. Note that there are buried subhypotheses in H_2.

Problem 16

3. See, in particular, Hayes, Dr. Richard E. (1975) "A Guide to Network Construction and Utilization." CACI, Inc.-Federal, November. Arlington, Va. AD A115831.

Problem 17

4. As we noted in problem 8iva, this is one time we prefer a straight systemic approach over an approach dealing with systemic forces influencing a future decisional situation - a reversal of our usual preferences.

Problem 18

5. Drawn from a contractor proposal and modified by me.

Problem 19

6. Student Michael Burke pointed out that one exception would be in the case of attempts at a medical diagnosis of some problem. This is, of course, not a political/military problem, as a rule. (An obvious exception is the state of health of the Soviet leaders.)

Problem 20

7. SCHWIETZER (1978), p. 29.

Problem 22

8. SCHWEITZER (1978), p. 28.

9. *Ibid*.

10. *Ibid*.

7
Additional Problems Involving Evidence

In this chapter we will deal with more problems involving evidence, including the problems of whether to assign probabilities to the occurrence of reports or to the contents of those reports, of how many items of evidence to abstract from a given report, of possibly overweighting an item of evidence by assessing it each time it is reported, of how to handle different characterizations of the true data state, of how often to count "negative" evidence (non-occurrences of an event in a particular time period), of how to deal with the lack of data about an event for a certain period of time, and of how to deal with the non-observation of a particular event in a given time period.

PROBLEM 23. The probability of reports *vs*. the probability of evidence: the problem of deciding when to ask the conditional probability that the *report* (of certain evidence) *will occur* rather than asking the usual conditional probability that the *evidence* (contained in the report) *will occur*.

The problem: In assigning our conditional probabilities, we need to distinguish between the two probabilities we can assign and decide which one we actually want to use. We either can assign the conditional probability that the *report* itself (of certain evidence) *will occur* i.e., P(report of evidence/H_j& previous evidence), *or* assign the conditional probability that the *evidence* (contained in the report) *will occur* i.e., P(evidence/H_j & previous evidence). *Normally*, it is the latter probability that we will want to assign. The problem is deciding when we want to ask the former, i.e., when we want to assign the conditional probability that the *report itself* (of certain evidence) will occur.

The solution: Normally, as will be discussed later, only in the case when data on an event has not been received for a period of time will we ask the probability of the report itself (not) occurring.

Discussion: Normally, we are interested in the conditional probability of an *event* rather than the conditional probability of the report of the event (evidence).[1] There are a number of problems that we encounter when we ask not the probability of the evidence (i.e., event) reported on but instead ask the probability that the report itself would occur. For one thing, the probability of the report occurring involves the likelihood of the observation of the event, and the likelihood of the observation might depend upon (1) an observer being in the right place at the right time, (2) the abilities of the observer, etc.

PROBLEM 24. How many events to abstract: the problem of how many events, i.e., items of evidence, to abstract from a given report if we decide that it is units of evidence for which we wish to assess probabilities rather than reports.

The problem: As stated.

The solution: It is a matter of judgment. An events data approach could be used, Dr. Hopple notes, "to define items in terms of discrete items referring to specific (verbal/physical) behavior."

PROBLEM 25. "Double-counting" evidence: the problem of overweighting items of evidence by assessing the same item each time it is reported.

The problem: Student Sylvia Smith put it nicely when she said that we want to insure that the revised probabilities of hypotheses will reflect the different events (items of evidence) that have occurred rather than the volume of reporting on the events.

The solution: The basic assumption in this problem, as it is discussed in the Bayes literature, is that the conditionally independent form of Bayes is being used. If this assumption is true, then the solution is to include each distinguishable event (item of evidence) only once in the Bayesian analysis. If the conditionally dependent form of Bayes is being used, the same event can be included each time it is reported, and no harm will be done. However, the same results will come about if we adopt the same rule as for conditional independence: include each item only once in our Bayesian analysis.

(This problem is related to Zlotnick's suggestion to segment evidence as much as possible, i.e., to group items of evidence which are closely related into a single item of evidence. He wanted to do this to achieve conditional independence of grouped items; here, the idea is to avoid overweighting evidence on which the volume of reporting is high.)

Discussion: Unfortunately, this problem is not clearly discussed in the literature on Bayesian analysis. The problem is incorrectly stated as being the problem of "double-counting" and overweighting evidence by taking reports as the units of evidence rather than distinguishable events; the solution then provided is to take distinguishable events as the units of evidence rather than reports. The reason for asserting that this portrayal of the problem (and solution) is incorrectly stated is that it confuses assessing the probability that a _report_ (of certain evidence) _will occur_ with assessing the probability that the _evidence_ (contained in the report) _will occur_. When the explanations in the literature are examined, it seems that they should be re-phrased as follows: there is a problem of double-counting and overweighting evidence if each item of evidence is included in a Bayesian analysis each time it is reported. The solution is to count each item of evidence only once, not each time it is reported. We can expand a little on this revised wording of the problem as follows. If we count each item of evidence each time it is reported, then we could receive one report on a particular event, call the event E_1, make our assessments $P(E_1/H_1)$, $P(E_1/H_2)$, $P(E_1/H_3),\cdots$, $P(E_1/H_m)$, and subsequently consider the evidence contained in another report (which might have been on hand or received subsequently). If the second report says essentially the same thing as the first, and we make an input E_2 based on this additional (perhaps later) report, people argue then we are engaged in double-counting the same event (i.e., same unit of evidence) and overweighting the impact of the particular event (unit of evidence). (There could be further overweighting, of course, if there were additional, perhaps subsequent, reports again saying the same thing about the event, if we made inputs E_3 etc. based on them.)

Thus, if we count each item (unit) of evidence each time it is reported, we will overweight events on which the volume of reporting is high, and underweight possibly more significant events on which it is low. Hence, the advice is to count distinct items of evidence, distinguishable events that is, only once and not each time they are reported, so that each item of evidence is included only once in a Bayesian analysis.

Now it can be asserted that this is good advice, even though the basic argument in the literature leading to the advice still is flawed. The reason for saying it still is a flawed argument is that the argument seems to assume that the conditionally independent form of Bayes is being used. That is, if we have an E_1 based on a report and make the assessments $P(E_1/H_1)$, $P(E_1/H_2)$, $P(E_1/H_3),\cdots$ $P(E_1/H_m)$, and then for an E_2 essentially the same as E_1 we enter in the conditionally independent assessments $P(E_2/H_1)$, $P(E_2/H_2)$, $P(E_2/H_3)$, $\cdots$ $P(E_2/H_m)$, then we can expect to have $P(E_2/H_1) = P(E_1/H_1)$, $P(E_2/H_2) = P(E_1/H_2)$, etc. and we have in essence double-counted E_1. (Perhaps this is clearer in the odds-likelihood formulation. Suppose that we have only two hypotheses; then

$$\frac{P(H_1/E_1\&E_2\&\cdots\&E_n)}{P(H_2/E_1\&E_2\&\cdots\&E_n)} = \frac{P(H_1)}{P(H_2)} \cdot \frac{P(E_1/H_1)}{P(E_1/H_2)} \cdot \frac{P(E_2/H_1)}{P(E_2/H_2)} \cdots \frac{P(E_n/H_1)}{P(E_n/H_2)}.$$

Then, if E_2 is essentially the same as E_1, the first two likelihood ratios would be the same and we would indeed have double-counted E_1.)

But, as noted, this assumes we are using the conditionally independent form of Bayes. If we use the conditionally dependent form of Bayes, then for an item of evidence E_2 (based on another report), essentially the same as an item of evidence E_1 (from a previous report), we will find that $P(E_2/H_1\&E_1) = 1$, $P(E_2/H_2\&E_1) = 1$, $P(E_2/H_3\&E_1) = 1$, etc. (See below for proof.) (Or, in the case of the odds-likelihood form of Bayes, the likelihood ratio for E_2 would be

$$\left. \frac{P(E_2/H_1\&E_1)}{P(E_2/H_2\&E_1)} = \frac{1}{1}. \right)$$

Because these values are "1," the problem of double-counting disappears. If we enter in an item of evidence essentially the same as one reported earlier, the conditionally dependent probability of its occurrence for each hypothesis will be "1" and it will have no diagnostic impact. E.g., if E_2 and E_1 are "Brezhnev is in Prague talking to Dubcek," then

$$P\left(\begin{array}{l}\text{"Brezhnev is in Prague}\\ \text{and talking to Dubcek"}\end{array} \middle/ H_j\& \begin{array}{l}\text{"Brezhnev is in}\\ \text{Prague and talking}\\ \text{to Dubcek"}\end{array}\right) = 1 \quad \text{for each } H_j.$$

(The mathematical proof of this is as follows:

$$P(E_2/H_j\&E_1) = \frac{P(E_2\&H_j\&E_1)}{P(H_j\&E_1)}$$ by the conditional probability formula.

But if $E_1 = E_2$, then $E_2\&H_j\&E_1 = E_1\&H_j\&E_1 = H_j\&E_1$.

Hence $P(E_2/H_j\&E_1) = \frac{P(H_j\&E_1)}{P(H_j\&E_1)} = 1$ (for each H_j) .)[2]

Thus, it can be concluded that there is no real double-counting problem mathematically, <u>if</u> the conditionally dependent form of Bayes is used. There is <u>a</u> problem only if the conditionally independent form is used. As has been suggested above, it would appear to be better to use the conditionally dependent form of Bayes.

However, with respect to cutting down on busywork, it probably is a good idea to follow the advice given above about counting discrete units of evidence (distinguishable events, that is) only once, rather than each time reported. That is, to avoid the busywork of having to input items of evidence which are essentially the same as items of evidence previously considered, it is better simply to include the evidence once. (Moreover, as the above argument shows, if the conditionally independent form of Bayes is used, this advice is mandatory to avoid doublecounting.)

Now, all of the above discussion tended to ignore the issue of the time of receipt of the reports in which the same distinguishable event is described. If there are a number of reports <u>on hand</u> and there is a distinguishable event which is described essentially the same way in each of the reports, it is an easy matter to just input this event once into Bayes.

If <u>subsequently</u> another report is received, which again describes essentially the same distinguishable event in the same way, we either (1) can choose to say that we have already considered the event and hence need not further deal with it or (2) if we are using the conditionally dependent form of Bayes, we can include it by assigning conditional probabilities of "1" for its occurrence. (We <u>must</u> do (1) if we are using conditionally independent form of Bayes.)

(See below for the problem of a later report on the same distinguishable event which alters the description of that event.)

<u>PROBLEM 26</u>. Different characterizations of the true data state: <u>the problem of assessing the probability of an event when there are multiple reports on the event which do not characterize the event the same, but rather characterize the event in a way that is either complementary (supplementary) or contradictory in nature; in short, the problem of how to handle different characterizations of the true data state.</u>

<u>The problem</u>: The previous problem, in essence, was the problem of what to do if the same, identical event (item, unit of evidence), was reported more than once. The problem here, however, is that, when an event is reported on more than once, the later reports either add to the description of the event in a complementary, supplementary fashion <u>or</u> actually contradict the earlier description of the event.

<u>The solution</u>: When we have collated all the new information and arrive at a new characterization of the event, we should reassess our probabilities for the event (and the subsequent events if we are using the conditionally dependent form of Bayes).

<u>Discussion</u>: This might be particularly troublesome in a crisis situation, where we could get a report, yielding a distinguishable event E_i, put in our $P(E_i/H_j\&E_1\&\cdots\&E_{i-1})$ values for each H_j, and then get an additional report later which altered our description of E_i. (If our view of E_i was not altered, the remarks made in the previous "double-counting" problem apply.)

We really have dealt with this problem in our discussion of conditional independence and dependence. There seems little choice but to <u>reassess</u> the probabilities for $P(E_i/H_j\&E_1\&\cdots\&E_{i-1})$ (for all H_j), for $P(E_{i+1}/H_j\&E_1\&\cdots\&E_i)$ (for all H_j), etc. up to the latest event where the "E_i" now inputted is based on the latest data. Admittedly, there is a severe "contamination" problem in the sense that events subsequent to E_i, i.e., E_{i+1}, etc. are known when the reassessments are made. Hence, the probabilities assigned may (improperly and subconsciously) reflect data subsequent - rather than just prior - to each event whose probability is being reassessed.

If the reports are supplementary (complementary and not contradictory), there would appear at first to be an alternate solution if the conditionally dependent form of Bayes is used. Each could be entered as a separate item of evidence, and the usage of the conditionally dependent form of Bayes would tend to decrease problems of overweighting the evidence. (We could use Zlotnick's exponent (described below in problem 30), if we wanted to, for each item.)

However, there is some difficulty in practice even if the conditionally dependent form of Bayes is used. If E_2 fills in the picture more completely about the underlying event that led to the description embodied in the statement of E_1, then assessing $P(E_2/H_j\&E_1)$ for each H_j may not be all that easy. Since E_2 is not identical to E_1, we cannot say that $P(E_2/H_j\&E_1) = 1$ for each H_j, as we did in the last problem. Assigning $P(E_2/H_j\&E_1)$ is then not too easy. The best solution to the problem, as we have suggested, is to assemble all the reports, arrive at one detailed description of the event, and reassess. (If this is done, however, we can't use Zlotnick's exponent - there is no single source for the amalgamated description of the event.)

(There are some highly technical alternate ways of handling this problem but they probably are too difficult for normal use.)[3]

If the reports are in conflict with one another, that is, if separate reports talk about the same event, but provide meaningfully different and conflicting descriptions of that event, we should proceed in the same fashion by assembling all the reports, arriving at one detailed description of the event, and reassessing. (Again, the highly technical, alternate ways of dealing with this problem probably are too difficult for normal use.)

The problem of complementary or supplementary characterizations of the same event becomes even more complicated if we want to consider the reliability of the different sources.

(My thanks to student Bob Wright for helpful suggestions on this section.)

PROBLEM 27. How often to count "negative" evidence: the problem of how often to input a particular item (event) of "negative" evidence.

The problem: If we include non-occurrences in our Bayesian analysis, one problem is how often to include them and how to avoid overweighting the non-occurrences. (We are assuming here that we are definitely sure that the event has not occurred - we are assuming perfect observation and reporting.)

The solution: Not obvious. Suggestions by readers are solicited.

Discussion: The question here is how many times we can count the non-occurrence of an event without overweighting the fact of non-occurrence. Just how often should non-occurrences be included? After all, something can fail to happen this hour, the next five minutes, the next hour, the rest of the day, the next day, the next five days, next week, etc. When we look back, do we want to make an assessment for every hour, half-day, day, week, month or what with regard to the latest time period when the particular event failed to happen? In short, what is the time period in which we wish to consider the non-occurrence of the event?

There is no definitive answer to this question. Technically, it can be argued that the analysis could be the same (or, more exactly, yield the same results) no matter what the choice of time intervals is, if the conditionally dependent form of Bayes is used. (Compare the analysis below to the previous analysis of the "double-counting" problem, where it was concluded that it is best to include distinct events once, rather than including them each time they are reported.) That is, if we want to ask the probability that an event E_i would fail to occur yesterday given (i) a particular hypothesis, (ii) information that the event did not occur the day before yesterday, and (iii) other previous data, i.e.,

$$P\left(\begin{array}{l}E_i \text{ did not occur} \\ \quad\text{yesterday}\end{array} \middle/ \; H_j \& E_i \begin{array}{l}\text{did not occur} \\ \text{the day before} \\ \text{yesterday}\end{array} \;\&\text{ other previous data}\right)$$

then we will get probabilities for the different H_j's that are closer to one another (and hence less diagnostic) and higher than those when we ask

$$P\left(\begin{array}{l}E_i \text{ did not occur} \\ \quad\text{last week}\end{array} \middle/ \; H_j \& E_i \begin{array}{l}\text{did not occur} \\ \text{the week before} \\ \text{last}\end{array} \;\&\text{ other previous data}\right).$$

The latter probabilities for the different H_j's will tend to diverge more from one another (and hence be more diagnostic) and be lower than the former. The reasoning behind these assertions is that, as we compress the latest time interval, the probabilities will tend to come closer together (and hence not be very diagnostic) and become larger because, if something did not occur fairly recently, then it is more likely not to have occurred even more recently, regardless of the particular hypothesis and regardless of other previous data. For example, probability values of

$$P\left(\begin{array}{l}E_i \text{ did not occur} \\ \quad\text{yesterday}\end{array} \middle/ \; H_j \& E_i \begin{array}{l}\text{did not occur} \\ \text{the day before} \\ \text{yesterday}\end{array} \;\&\text{ other previous data}\right)$$

will tend to be higher and closer to one another for the different H_j's than values of

$$p\left(\begin{matrix}E_i \text{ did not occur} \\ \text{last week}\end{matrix} \Big/ H_j \& E_i \begin{matrix}\text{did not occur} \\ \text{the week before} \\ \text{last}\end{matrix} \text{ \& other previous data}\right)$$

simply because if E_i did not occur the day before yesterday then it is also more likely not to have occurred yesterday (regardless of the particular hypothesis and the other previous data) -- more likely in the sense of being higher than the probability we would calculate for E_i not occurring last week if it had not occurred the week before last.

The shorter the time interval, then, the more often the non-occurrence of the event is counted and the more undiagnostic the probabilities are; the longer the time interval, the less often the non-occurrence of the event is counted and the more diagnostic the probabilities are.

A more extreme example may illustrate this idea better. We would expect that in general values of

$$P\left(\begin{matrix}\text{War did not break} \\ \text{out in the last} \\ \text{hour}\end{matrix} \Big/ H_j \& \begin{matrix}\text{War did not break} \\ \text{out in the hour} \\ \text{before last}\end{matrix} \text{ \& other previous data}\right)$$

not only would be higher but also closer together for the different H_j's than values of

$$P\left(\begin{matrix}\text{War did not break} \\ \text{out last week}\end{matrix} \Big/ H_j \& \begin{matrix}\text{War did not break} \\ \text{out the week before} \\ \text{last}\end{matrix} \text{ \& other previous data}\right).$$

Briefly put, shorter time intervals generally should yield higher conditionally dependent probability judgments that are closer together and hence less diagnostic. So, including a non-occurrence often simply could yield offsetting, rather undiagnostic probabilities -- there could be no overweighting from frequent inclusion of non-occurrence. On the other hand, as the time intervals are lengthened, the probabilities should be lower and tend to diverge more -- the evidence should be more diagnostic. The final results could be the same regardless of how often the non-occurrence is counted.

This sort of off-setting of results could mean, in effect, that if we used the conditionally dependent form of Bayes that there would be no overweighting (or underweighting) no matter how often we included the non-occurrence of an event. This mathematical argument really does not justify, however, a definite conclusion that off-setting will be so exact that it does not matter how often we include the non-occurrence of an event. Moreover, all this assumes that we are, in some sense, highly internally consistent, i.e., almost human computers. Only a human computer could be expected to be so consistent that it would make no difference whether s(he) included the non-occurrence of an event often (assigning less diagnostic values) or less often (assigning more diagnostic values) -- only a human computer would have such internal consistency that the revised probabilities of hypotheses would be the same regardless of how often the non-occurrence of the event was counted.

But, since we are not human computers, inconsistencies can be expected. It is not immediately obvious what the direction of the inconsistencies will be. This line of argument, then, tends to erode the previous (rosy) purely mathematical argument which implied that we might have nothing to worry about. We are back to the original question, even if we do use the conditionally dependent form of Bayes, of asking how often to include the non-occurrence of an event. Simple answers are not obvious.[4]

If we try to find a way out by using the conditionally independent form of Bayes, we are even worse off. In this case, overweighting clearly could be a severe problem indeed -- the probabilities would not be assigned taking into consideration the closeness of the previous period when there was a non-occurrence.

This is a fairly strong argument for using the conditionally dependent form of Bayes. A further argument for the dependent form resides in the term "other previous data" in

$$p\left(\begin{array}{l}E_i \text{ has not occurred} \\ \text{in the last time} \\ \text{interval (whatever} \\ \text{it is)}\end{array} \Bigg/ \begin{array}{l}H_j \& E_i \text{ did not occur} \\ \quad \text{in the time interval} \\ \quad \text{(whatever it was)} \\ \quad \text{prior to the last} \\ \quad \text{time interval}\end{array} \begin{array}{l}\& \text{ other previous} \\ \quad \text{data}\end{array}\right).$$

The "other previous data" presumably includes previous non-occurrences. We really need to consider the <u>pattern</u> of non-occurrences in assessing the latest non-occurrences. Just asking the conditionally independent probability of

$$p\left(\begin{array}{l}E_i \text{ has not occurred} \\ \text{in the last time} \\ \text{interval}\end{array} \Bigg/ \quad H_j\right)$$

is to ask for a relatively meaningless probability and can indeed lead to overweighting Bayes if this probability is assessed often.

If an indicator list approach is used, where the non-occurrence of a particular event is one possible setting of the indicator for the event, then, if we included information on the previous settings of the indicator and considered this data in assessing our probabilities, we still would expect different results for different choices as to how often to set the indicators.

Schweitzer's suggested solution to the problem of how to include negative evidence, which he says "at least partially" solves the problem, is to include "the following item whenever it appears necessary: 'How likely is it that only these events would occur (and be seen) if hypothesis 1 is true?', '... if hypothesis 2 is true,' etc."[5] There appear to be a number of problems with this proposed solution. First there is the question of how to determine "whenever it appears necessary." Overweighting by overcounting still could occur. Second, the probability he suggests we assess seems to be very difficult to assess - it focuses not on a particular item of negative evidence but on the items which have occurred. When we ask how likely is it that "only" a long list of events would occur, we are asking the probability of the joint occurrence of perhaps a large number of events - a very difficult thing to assess. We only compound the difficulty by asking the probability that the events would be seen as well. All in all, his suggestion seems to be not helpful.

PROBLEM 28. The absence of data: the problem of how to deal with the absence of data about a particular event for a certain period of time.

The problem: One type of situation which loosely might be considered an example of negative evidence, and which is particularly difficult to deal with, is the time interval between data, or, more exactly, the absence of data about a particular event for a certain period of time. There are several difficulties. First, there is a legitimate question as to whether this really is an example of the non-occurrence of an event. Second, there is the question of how to consider the reasons for not receiving data when we do our Bayesian analysis. Third, we have the question of how often to include the item of not receiving data, which involves our choice of the length of the time period (interval) in which we consider the receipt or non-receipt of data. We want to avoid biasing our analysis by overcounting the absence of data.

The solution: The first difficulty is a matter of categorization, and really does not need to be worried about greatly. Analytically, the problem does seem, in a minor way, to be different from the negative evidence problem. The solution to the second difficulty appears to be the use of hierarchical Bayesian analysis. The third difficulty, however, is not solved by hierarchical Bayes, and no solution is obvious using normal Bayes. Overloading is possible using either technique, but is almost assured if we use hierarchical Bayes.

On balance, the use of the conditionally dependent form of normal Bayes seems desirable. Its usage does not solve the second difficulty, but seems to avoid the problems to which hierarchical Bayes leads if the latter technique is used to try to solve the third difficulty.

Discussion: The lack of data about a particular event for a certain period of time may itself be a sign that "something is up," the "lull before the storm syndrome," Dr. Hopple notes.[6]

Turning to the first difficulty, it is debatable whether or not we wish to consider the absence of data for a certain period of time to be an example of the non-occurrence of an event. Student Timothy Laur argued that

> this is not the same as the problem of the non-occurrence of an event. The non-occurrence of an event is the event "not happening," [e.g.] the Soviets not invading Poland. The lack of data mentioned here is [the situation of] not receiving data to say if or if not an event occurred. [Thus,] this problem refers to the situation where, for some reason or another, data on a particular situation stop being received. The event relating to a hypothesis (i.e., an invasion) may or may not have occurred; we don't know because information has stopped. As in the "non-occurrence" problem, the "non-data" problem could in itself be a datum. Is the lack of reporting due to a definitive act by a nation (radio silence), or physical (atmospherics, etc.), or mechanical (radio outage) problems? Or, problems on our side may preclude timely data receipt. What is an acceptable interval of "non-data?"

Note that Laur's discussion really is bringing us back to a previous question of whether we want to assess the probability of the occurrence of an event which was reported or the probability of occurrence of a report of an event. Here, it seems that when we are talking about the problem of lack of data about an event, we really are talking about the probability of not receiving reports on a particular event (given the particular hypothesis). Thus, this problem is, in what might be considered a minor way, actually different from the negative evidence problem, where we were talking about the probability of non-occurrences of events.

The second difficulty is how to consider the reasons for not receiving data when we do our Bayesian analysis. It seems that we cannot be sure, if data are not reported for a certain period of time, whether it is because (i) the event actually has not occurred or (ii) although it has occurred, it has not been observed, or (iii) although it has occurred and has been observed, the report has not been received (for a variety of possible reasons).

Although there are these three basic types of reasons for the lack of data, in using the conditionally dependent form of normal Bayesian analysis all we can do is ask, for each H_j,

$$P\left(\begin{array}{l}\text{data on the particular}\\ \text{event has not been}\\ \text{received in the latest}\\ \text{time period}\end{array} \Big/ H_j \text{ \& previous items of evidence}\right).$$

Since there are these various reasons for not receiving data, it is not particularly easy to assign such probabilities.

If hierarchical Bayes is used, however, we assign values at the lowest level of the hierarchy as follows:

Event	Possible Reasons for Event		
	Event did not occur	Event did occur but was not observed	Event did occur and was observed, but a report has not been received
data on an event has not been received in the latest time period	P(data on an event has not been received in the latest time period / Event did not occur)	P(data on an event has not been received in the latest time period / Event did occur but was not observed)	P(data on an event has not been received in the latest time period / Event did occur and was observed, but a report has not been received)

Note that the probabilities that are assigned are conditionally independent (this essentially must be the case if we use hierarchical Bayes) and that the entries in the matrix are the probability of the row heading given the column heading.

These probabilities still are not easy to assign, but are somewhat more meaningful than the probability judgments we must make in using normal Bayesian analysis, since the various reasons for the lack of data now are being explicitly considered. On the other hand, the necessary assumption of conditional independence means we can't consider the patterns of periods of lacking data. That is, we can't look at, for example

P(data on the particular event has not been received in the latest time period / H_j & data on the particular event was not received in the time period prior to the latest one & other previous data)

The probability here could be influenced greatly by the pattern of non-receipt of data in earlier time periods - but we can't take this into account if we use hierarchical Bayes.

The third difficulty is the question of how often to include the item of not receiving data, which involves our choice of the length of the time period (interval) in which we consider the receipt or non-receipt of data. We want to avoid the overcounting of the absence of data which would result in biasing our analysis.

Now, it should be clear that this difficulty is quite analagous to the problem of how often to count negative evidence, i.e., how often to count the non-occurrence of an event. We found in our treatment of this latter question that there was no happy solution, but that conditional dependence is better than conditional independence. We would reach the same conclusion if we carried out a parallel analysis here. It is better, in short to be able to ask

$$P\left(\begin{array}{l}\text{data on the}\\ \text{particular}\\ \text{event has}\\ \text{not been}\\ \text{received}\\ \text{in the}\\ \text{latest time}\\ \text{period}\end{array}\middle/ H_j \ \&\ \text{previous data}\right)$$

or ask

$$P\left(\begin{array}{l}\text{data on the}\\ \text{particular}\\ \text{event has}\\ \text{not been}\\ \text{received}\\ \text{in the}\\ \text{latest}\\ \text{time}\\ \text{period}\end{array}\middle/ H_j \ \&\ \begin{array}{l}\text{data on the}\\ \text{particular}\\ \text{event was}\\ \text{not received}\\ \text{in the time}\\ \text{period prior to}\\ \text{the last time}\\ \text{period (either)}\end{array} \ \&\ \text{other previous data}\right)$$

which shows the patterns rather than to be limited to asking

$$P\left(\begin{array}{l}\text{data on the}\\ \text{particular}\\ \text{event has}\\ \text{not been}\\ \text{receieved in}\\ \text{the latest}\\ \text{time period}\end{array}\middle/ H_j\right).$$

But, we can't make the conditionally dependent judgments if we use the hierarchical form of Bayes as we saw above (in this problem) at the end of our discussion of the second difficulty! Moreover, use of hierarchical Bayes very easily could lead to overcounting the lack of data, since we would have to input an entry at the bottom level of the matrix for the lack of data for each time period in which there were no data. There is no way of discounting via use of conditional dependence if we use hierarchical Bayes.

Because use of hierarchical Bayes could lead to overweighting the lack of data and does not permit the pattern of lack of data to be taken into account, I suggest, at this point, the use of the conditionally dependent form of normal Bayes. This is hardly a happy choice because it does not permit us to consider directly the reasons for the lack of data, i.e., we solve the third difficulty but not the second.

PROBLEM 29. No observations of an event: the problem of how to deal with the item of evidence that a particular event has not been observed in a particular time period.

The problem: This problem is different from the preceding problem in that we do not lack data for a certain period of time. We do receive data, namely a report of the item of evidence that a particular event has not been observed in a particular time period. This also differs from the non-occurrence problem because, in that problem, we were sure that the event definitely had not occurred in a particular time period. In this problem, the report says that a particular event has not been observed in a particular time period and we are not sure if (i) the particular event has not occurred or (ii) the particular event has occurred, but has not been observed. Analagous to the last problem, we face the difficulties of how to incorporate the reasons for the lack of an observation in a given time period into our Bayesian analysis, and the choice of an appropriate time interval to avoid overcounting the non-observations of a particular event.

The solution: Analagous to the last problem, we find that use of hierarchical Bayes will allow us to incorporate the reasons for the item of evidence of not observing the occurrence of a particular event. However, use of hierarchical Bayes almost insures overcounting the non-observations of a particular event, so we probably are better off using the conditionally dependent form of normal Bayes. Use of normal Bayes also has the advantage, which hierarchical Bayes does not, that we can take into account the pattern of non-observations of the event.

Discussion: A matrix may help to show the distinctions we have made among the problems of the non-occurrence of an event in a particular time period, the lack of data about an event for a certain period of time, and the non-observation of a particular event in a given time period.

Name of the problem	What has happened	The possible reasons for what has happened and what probability we can assess in normal Bayes
Negative evidence: non-occurrence of an event in a given time period (problem 27)	An event definitely has not occurred; we have perfect observation and reporting	The event definitely has not occurred; we assess $P\left(\text{event has not occurred in the latest time time period} \,/\, H_j \text{ \& previous data}\right)$

Lack of data on a particular event: absence of data for a certain period of time (problem 28)	<u>Reports</u> regarding an event have not been received for a certain period of time	Either (i) the event actually has not occurred or (ii) the event has occurred but has not been reported or (iii) the event has occurred and has been observed but the report has not been received; we can assess only $P\left(\text{data on the particular event has not been received in the latest time period} \mid H_j \,\&\, \text{previous data}\right)$
Non-observation of an event in a given time period (problem 29)	An item of evidence of the lack of the observation of a particular event (in a given time period) has occurred	Either (i) the particular event has not occurred or (ii) it has occurred but has not been observed; we can assess only $P\left(\text{the particular event has not been observed in the latest time period} \mid H_j \,\&\, \text{previous data}\right)$

The second column for the last problem entry is difficult to formulate. We do receive a report, and the content of the report is the item of evidence that there has been no observation of a particular event in a given time period. This distinction is arguable, however, and a case could be made that the content of the report is the <u>report</u> that there has been no observation of a particular event in a given time period. The distinction does not appear to be critical.

As before, normal Bayes has no good way of taking into account the reasons behind this particular unit of evidence (whether we wish to characterize it as an item of evidence or a report). Once more, if we are willing to give up conditional dependence, hierarchical Bayes can be a help. At the lowest level of the hierarchy we could input

Evidence	Possible Reasons for Evidence: Event did not occur in a given time period	Possible Reasons for Evidence: Event occurred but was not observed in a given time period
a particular event has not been observed in a given time period	$P\left(\text{event has not been observed in a given time period} \mid \text{event did not occur in a given time period}\right)$	$P\left(\text{event has not been observed in a given time period} \mid \text{event occurred but was not observed in a given time period}\right)$

We ask the probability of the row entry given the column heading. At the next highest level, we might have (where the two column headings are indicator states)

Possible Reasons for Evidence	I_1^1	I_2^1
Event did not occur in a given time period	P(Event did not occur in a given time period / I_1^1)	P(Event did not occur in a given time period / I_2^1)
Event occurred but was not observed in a given time period	P(Event occurred but was not observed in a given time period / I_1^1)	P(Event occurred but was not observed in a given time period / I_2^1)

In the end, the probabilities of the event not being observed, given the various hypotheses, are calculated.

As we saw in the previous problems, however, use of the hierarchical form of Bayes is not a cure-all. Its use means we can't assess the conditionally dependent probabilities

P(there has been no observation of a particular event in the latest time period / H_j & there was no observation of the event in the time period prior to the last one & other previous evidence).

The probability assessed here could depend greatly on the lack of an observation in the previous time period.

Also, analagous to what we saw before in the previous problems, the use of hierarchical Bayes can result in overweighting the non-observations of an event since we must input an entry for each given time period in which there was no observation of the particular event. We probably are better off, on balance, using the conditionally dependent form of normal Bayesian analysis, where we can take into account the pattern of non-observations even if we can't explicitly consider the underlying reasons for (or categories of) the non-observations.

NOTES

Problem 23

1. This is a much deeper issue than the analysis here indicates. I am trying to deal with basic problems of Bayesian analysis and readily understandable solutions in this book. This issue raises highly complex issues with very sophisticated analyses that are

too complex for the level at which I have endeavored to keep this book. The reader wishing to pursue the issue can see, for example, the references in f.n. 3 of Problem 26 below.

Problem 25

2. Note again that we are not saying that the reports that led to E_1 and E_2 are the same, but instead that the contents of E_1 and E_2 of the reports are the same.

Problem 26

3. See the following, for example:

SCHUM, DAVID A. and DU CHARME, WESLEY M. (1971) "Comments on the Relationship Between the Impact and Reliability of Evidence." Organizational Behavior and Human Performance 6: 111-131.

GETTYS, CHARLES F. and WILKE, T.A. (1969) "The Application of Bayes's Theorem When the True Data State is Uncertain." Organizational Behavior and Human Performance 4: 125-141.

SCHAEFER, RALF E. and BORCHERDING, KATRIN (1973) "A Note on the Consistency between Two Approaches to Incorporate Data from Unreliable Sources in Bayesian Analysis." Organizational Behavior and Human Performance 9: 504-508.

SCHUM, DAVID A. and KELLEY, CLINTON W. (1973) "A Problem in Cascaded Inference: Determining the Inferential Impact of Confirming and Conflicting Reports from Several Unreliable Sources." Organizational Behavior and Human Performance 10: 404-423.

Problem 27

4. "Also relevant," Dr. Hopple notes, "is the time frame - current vs. estimative, long vs. short term, strategic assumptions vs. tactical indicator, etc."

5. SCHWEITZER (1976), Appendix 2, p. iv. (Almost identical wording appears in SCHWEITZER (1978), p. 29.)

Problem 28

6. See problem 55.

8
Problems of Source Reliability and the Credibility of Evidence

In this chapter we will discuss problems involving source reliability and the credibility of items of evidence.

PROBLEM 30. Source reliability: the problem of how to take into the account the reliability of a source of information.

The problem: As stated.

The solution: The solutions available in the literature[1] seems to me to be too complex to be used by analysts. The best "solution" simply may be to ignore the problem, but revise our description of events when necessary and then reassess.

Discussion: This problem can be distinguished analytically from the next problem, which is how to deal with evidence which does not appear believable.

In doing Bayes, we normally take the evidence as "true." We ask the probability the event will happen, given the particular hypothesis and the previous events. There is no expression of doubt about the event as it is described. Nevertheless, there might be some doubt about the reliability of the source and hence about whether the event occurred the way it was reported or even happened at all, and we might want to consider ways of dealing with this problem of source reliability. Student Norman Allred pointed out the problem quite clearly when he said that "If some particularly diagnostic data comes from a questionably reliable source, one may be left with the dilemma of not using the data, using it in its questionable state or using it and trying to make some kind of an appropriate judgment value or adjustment to the answer received."

A number of ways of trying to deal with this problem have been employed in the intelligence community. None is totally satisfactory. Perhaps the most common way of dealing with the problem simply has been to ignore it. This is not really a bad approach. We simply can charge ahead and make assessments based on the description of the event. If, later, it is found that the event was different than described by the source, we can make the necessary reassessments based on the properly revised description of the event. If we feel that this sort of revision will not be required very often, then this approach to the problem of source reliability does not appear to be a bad one.

A second, much less defensible way of dealing with the problem, a way that is used often according to some observers, is to intuitively discount an item of dubious evidence by the weights (i.e., conditional probabilities) assigned to it. This appears to be a most unfortunate approach because of the analyst's inability to justify the degree of discounting. There is, in short, no specified discounting rule. Once the analyst makes the subjective conditional probability judgments required, (s)he intuitively discounts the weights (conditional probabilities) assigned, or does both weighting and discounting simultaneously. The direction of discounting is clear only in the sense that the closer the analyst makes the conditional probabilities for any E_i, the less the impact of that E_i will be on the revised probabilities of the hypotheses. I.e., the closer the analyst makes

$$P(E_i/H_1\&E_1\&\cdots\&E_{i-1})$$

$$P(E_i/H_2\&E_1\&\cdots\&E_{i-1})$$

$$\vdots$$

$$P(E_i/H_m\&E_1\&\cdots\&E_{i-1})$$

to one another, the less the impact of that E_i will be. At the extreme, if the analyst has no faith in the reliability of the source and wants to totally discount an item of evidence, (s)he simply can make

$$P(E_i/H_1\&E_1\&\cdots\&E_{i-1}) = P(E_i/H_2\&E_1\&\cdots\&E_{i-1}) = \cdots = P(E_i/H_m\&E_1\&\cdots\&E_{i-1})$$

and then E_i will have no diagnostic impact at all. (We can see this more clearly in the odds-likelihood form of Bayes where the likelihood ratio for E_i is

$$\frac{P(E_i/H_1\&E_1\&\cdots\&E_{i-1})}{P(E_i/H_2\&E_1\&\cdots\&E_{i-1})}.$$

If we make

$$P(E_i/H_1\&E_1\&\cdots\&E_{i-1}) = P(E_i/H_2\&E_1\&\cdots\&E_{i-1})$$

then the likelihood ratio becomes "1" and then E_i will have no diagnostic impact.)

A third approach, suggested some years ago by Zoltnick, can be dubbed Zlotnick's (source discounting) exponent. It is best illustrated in the odds-likelihood format. We will provide an example.

To make the example easier to follow, suppose that $P(H_1) = 0.5$ and $P(H_2) = 0.5$ and that for a given E_1 the assessment is made that

$$\frac{P(E_1/H_1)}{P(E_1/H_2)} = \frac{2}{3}.$$

Then

$$\frac{P(H_1/E_1)}{P(H_1/E_2)} = \frac{P(H_1)}{P(H_2)} \cdot \frac{P(E_1/H_1)}{P(E_1/H_2)} = \frac{0.5}{0.5} \cdot \frac{2}{3} = \frac{2}{3}.$$

Then, $P(H_1/E_1) = \frac{2}{2+3} = \frac{2}{5} = 0.4$ and $P(H_2/E_1) = 1\text{-}0.4 = 0.6$.

The pictorial view of this illustrates the facts that the probability of H_1 has moved from 0.5 to 0.4, while the probability of H_2 has moved from 0.5 to 0.6. Figure 8.1 provides the illustration.

FIGURE 8.1
Diagram of the Change in the Probabilities of Two Hypotheses after One Item of Evidence

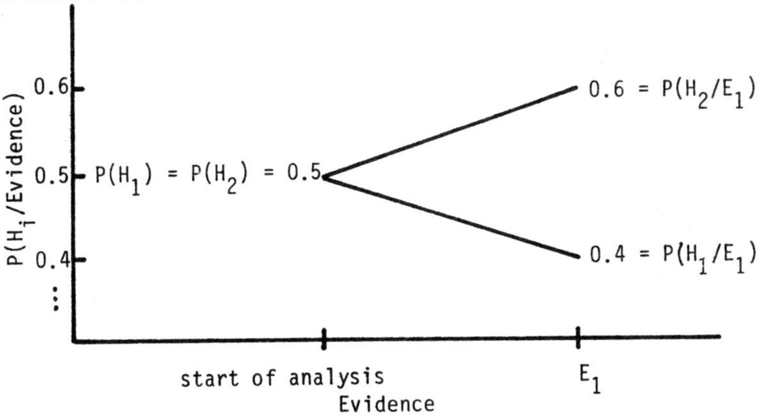

Suppose that it is decided that the source is only 80 percent reliable. This may be a subjective judgment, or it may be based on frequency counts.

Zlotnick's idea was to make source reliability an exponent on the entire likelihood ratio. This clearly makes sense at the extremes. That is, if a source is considered to be 100 percent reliable, Zlotnick's idea makes sense because the exponent is "1," and the likelihood ratio raised to first power ("1") is just the likelihood ratio - it "counts completely." (E.g., $(2/3)^1 = 2/3$.) On the other hand, if the source is considered to be completely unreliable, Zlotnick's idea again makes sense because the exponent is "0," and the likelihood ratio raised to the zero power ("0") is the fraction 1/1, so it is as if the likelihood ratio had never been entered into the equation -- it has no diagnostic impact at all, it doesn't count at all. (E.g., $(2/3)^0 = 1/1$.)

For a reliability rating between 0 and 1, it should be the case that Zlotnick's procedure would yield a likelihood ratio that would not lead to as much revision of probabilities as if the source were completely reliable, but would have more diagnostic impact than if the source were completely unreliable. This, fortunately, is just the sort of result obtained by using Zlotnick's procedure.

In the above example, if the source reliability is 0.80, we don't want the probability of H_1 to move all the way down from 0.5 to 0.4, or for the probability of H_2 to move all the way up from 0.5 to 0.6. But, since 0.8 is a high reliability rating, we would expect the revised probability of H_1 to be slightly greater than 0.4, and the revised probability of H_2 to be slightly less than 0.6.

When the required calculation is carried out, the results are what we said they should be. That is, using Zlotnick's exponent, we have

$$\frac{P(H_1/E_1)}{P(H_2/E_1)} = \frac{P(H_1)}{P(H_2)} \cdot \left[\frac{P(E_1/H_1)}{P(E_1/H_2)}\right]^R = \frac{0.5}{0.5} \cdot \left(\frac{2}{3}\right)^{0.8}$$

$$= \frac{1}{1} \cdot \left(\frac{0.66}{1}\right)^{0.8} = \frac{0.7224027}{1},$$

$$\text{so } P(H_1/E_1) = \frac{0.7224027}{0.7224027+1} = 0.42 \text{ and } P(H_2) = 1-0.42 = 0.58.$$

This can be shown on a diagram as follows (Figure 8.2).

Note, by the way, that multiplying by the reliability doesn't work! For example, using the same numbers as before, if we were to calculate

$$\frac{P(H_1/E_1)}{P(H_2/E_1)} = \frac{P(H_1)}{P(H_2)} \cdot \frac{P(E_1/H_1)}{P(E_1/H_2)} \cdot R = \frac{0.5}{0.5} \cdot \frac{2}{3} \cdot \frac{0.8}{1} = \frac{1.6}{3},$$

$$\text{then } P(H_1/E_1) = \frac{1.6}{4.6} = 0.35,$$

a greater revision downward than before. This is a nonsense answer, because it is saying that, when we take reliability into account, the probability revision downward for $P(H_1)$ is 0.15, i.e., from 0.5 to 0.35 for a source of 80 percent reliability while, when we ignore reliability (which is equivalent to a reliability of R = 100 percent), the downward revision is only 0.10 i.e., from 0.5 to 0.4. So, the lesser reliability yields a greater revision -- which deserves to be called a nonsense result. Multiplying by R just does not work.

FIGURE 8.2
Diagram of the Change in the Probabilities of Two Hypotheses after One Item of Evidence when Zlotnick's Exponent is Used

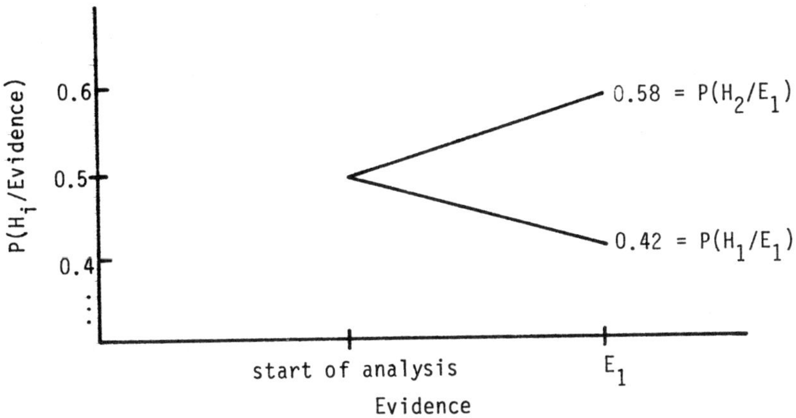

Clearly, Zlotnick's exponent procedure can be used for multiple hypotheses by applying the exponent to all entries in the column for a particular event whose source is being considered. For example, suppose we have the situation shown in Table 8.1.

TABLE 8.1
Tabular Array

j	$P(H_j)$	$P(E_1/H_j)$	Row Product	Revised Probability of H_j
1	0.4	0.7	0.28	0.667
2	0.5	0.1	0.05	0.119
3	0.1	0.9	0.09	0.214
	1.0		0.42	1.000

If we consider the source for E_1 to be only 80 percent reliable, and we want to use Zlotnick's exponent solution, we can proceed as in Table 8.2.

TABLE 8.2
Tabular Array Using Zlotnick's Exponent

j	$P(H_j)$	$(P(E_1/H_j))^{0.8}$	Row Product	Rev Prob of H_j	Rev Prob of H_j not using Zlotnick's exponent
1	0.4	$(0.7)^{0.8}$ (=0.751759)	0.300704	0.637	0.667
2	0.5	$(0.1)^{0.8}$ (=0.158489)	0.079244	0.168	0.119
3	0.1	$(0.9)^{0.8}$ (=0.919166)	0.091917	0.195	0.214
	1.0		0.471865	1.000	1.000

There appear to be two major problems with Zlotnick's approach. The first is how to assess validly the exponent (see the next problem). The second is that there is no obvious mathematical justification for assigning such an exponent - it is equivalent to putting exponents on the probabilities attached to tree branches in a probability diagram. There is no obvious justification for doing this; the whole procedure seems to be one of introducing a "fudge" factor. It is, however, quite ingenious and should not be rejected out of hand. It does work, if we can justify the numerical exponent.

A fourth solution to the problem is a variation on the first solution of ignoring the whole issue. We would do this as long as the data either were not very diagnostic or came from a fairly reliable source. If an item was very diagnostic and came from a very unreliable source, we could run parallel Bayesian analyses. In general, we would run one analysis with all the evidence, and a parallel analysis deleting all items which were highly diagnostic but were reported by very unreliable sources. Both analyses should be provided to the consumer if this fourth "solution" is used. This use of parallel analyses seems to be a fairly good way of handling the dilemma that not using data that would provide a significant input (because of reliability questions) would bias the results, while using untrue data also would skew the results.

Student Norman Allred pointed out that a practical problem with all of this may be that the analyst may be denied information on the source of the data (and therefore may have to adopt the first solution).

Student James McGowan's work suggests that a solution to both the reliability problem and "old" data problem (problem 52 below) would be to use an exponent of $R\cdot(1/T)$, where T is measured, say, in months.

The several highly technical works of footnote 1 give much more sophisticated solutions to this problem. The techniques are, I think, too complicated for normal usage.

PROBLEM 31. Assessing source reliability: the problem of how to assess the Zlotnick exponent.

The problem: As stated.

The solution: No good "objective" solution is obvious, i.e., no formula or counting rule is obvious.

Discussion: One student (I don't know which) pointed out that it may *not* be valid to try to base the analyst's reliability rating on frequency counts. The student pointed out that reliability is a function of the source's access, which in turn is a function of the importance of the information. The source might be quite good (reliable) on mundane subjects, but poor (unreliable) on important subjects. This, then, would mean that it would not be valid to say, even if the source has been reporting over some length of time, that his (her) reliability equals the number of verified, correct reports, divided by the total number of reports.

The absence of a good, objective rule could allow for a significant amount of analyst bias to enter the process, student Gerald Shea pointed out. He went on to note that institutional bias could enter in - organizations enamored of technical collection would want exponents of "1" for their information, for example.

PROBLEM 32. New reports and source reliability: the problem, if Zlotnick's exponent is used, of what to do when new reports reflecting the true data state are received.

The problem: As stated.

The solution: Not obvious.

Discussion: An example of this problem may make it clearer. Suppose we have used a Zlotnick exponent on an event reported by a human intelligence (HUMINT) source, and get another report on the same event from photography. Whether the report from photography confirms or denies the HUMINT report, do we go back and change the exponent for the HUMINT report? What other alternatives are there? Suggestions by readers for non-complex solutions are invited.

PROBLEM 33. Multiple reports and source reliability: the problem, if there are multiple reports on the true data state, of whether to include the source reliability for each report and, if so, how to do so.

The problem: Regardless of whether a series of reports, each characterizing a particular event, are identical *or* supplementary *or* contradictory in their characterization of the event (i.e., their description of the true data state), we might want to include the reliability of each source in our analysis. The question is how to do so, and if we do, what characterization of the event is to be taken if we continue to use the conditionally dependent form of Bayes.

The solution: The solutions available in the literature[2] seems to me to be too complex to be used by analysts. Suggestions by readers for more elementary solutions would be welcomed.

PROBLEM 34. Unbelievable evidence: the problem of how to deal with evidence which does not appear believable.

The problem: This problem can be most severe. The problem is what to do about reported events which we find simply not believable, regardless of our feeling about the source's reliability. Reports on aliens emerging from UFOs or (circa early 1962) reports on Soviet ballistic missiles in Cuba fall into this category. Such "evidence" may be felt to be too dubious and possibly misleading to be included in our analyses.

There is a great danger in leaving this sort of reported event out of a Bayesian analysis. That is, there is a great danger of ignoring evidence that seems totally incompatible with the hypothesis in which the analyst believes (e.g., the Soviets would never put ballistic missiles into Cuba; there are no such things as alien-operated UFOs). Ignoring such evidence subverts Bayes, because such a judgment about credibility is based on the analyst's preconceptions (prejudgment) about the prior probabilities and relative merits of the hypotheses. If this sort of prejudgment occurs, the previously cited advantage -- of decoupling the analyst's ego from the assessments made -- simply disappears. Such evidence might be critical in a non-stationarity of hypotheses situation.

The solution: How, then, should we handle evidence which does not appear believable? The suggested solution is to run two Bayesian analyses, one leaving out the unbelievable event (evidence), and one including the unbelievable event, and present both analyses to decision makers.

Discussion: If we adopt the suggested solution, we have to double the number of judgments to be made. That is, for each event E_j subsequent to the dubious event E_i, we have to assess the probabilities for E_j leaving the event E_i out, and then reassess the probabilities for E_j where E_i is included. More exactly, we must assess

$P(E_j/H_1\&E_1\&\cdots\&E_{j-1})$ where E_i is not included,

$P(E_j/H_2\&E_1\&\cdots\&E_{j-1})$ where E_i is not included,

$\vdots$

$P(E_j/H_m\&E_1\&\cdots\&E_{j-1})$ where E_i is not included, and then assess

$$P(E_j/H_1\&E_1\&\cdots\&E_{j-1}) \quad \text{where } E_i \text{ is included,}$$

$$P(E_j/H_2\&E_1\&\cdots\&E_{j-1}) \quad \text{where } E_i \text{ is included,}$$

$$\vdots$$

$$P(E_j/H_m\&E_1\&\cdots\&E_{j-1}) \quad \text{where } E_i \text{ is included.}$$

We then can see the cumulative impact of leaving the dubious event out compared to the cumulative impact of leaving it in.

In addition, there could be a simultaneous Bayesian analysis where one hypothesis is that the dubious event is true, and the other hypothesis is that the dubious event is false. As the probabilities are revised in this supplementary Bayesian analysis, we can alter our view as to which of the two primary Bayesian analyses -- one leaving the dubious event out, the other including it -- has more value.

An additional technical point needs to be made. It has been suggested[3] that, when we include a dubious event E_i, a good way of considering it while at the same time admitting its questionable status would be to say that event E_i is that "unsubstantiated evidence that x is the case has appeared and been observed." We could then make assessments of

$$P\left(\begin{array}{l}\text{"unsubstantiated evidence that}\\ \text{x is the case would appear}\\ \text{and be observed"}\end{array} \Big/ \; H_j\&E_1\&\cdots\&E_{i-1}\right)$$

for each H_j. Verbally, we would be asking "What is the probability that unsubstantiated evidence that x is the case would appear and be observed if H_j is assumed to be true and if $E_1 \cdots E_{i-1}$ have occurred?" This approach does not seem useful, however, because the probability assigned is now dependent upon the <u>appearance</u> of the unsubstantiated evidence and upon the <u>observation</u> -- where the probability of observation might depend upon an observer being at the right place at the right time. We are then back to the problem of asking not the probability of the evidence reported on but instead asking the probability that the <u>report</u> itself of the event would occur. As has been argued before, we usually are interested in the probability of the event and <u>not</u> in the probability of the report, and this is the case here, too. Thus it is better not to speak of "unsubstantiated evidence appearing and being observed," but instead to just treat the evidence like any other piece of evidence -- i.e., as "true."

If there is more than one item of evidence which does not appear believable, it probably is best just to continue on with two parallel analyses, one including all the dubious evidence, and the other deleting all the dubious evidence.

NOTES

Problem 30

1. The same references as in f.n. 2 (for Problem 26) apply here. See also the following Schum references:

SCHUM, DAVID A. and PFEIFFER, PAUL E. (1973) "Observer Reliability and Human Inference." IEEE Transactions on Reliability R-22 (No. 3) (Aug), 170-176.

SCHUM, DAVID A. (1977a) "The Behavorial Richness of Cascaded Inference Models: Examples in Jurisprudence," pp. 149-173 in N. John Castellan *et. al.* (eds.) Cognitive Theory (Vol 2). Hillsdale, N.J.: Lawrence Erlbaum Associates.

SCHUM, David A. (1977b) "Contrast Effects in Inference: On the Conditioning of Current Evidence by Prior Evidence." Organizational Behavior and Human Performance 18: 217-253.

SCHUM, David A. (1980) "Current Developments in Research on Cascaded Inference Processes," pp. 179-210 in Thomas S. Wallsten (ed.) Cognitive Processes in Choice and Decision Behavior. Hillsdale, N.J.: Lawrence Erlbaum Associates.

Problem 33

2. The references mentioned in footnote 1 above apply.

Problem 34

3. By Zlotnick.

9
Problems Involving the Probability Numbers That Are Assigned

In this chapter we will discuss some of the problems involving the probability numbers that are assigned when we carry out a Bayesian analysis.

PROBLEM 35. Phrasing conditional probability judgments.

The problem: The problem is how to phrase expressions of the conditional probability of events, i.e., how to verbalize $P(E_i/H_j\&E_1\&E_2\&\cdots\&E_{i-1})$.

The solution: There are several possible ways to verbalize $P(E_i/H_j\&E_1\&E_2\&\cdots\&E_{i-1})$. We can say

a. The probability of event E_i given H_j and that E_1 through E_{i-1} have happened, or

b. The probability of E_i assuming that H_j is true and given that E_1 through E_{i-1} have happened, or

c. The probability that E_i would happen given H_j and that E_1 through E_{i-1} have happened, or

d. The probability that E_i would happen assuming that H_j is true and given that E_i through E_{i-1} have happened, or

e. The % (percent) (or proportion) of the time E_i would happen given H_j and that E_1 through E_{i-1} have happened, or

f. The % (or proportion) of the time E_i would happen assuming that H_j is true and given that E_1 through E_{i-1} have happened, or

g. The % (or proportion) of the time E_i would happen given H_j and that E_1 through E_{i-1} have happened, compared to the % (or proportion) of the time E_i would not happen given H_j and that E_1 through E_{i-1} have happened, or

h. The % (or proportion) of the time E_i would happen assuming H_j to be true and given that E_1 through E_{i-1} have happened compared to the % (or proportion) of the time E_i would <u>not</u> happen assuming H_j to be true and given that E_1 through E_{i-1} have happened.

<u>Discussion</u>: Alternatives (e) and (f) may be particularly useful at times. An E_i which otherwise might not make much sense with respect to assessing $P(E_i/H_j\&E_1\&E_2\&\cdots\&E_{i-1})$ may make sense if this terminology is employed.

Another, rather rare way of looking at $P(E_i/H_j\&E_1\&E_2\&\cdots\&E_{i-1})$, stems from noting that

$$P(E_i/H_j\&E_1\&E_2\&\cdots\&E_{i-1}) + P(\text{not } E_i/H_j\&E_1\&E_2\&\cdots\&E_{i-1}) = 1,$$

as this portion of the tree illustrates (Figure 9.1).

FIGURE 9.1
Portion of a Probability Tree Diagram for Hypothesis H_j and Event E_i

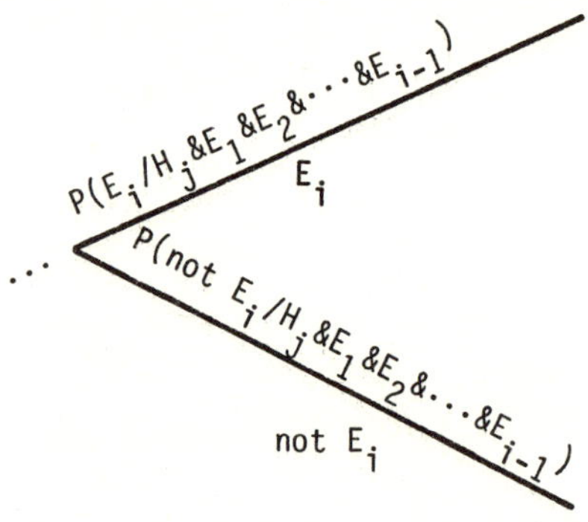

We then can say that $P(E_i/H_j\&E_1\&E_2\&\cdots\&E_{i-1})$ is the percentage (or proportion) of the time that E_i would happen, given H_j and that E_1 through E_{i-1} have happened, compared to the percentage (or proportion) of the time that E_i would <u>not</u> happen, given H_j and that E_1 through E_{i-1} have happened. Alternatively, we can say that $P(E_i/H_j\&E_1\&E_2\&\cdots\&E_{i-1})$ is the percentage (or proportion) of the time that E_i would happen, assuming H_j to be true and given that E_1 through E_{i-1} have happened, compared to the percentage (or proportion) of the time E_i would <u>not</u> happen, assuming H_j to be true and given that E_1 through E_{i-1} have happened. Note that the notion of phrasing $P(E_i/H_j\&E_1\&E_2\&\cdots\&E_{i-1})$ as the probability that

E_i would "precede" H_j was dealt with and rejected in the section on what sort of questions Bayes can answer (i.e., in problem 8).

PROBLEM 36. "Blow-up": the problem of an unwarranted rapid increase or decrease in the revised probabilities of some hypotheses.

The problem: If we even are slightly careless or sloppy about assigning, for each E_i, the probabilities $P(E_i/H_j\&E_1\&E_2\&\cdots\&E_{i-1})$, the net result can be an unwarranted - and therefore incorrect - rapid increase or decrease in the revised probabilities of some hypotheses. This can be termed the "blow-up" problem.

The solution: The way to avoid the "blow-up" problem is to be conservative, for each E_i, in assigning the set of probabilities

$$P(E_i/H_1\&E_1\&E_2\&\cdots\&E_{i-1})$$
$$P(E_i/H_2\&E_1\&E_2\&\cdots\&E_{i-1})$$
$$\vdots$$
$$P(E_i/H_m\&E_1\&E_2\&\cdots\&E_{i-1}),$$

where being "conservative" means that we should be reluctant to assign even slightly different probabilities unless there is good reason for doing so.

Discussion: The last advantage in the chapter on advantages of Bayesian analysis was that the revised probabilities of the hypotheses can change very quickly - either upward or downward. Even a few items of diagnostic evidence (or just one) can lead to this "blow-up" situation. As was pointed out, however, this is an advantage only insofar as we have been extremely careful in making the probability assessments for each given datum.

The reason, then, for saying there is a "blow-up" problem is that, if we are careless or sloppy in making probability assessments, the net result can be an unwarranted blow-up situation. The estimates then provided to decision makers can be greatly erroneous, with possible harm to the country as a result.

The way to avoid the blow-up problem is to be conservative in assigning conditional probabilities. We probably should be reluctant when we assign the set of values for each E_i of

$$P(E_i/H_1\&E_1\&E_2\&\cdots\&E_{i-1})$$
$$P(E_i/H_2\&E_1\&E_2\&\cdots\&E_{i-1})$$
$$\vdots$$
$$P(E_i/H_m\&E_1\&E_2\&\cdots\&E_{i-1}),$$

to assign values that are even slightly different unless there is good reason for doing so. (As always, a paragraph or so is needed to justify the numbers assigned.) If, for example, in a tabular array we assign the probability values shown in Table 9.1, we should examine the conditionally dependent probabilities assigned for E_6 to see if the probabilities can be justified in light of the blow-up problem. If there is any doubt about the line of argument used to justify even this "minor" 0.05 difference for $P(E_6/H_1\&E_1\&E_2\&\cdots\&E_5)$, we would be wise to make the column entries for $P(E_6/H_j\&E_1\&E_2\&\cdots\&E_5)$ all 0.45, rather than run the risk that this item of evidence in combination with others will result in an unwarranted increase in the revised probability of H_1. (As we will see in more detail later, it is the ratios of the probability values that are assigned that determines the impact of the item of evidence.)

TABLE 9.1
Tabular Array for Revising Probabilities of Hypotheses Upon Receipt of a Sixth Item of Evidence

j	Latest Row Product	$P(E_6/H_j\&E_1\&E_2\&\cdots\&E_5)$	New Row Product	Latest Revised Probability of H_j
1	0.078	0.50	0.03900	0.354
2	0.065	0.45	0.02925	0.266
3	0.093	0.45	0.04185	0.380
			0.11010	1.000

A question that can be raised at this point is whether this conservatism may not reintroduce the conservatism people have in revising probabilities of hypotheses -- the conservatism that Bayesian analysis supposedly overcomes. There appears to be some truth to this claim, but it seems worth the price to avoid a blow-up situation. Moreover, if items of evidence appear that lead to the assessment of diagnostic probabilities, i.e., different probabilities for different hypotheses, the revised probabilities that will be calculated can be expected to be ahead of intuitive revisions. Hence, we still can expect that conservatism with respect to revising probability estimates of hypotheses will be overcome.

PROBLEM 37. "Error-bounding": the problem of the high sensitivity of the revised probabilities of hypotheses to minor variations in the assigned probabilities $P(E_i/H_j\&E_1\&E_2\&\cdots\&E_{i-1})$.

The Problem: Quite often in numerical problems, we want to get some feeling for how sensitive the final answer is to minor variations in the values of the influencing variables. We hope that such minor variations in the influencing variables will lead to only minor changes in the final answer. Unfortunately, this is not the case in Bayesian analysis. The revised probabilities of hypotheses are highly sensitive to minor variations in the assigned probabilities of $P(E_i/H_j\&E_1\&E_2\&\cdots\&E_{i-1})$.

The solution: It can be argued reasonably that the method presented here to calculate the variations in the hypotheses is not the usual method for assessing the sensitivity of results and therefore may overstate the case somewhat. However, in light of what we know about the sensitivity of the final results when Bayesian analysis is used, it is suggested that we put more emphasis on the trend line we plot for our best estimates of revised probabilities of hypotheses and put less emphasis on the exact numbers themselves.

Discussion: Besides the "blow-up" problem, where the revised probabilities of some hypotheses improperly go up or down very rapidly, there is an analytically distinguishable problem which might be labeled the "error-bounding" problem. The point is that, especially in numerical analyses of problems, we want some feeling about how sensitive the final answer is to minor (small) variations in the input data. For example, if we find that the revised probability of a given H_j, i.e., $P(H_j/E_1\&\cdots\&E_n)$, is 0.65, then we would hope that, if we were to make minor, reasonable changes in the input probabilities (reflecting the uncertainty about the preciseness of those assessments which we had made), the resulting range of values for H_j then might be, say, from 0.55 to 0.75. This would be a fairly comforting range of values in the sense that it would say that, while the best estimate for $P(H_j/E_1\&E_2\&\cdots\&E_n)$ is 0.65, this revised probability might be as low as 0.55 or as high as 0.75 - a fairly narrow (i.e., 0.20 or less) bandwidth about the best estimate of 0.65.

Unfortunately, such a fairly narrow bandwidth is not at all what we typically get. Bayes is highly sensitive to minor variations in the input probabilities in the sense that such minor variations lead to a wide range of values for the revised probabilities of hypotheses and therefore, in the eyes of some analysts, there are grave doubts about the reliability of those revised probabilities. Instead of a bandwidth from 0.55 to 0.75 about the best estimate of 0.65, a fairly typical bandwidth after five or six items of evidence is from 0.05 to 0.95. This sort of "error interval" is so wide as to undermine confidence in the best estimate. Before addressing this problem further, we will show how to go about establishing the error interval. An example will illustrate the approach when we have three hypotheses.

Suppose we have the tabular array of Table 9.2.

TABLE 9.2 Tabular Array for Revised Probabilities of Hypotheses after Three Items of Evidence

j	$P(H_j)$	$P(E_1/H_j)$	$P(E_2/H_j\&E_1)$	$P(E_3/H_j\&E_1\&E_2)$	Row Product	Revised Probability of H_j
1	0.4	0.7	0.2	0.6	0.0336	0.351
2	0.2	0.9	0.2	0.3	0.0108	0.113
3	0.4	0.4	0.4	0.8	0.0512	0.536
	1.0				0.0956	1.000

Thus, $P(H_1/E_1\&E_2\&E_3) = 0.351$
$P(H_2/E_1\&E_2\&E_3) = 0.113$
$P(H_3/E_1\&E_2\&E_3) = 0.536.$

To see how <u>large</u> $P(H_1/E_1\&E_2\&E_3)$ reasonably might be, we take the original array, and for each event <u>column</u> make the value for j = 1 as <u>large</u> as it reasonably could be and the values for j = 2, j = 3 as <u>small</u> as they reasonably could be.

For the first event column, we might feel that $P(E_1/H_1)$ reasonably might be as large as 0.8, while $P(E_1/H_2)$ reasonably might be as low as 0.8 and $P(E_1/H_3)$ reasonably be as low as 0.3. For the second event column, we might feel that $P(E_2/H_1\&E_1)$ reasonably might be as large as 0.3, while $P(E_2/H_2\&E_1)$ reasonably might be as low as 0.1, and $P(E_2/H_3\&E_1)$ reasonably might be as low as 0.3. For the third event column, we might feel that $P(E_3/H_1\&E_1\&E_2)$ reasonably might be as large as 0.7, while $P(E_3/H_2\&E_1\&E_2)$ reasonably could be no lower than its present value of 0.3, and $P(E_3/H_3\&E_1\&E_2)$ reasonably could be as low as 0.75. Table 9.3 shows the revised tabular array for finding the largest reasonable value for H_1.

TABLE 9.3
Revised Tabular Array to find the <u>Largest</u> Reasonable Revised Probability of H_1 after Three Items of Evidence

j	$P(H_j)$	$P(E_1/H_j)$	$P(E_2/H_j\&E_1)$	$P(E_3/H_j\&E_1\&E_2)$	Row Product	Revised Probability of H_j
1	0.4	0.8	0.3	0.7	0.0672	0.679
2	0.2	0.8	0.1	0.3	0.0048	0.048
3	0.4	0.3	0.3	0.75	0.0270	0.273
	1.0				0.0990	1.000

Thus, the <u>largest</u> reasonable value for the revised probability of H_1, after three items of evidence, is 0.679, a value almost twice as large as the originally calculated value of 0.351.

To see how small $P(H_1/E_1\&E_2\&E_3)$ reasonably might be, we take the original array, and for each event column make the value for j = 1 as small as it reasonably could be, and the values for j = 2, j = 3 as large as they reasonably could be. We might, then, end up with Table 9.4.

TABLE 9.4
Revised Tabular Array to find the Smallest Reasonable Revised Probability of H_1 after Three Items of Evidence

j	$P(H_j)$	$P(E_1/H_j)$	$P(E_2/H_j\&E_1)$	$P(E_3/H_j\&E_1\&E_2)$	Row Product	Revised Probability of H_j
1	0.4	0.6	0.1	0.5	0.0120	0.097
2	0.2	0.9	0.3	0.4	0.0216	0.175
3	0.4	0.5	0.5	0.9	0.0900	0.728
	1.0				0.1236	1.000

Thus, the smallest reasonable value for the revised probability of H_1, after three items of evidence, is 0.097. The error-interval" about our first value of 0.351 therefore ranges from 0.097 to 0.679. Rounding off a bit, we can say that our revised probability for H_1 after three items of evidence could be as low as 0.10 or as high as 0.68, with the most "reasonable" value being 0.35. A range of 0.58 (0.68-0.10) is somewhat disquieting.

To find the largest reasonable revised probability of H_2, we take the original array and for each event column make the value for j = 2 as large as it reasonably could be, and the values for j = 1 and j = 3 as small as they reasonably could be. (Note that all these values were assigned in the course of dealing with H_1.) Table 9.5 displays this tabular array.

TABLE 9.5
Revised Tabular Array to find the Largest Reasonable Revised Probability of H_2 after Three Items of Evidence

j	$P(H_j)$	$P(E_1/H_j)$	$P(E_2/H_j\&E_1)$	$P(E_3/H_j\&E_1\&E_2)$	Row Product	Revised Probability of H_j
1	0.4	0.6	0.1	0.5	0.0120	0.198
2	0.2	0.9	0.3	0.4	0.0216	0.356
3	0.4	0.3	0.3	0.75	0.0270	0.446
	1.0				0.0606	1.000

So, the largest reasonable value for the revised probability of H_2 after three items of evidence is 0.356, over three times as large as the originally calculated value of 0.113.

To find the smallest reasonable value for the revised probability of H_2, we take the original array and for each event column make the value for j = 2 as small as it reasonably could be and the values for j = 1 and j = 3 as large as they reasonably could be. (Again, all these values were assigned in the course of dealing with H_1.)

TABLE 9.6
Revised Tabular Array to find the Smallest Reasonable Revised Probability of H_2 after Three Items of Evidence

j	$P(H_j)$	$P(E_1/H_j)$	$P(E_2/H_j\&E_1)$	$P(E_3/H_j\&E_1\&E_2)$	Row Product	Revised Probability of H_j
1	0.4	0.8	0.3	0.7	0.0672	0.415
2	0.2	0.8	0.1	0.3	0.0048	0.030
3	0.4	0.5	0.5	0.9	0.0900	0.556
	1.0				0.1620	1.001

So, the smallest reasonable revised probability of H_2 after three items of evidence is 0.03. Thus, the error interval about our first value of 0.113 ranges from 0.030 to 0.356. Rounding off a bit, we can say that our revised probability for H_2 after three items of evidence could be as low as 0.03 or as high as 0.36, with the most reasonable value being 0.11. A range of 0.33 also is somewhat disquieting.

The procedure to find the largest and smallest reasonable revised probabilities of H_3 should be clear. Our tabular arrays are shown in Tables 9.7 and 9.8.

TABLE 9.7
Revised Tabular Array to find the Largest Reasonable Revised Probability of H_3 after Three Items of Evidence.

j	$P(H_j)$	$P(E_1/H_j)$	$P(E_2/H_j\&E_1)$	$P(E_3/H_j\&E_1\&E_2)$	Row Product	Revised Probability of H_j
1	0.4	0.6	0.1	0.5	0.0120	0.112
2	0.2	0.8	0.1	0.3	0.0048	0.045
3	0.4	0.5	0.5	0.9	0.0900	0.843
	1.0				0.1068	1.000

Thus, the smallest reasonable value for the revised probability of H_3, after three items of evidence, is 0.233, and the largest reasonable value 0.843. Rounding off a bit, we can say that our revised probability for H_3 after three items of evidence could be as low as 0.23 and as high as 0.84, with the most "reasonable" value being 0.54. A range of 0.61 (0.84-0.23) is rather disquieting.

TABLE 9.8
Revised Tabular Array to find the Smallest Reasonable Revised Probability of H_3 after Three Items of Evidence

j	$P(H_j)$	$P(E_1/H_j\&E_1)$	$P(E_2/H_j\&E_1)$	$P(E_3/H_j\&E_1\&E_2)$	Row Product	Revised Probability of H_j
1	0.4	0.8	0.3	0.7	0.0672	0.580
2	0.2	0.9	0.3	0.4	0.0216	0.187
3	0.4	0.3	0.3	0.75	0.0270	0.233
	1.0				0.1158	1.000

Putting all our results together, we have Table 9.9.

TABLE 9.9
Summary of Results

	Low Value	Most Reasonable Value	High Value	Range
$P(H_1/E_1\&E_2\&E_3)$	0.10	0.35	0.68	0.58
$P(H_2/E_1\&E_2\&E_3)$	0.03	0.11	0.36	0.33
$P(H_3/E_1\&E_2\&E_3)$	0.23	0.54	0.84	0.61

If we were to present these results to a decision maker, and say that, on the basis of three items of evidence, the probability of H_1 had moved from an initial value of 0.4 to a value of 0.35, but that, given the reasonable range of uncertainties in the assigned probabilities, the revised probability of H_1 could be as low as 0.10 or as high as 0.68, the decision maker might well lack faith in the estimate provided. Once similar comments about H_2 and H_3 had been provided, the decision maker's faith might be even less. Moreover, we find that as we increase the number of events, the range of values increases. At five items of evidence, ranges of from 0.05 to 0.95 are quite common.

There is no doubt that this very wide "error interval" is quite disturbing. In essence, it shows that a sensitivity analysis of the sort described here yields "error intervals" that are so wide as to be meaningless and that therefore there is no point in doing a sensitivity analysis of this type. Beyond this, however, there is the question of what faith we can have in Bayesian analysis if the "error intervals" are so terribly wide. One answer obvious at this point is as follows: We can say that we have followed conservative procedures in arriving at the "best estimate" revised probability of the hypotheses (assuming we are doing what was suggested in the discussion of the blow-up problem), and therefore we have some confidence in the revised probabilities provided. We are prepared to defend our assessments of all the input conditional probabilities we assigned.

A second answer is that the type of procedure used in doing the sensitivity analysis really was a worst case procedure for both the upper and lower limits for $P(H_j/E_1\&E_2\&\cdots\&E_n)$. Every new assessment made, that is, was a worst case assessment, and it can be argued that in real life the "true" (in the sense of being known only by "God") revised probability probably would be much closer to the analyst's "best estimate" revised probability than the worst case high or low values. This line of argument may alleviate partially our distress about the implications of the terribly large "error intervals" that result from doing a sensitivity analysis using the method described.

A third answer to the question of what faith we can have in Bayesian analyses if the "error intervals" are so terribly wide is that the described method of finding those "error intervals" is not the usual sort of method and actually is a distortion of the "normal" process. Several engineers[1] have pointed out in class that, in a sensitivity analysis, it is usual to only vary the values of influencing variables one at a time instead of varying all the values as the described method does. They point out that the method used essentially guarantees large "error-intervals" and would do so in many analyses. They argue that the procedure used is therefore misleading and that the sensitivity problem as described is exaggerated.

I find these lines of argument reasonable, but still feel that in light of what we know about the sensitivity of the final results when using Bayes that we should put more emphases on the trend line we plot for our best estimates of revised probabilities of hypotheses and put less emphasis on the exact numbers themselves.

PROBLEM 38. The importance of prior probabilities: the problem of the impact of the prior probabilities of the hypotheses if there are few items of evidence.

The problem: Writers on Bayes often indirectly address this problem by stating that, in order to use Bayesian analysis, we need a "lot" of evidence. This statement is the solution; the problem is that, if there are only a "few" items of evidence, they may not be very diagnostic; in this case, then, the prior probabilities of the hypotheses (which may be very subjective and certainly are inductively assessed) will largely determine the revised probabilities of the hypotheses rather than the weight of the deductively assessed evidence.

The solution: Although the problem as stated appears to be a fact, and in a way cannot be "solved," student Kenneth Boll noted that one approach clearly is to search diligently for and collect evidence as quickly as we can "so that diagnostic evidence is applied sufficiently to reflect the true state of nature."

As we saw in the discussion in the chapter on the advantages of Bayesian analysis, a few items of very diagnostic evidence (or even a number of items of evidence of low diagnostic value) can lead to a radical revision of the probabilities of the hypotheses. It also turns out that the prior probabilities are quickly "swamped" by the evidence, i.e., it almost doesn't matter what the priors are if we have a "fair" amount of diagnostic evidence. What "fair" means is somewhat ambiguous but, as a rough rule of thumb, it can be argued that if we have 20 items of evidence (all favoring one particular hypothesis), even if they are of low diagnostic value the prior probabilities will tend to be swamped.

However, if we are asked for the revised probabilities of the hypotheses after only a few items of not very diagnostic evidence have been received, the trend line we have plotted will show the decision maker that the evidence has not had much impact. (The prior probabilities of the hypotheses should be assigned therefore with the greatest of care.) We may wish to carry out and provide to the decision maker a quick sensitivity analysis of what the trend would be if our initial - and highly determining - probabilities of hypotheses were slightly different.[2]

Discussion: Often, writers on Bayes stress the need for a "lot" of evidence. The reason is that, if there is only a little evidence, it may not be very diagnostic and the prior probabilities of hypotheses, then, will be the major determinant of the revised probabilities of the hypotheses. Thus, the initial, subjective judgments, rather than the weight of the evidence, will be the major determinant of the revised probabilities. This is not a very good situation, particularly if the starting probabilities either are (1) subconsciously "loaded" in favor of the hypothesis the analyst may support or (2) assigned conservatively in a purportedly neutral fashion - by making $P(H_1) = P(H_2) = \cdots = P(H_m)$. This latter way of assigning probabilities may be "justified" by saying that, by making such a "hypothesis neutral" assignment, the "evidence" will be the determining factor with respect to the revised probabilities. But, if there is little evidence and it is not diagnostic, the so-called "conservative" or "hypothesis-neutral" assignment will be what determines the revised probabilities and this can be very misleading. We always should do the best political/military analysis possible in order to assign our initial probabilities of hypotheses, for these initial probabilities may indeed be critical until we get a "fair" amount of evidence.[3]

Of course, once we do get a good amount of data, the previously analyzed high sensitivity of Bayesian analysis to even a little data that is mildly diagnostic means that, whatever the prior probabilities are, they soon will be "swamped" by the evidence. The evidence, rather than the prior probabilities, then will be the critical factor in determining the revised probabilities. Hence, the call in the literature for a "lot" of evidence. I prefer the phrase "a fair amount," and feel that a reasonable rough rule of thumb is that if we have 20 items of evidence (even of low diagnostic value), the prior probabilities will tend to be swamped.

An example follows where the priors are not "quickly" swamped, because of the extreme ratio of the prior probabilities, but they still are swamped after 20 items of evidence. The array for the first 10 items of evidence is presented in Table 9.10.

TABLE 9.10
Tabular Array Using Original Priors (9 to 1) after 10 Items of Evidence

j	$P(H_j)$	$P(E_1/H_j)$	$P(E_2/H_j\&E_1)$	···								Row Product
1	0.9	0.8	0.6	0.4	0.9	0.6	0.4	0.7	0.9	0.5	0.3	0.0035271
2	0.1	0.6	0.5	0.3	0.7	0.5	0.3	0.6	0.8	0.4	0.2	0.0000362
												0.0035633

Revised probability of H_1 = 0.9898
" " " H_2 = 0.0102.

Reversing the prior probabilities, we have Table 9.11.

TABLE 9.11
Tabular Array Using Reversed Priors (i.e., 1 to 9) after 10 Items of Evidence

j	$P(H_j)$	$P(E_1/H_j)$	$P(E_2/H_j\&E_1)$	···								Row Product
1	0.1	0.8	0.6	0.4	0.9	0.6	0.4	0.7	0.9	0.5	0.3	0.0003918
2	0.9	0.6	0.5	0.3	0.7	0.5	0.3	0.6	0.8	0.4	0.2	0.0003265
												0.0007183

Revised probability of H_1 = 0.5455
" " " H_2 = 0.4545.

These probabilities differ significantly from those above in the first tabular array (Table 9.10) for the original (9 to 1) priors. So, the prior probabilities are not swamped yet. The reason is their extreme ratio, 9 to 1 (or, reversed, 1 to 9). On the other hand, if we have, say, another 5 items of evidence favoring H_1, the trend is upward for the revised probabilities in the array for reversed priors (i.e., 1 to 9 priors), as shown in Table 9.12.

TABLE 9.12
Tabular Array Using Reversed Priors (1 to 9) after 15 Items of Evidence

j	Previous Row Product	···					Latest Row Product
1	0.0003918	0.6	0.8	0.4	0.9	0.7	0.0000473
2	0.0003265	0.5	0.7	0.3	0.8	0.6	0.0000163
							0.0000636

Revised probability of H_1 = 0.744,
" " of H_2 = 0.256.

Given another 5 items of evidence, the trend is dramatic.

TABLE 9.13
Tabular Array Using Reversed Priors (1 to 9) after 20 Items of Evidence

j	Previous Row Product	...					Latest Row Product
1	0.0000473	0.4	0.8	0.9	0.7	0.5	0.0000047
2	0.0000163	0.3	0.7	0.8	0.6	0.4	0.0000006
							0.0000053

Revised probability of H_1 = 0.89,
" " of H_2 = 0.11.[4]

A little calculation will show that in the tabular array for the original priors (9 to 1), the revised probability of H_1 essentially is "1," and the revised probability of H_2 essentially is "0".

After 20 items of evidence, then, there is not a great deal of difference in the revised probabilities, whether the prior probabilities were

$P(H_1) = 0.9$
$P(H_2) = 0.1$

or the reversed (opposite), $P(H_1) = 0.1$,
$P(H_2) = 0.9$.

Given the high ratio (9 to 1) of the prior probabilities (and 1 to 9 for the reversed priors), it is not surprising that 20 or so items of mildly diagnostic data were needed to "swamp" the prior probabilities.

Note that this rule of thumb is very rough. If the ratio of the original probabilities was not as extreme as 9 to 1 or 1 to 9, then fewer items of evidence could "swamp" the priors. On the other hand, since in the example all the items of evidence were evaluated to favor the H_1 hypothesis, if the evidence was more "mixed" (with some data supporting H_1 and some supporting H_2), more than 20 items of evidence could be required to swamp the prior probabilities.

If we are doing a Bayesian analysis and are asked for the revised probabilities after only a few items of not very diagnostic evidence have been received, the trend line we have plotted will show the decision maker that the evidence has not had much influence. We also should do a quick sensitivity analysis to see what the trend would be if we changed our initial probabilities of hypotheses.

PROBLEM 39. The importance of ratios: the problem of the importance of the ratios when assigning the conditional probabilities of events occurring.

The problem: The ratios among the conditional probabilities assigned for an event, rather than their absolute values, are what determines how rapidly the revised probabilities of the hypotheses shoot up or down vis-a-vis each other. However, as student Beverly Lohre pointed out in class, the ratios resulting from the conditional probabilities that have been assigned may not correspond to the desired differences in weighting, especially when one or more small probabilities (i.e., <0.2) are assigned, and may be much more diagnostic than intended. That is, the highly sensitive nature of Bayes may yield badly misleading revised probabilities.

Also, when there are a series of events where, for each event, (i) none of the probabilities is small (i.e., all are >0.2), but (ii) all the probabilities are close to one another, and (iii) all the ratios are close to one, then carelessness in assigning the probability values can have a large cumulative effect.

The solutions: When we deal with an event where one or more small probabilities (i.e., <0.2) have been assigned, we should normalize all the assigned probabilities and see if the resulting set of ratios is really what we intended in terms of diagnosticity -- and, if not, we should then adjust the ratios until the set does seem to be correct.

When we deal with a series of events where, for each event, (i) none of the probabilities is small (i.e., all are >0.2), but (ii) all the probabilities are close to one another, and (iii) all the ratios are close to one, we should make the probabilities for each event the same unless there is very good reason to make them different.

Discussion: By this point, because of the number of examples presented in this book, the reader probably has become aware that it is the set of ratios of the conditional probabilities assigned to events, rather than their absolute values, that determines the revised probabilities of the hypotheses. For example, for an item of evidence E_i, it makes no difference if we assess

$$\frac{P(E_i/H_1\&E_1\&E_2\&\cdots\&E_{i-1})}{P(E_i/H_2\&E_1\&E_2\&\cdots\&E_{i-1})} = \frac{0.8}{0.4}$$

$$\text{or} \quad \frac{P(E_i/H_1\&E_1\&E_2\&\cdots\&E_{i-1})}{P(E_i/H_2\&E_1\&E_2\&\cdots\&E_{i-1})} = \frac{0.2}{0.1}$$

because the resulting ratio in each instance is $\frac{2}{1}$. If we are using multiple (i.e., greater than 2) hypotheses Bayes, the same principle would hold. Assigning values of, say,

$$P(E_i/H_1\&E_1\&E_2\&\cdots\&E_{i-1}) = 0.9$$
$$P(E_i/H_2\&E_1\&E_2\&\cdots\&E_{i-1}) = 0.6$$
$$P(E_i/H_3\&E_1\&E_2\&\cdots\&E_{i-1}) = 0.3$$

yields the same revised probabilities as the values of

$$P(E_i/H_1\&E_1\&E_2\&\cdots\&E_{i-1}) = 0.3$$
$$P(E_i/H_2\&E_1\&E_2\&\cdots\&E_{i-1}) = 0.2$$
$$P(E_i/H_3\&E_1\&E_2\&\cdots\&E_{i-1}) = 0.1$$

because the internal set of ratios in each case is 3 to 2 to 1. Unless we keep this fact in mind, however, severe and unintended problems can arise, particularly when we assign small values (i.e., values < 0.2). For example, for a given E_i, if we assign values of

$$\frac{P(E_i/H_1\&E_1\&E_2\&\cdots\&E_{i-1})}{P(E_i/H_2\&E_1\&E_2\&\cdots\&E_{i-1})} = \frac{0.20}{0.10}$$

this yields a ratio of 2 to 1, which is rather diagnostic and will definitely impact on the revised probability of H_1. For only a slightly different assessment of the denominator, however, say, for example,

$$\frac{P(E_i/H_1\&E_1\&E_2\&\cdots\&E_{i-1})}{P(E_i/H_2\&E_1\&E_2\&\cdots\&E_{i-1})} = \frac{0.20}{0.20}$$

the resulting ratio is $\frac{1}{1}$ and the data have no diagnostic impact. If we did not consider the ratio, we might be as likely to use the first set of values as the second.

Student Thomas McIntyre offered an example where an analyst felt that an event was highly unlikely to occur for either of two hypotheses and assigned conditional probabilities of 0.05 and 0.01. Unfortunately, although (s)he meant to convey the idea that there was a very slim chance of the event under the first hypothesis and even less of a chance of the event under the second hypothesis, the ratio is 5 to 1. This means that the revised probability of the first hypothesis will shoot up greatly. Even a minor shift in the probabilities assigned (e.g., 0.05 and 0.03) would yield a very different ratio and quite different revised probabilities.

Many other such examples of major differences in the ratios, resulting from very minor differences (on the order of 0.1) in the assigned probabilities when one or more of the probabilities is low to begin with, could be given.[5] The moral is clear. When dealing with events that yield one or more small conditional probabilities either (1) we should normalize (see below) all of the assigned probabilities and see if the resulting set of ratios is really what we intended -- and if not, we should adjust the ratios until the set does seem to be correct, or (2) we should directly assess the ratios in the beginning rather than assign (small) probabilities. The first solution probably is preferable, for reasons discussed in the next problem.

We can normalize easily. If, for example, we have made the assessments

$$P(E_i/H_1\&E_1\&E_2\&\cdots\&E_{i-1}) = 0.10$$
$$P(E_i/H_2\&E_1\&E_2\&\cdots\&E_{i-1}) = 0.05$$
$$P(E_i/H_3\&E_1\&E_2\&\cdots\&E_{i-1}) = 0.20$$
$$P(E_i/H_4\&E_1\&E_2\&\cdots\&E_{i-1}) = 0.15$$

then we could take the smallest probability (0.05) and divide each entry by this value, getting

$$\frac{\dfrac{P(E_i/H_1\&E_1\&E_2\&\cdots\&E_{i-1})}{P(E_i/H_2\&E_1\&E_2\&\cdots\&E_{i-1})}}{\dfrac{P(E_i/H_3\&E_1\&E_2\&\cdots\&E_{i-1})}{P(E_i/H_4\&E_1\&E_2\&\cdots\&E_{i-1})}} = \frac{\frac{2}{1}}{\frac{4}{3}}$$

or take the last probability in the list (0.15) and divide each entry by this value, getting

$$\frac{\dfrac{P(E_i/H_1\&E_1\&E_2\&\cdots\&E_{i-1})}{P(E_i/H_2\&E_1\&E_2\&\cdots\&E_{i-1})}}{\dfrac{P(E_i/H_3\&E_1\&E_2\&\cdots\&E_{i-1})}{P(E_i/H_4\&E_1\&E_2\&\cdots\&E_{i-1})}} = \frac{\frac{0.67}{0.33}}{\frac{1.34}{1}}.$$

Which normalizing procedure we select is somewhat arbitrary. Whichever schema allows us most comfortably to compare the resulting ratios is the one that should be used.

A slight modification, by the way, is to get rid of the decimal points. The last listing above can be converted to

$$\frac{\dfrac{P(E_i/H_1\&E_1\&E_2\&\cdots\&E_{i-1})}{P(E_i/H_2\&E_1\&E_2\&\cdots\&E_{i-1})}}{\dfrac{P(E_i/H_3\&E_1\&E_2\&\cdots\&E_{i-1})}{P(E_i/H_4\&E_1\&E_2\&\cdots\&E_{i-1})}} = \frac{\dfrac{67}{33}}{\dfrac{134}{100}}.$$

Indeed, in line with this idea of getting rid of the decimal points, another approach, which also is comfortable for some people, is to simply take the *original* probability assessment and multiply by the appropriate power of ten to get rid of the decimal points and then examine the resulting ratios. For the example above this yields

$$\frac{\dfrac{P(E_i/H_1\&E_1\&E_2\&\cdots\&E_{i-1})}{P(E_i/H_2\&E_1\&E_2\&\cdots\&E_{i-1})}}{\dfrac{P(E_i/H_3\&E_1\&E_2\&\cdots\&E_{i-1})}{P(E_i/H_4\&E_1\&E_2\&\cdots\&E_{i-1})}} = \frac{\dfrac{10}{5}}{\dfrac{20}{15}}.$$

Before turning to the next problem, another aspect of the problem of the importance of the ratios should be examined. This issue was raised by student Martin Joyce. It was noted earlier in footnote 5 that, when none of the probabilities for an event is small to begin with, minor differences in the values assigned will have much less impact on the revised probabilities of hypotheses than if one or more of the initial probabilities were small (0.20). This is true for the probability assessments for *one* event. However, consistent with what was found about the "blow-up" and "error-bounding" problems, it seems clear that if we have a *series* of events where, for each of the events, none of the probabilities is small but all of them are close to one another and have ratios close to one, then carelessness in assigning the values can lead to radically different revised probabilities.

That is, in such a case the impact of each event is little, regardless of the actual number assigned as a probability, because, in terms of odds, it's almost "even." However, a series of such events with *nearly* even, but *not* exactly even, odds in favor of one hypothesis over the others, will drive the revised probability up (or down). This can be seen as an advantage of Bayes, but it also can be its downfall because two different analysts may agree that the odds are *nearly* even, but one may cite them as 1 to 1.1, and the other might cite them as 1.1 to 1. The two analysts do not really disagree, but the revised probabilities which result after considering 12 pieces of evidence, for example, can differ greatly (e.g., 3 to 1 versus 1 to 3). Joyce suggested two solutions. One is to put aside those pieces of evidence where the odds are less than a given figure (e.g., 1.5 to 1) - a solution which he and I reject because it is "arbitrary and biased." It subverts the Bayes advantage of including all relevant data.

His second possible solution is to consider the evaluations by several analysts and to average their assigned probabilities. This solution may have some merit in a group project. "This may also have merit," Dr. Gerald Hopple suggests, "because the available evidence (admittedly rare) indicates that people disagree a lot on political/military judgments and probability assignments."

My advice is that, if for each of a series of events none of the probabilities is very small (where small means <0.20), but all are close to one another and the ratios are close to one, then the probabilities should be assigned with particular care and probably (for each event) should be made equal to one another unless there are strong justifications for differences. Otherwise, if there is even a little carelessness -- based on the mistaken belief that, "since for each event the probabilities are close together (and not small), it doesn't matter much if we assign one set of probabilities or another set of probabilities 'close' to the first" -- there can be radical differences in revised probabilities (the "blow-up" problem again).

PROBLEM 40. Probabilities or odds: the problem of whether to assign probabilities or odds.

The problem: The problem is whether we should assign probabilities for hypotheses and events or whether we should assign odds.

The solution: We should choose the approach with which we are most comfortable. However, use probabilities if possible. As noted in the previous problem, if we assign small probabilities to one or more events, we should normalize, i.e., convert to odds, as a check.

Discussion: The question of whether to use the conditionally independent or dependent form of Bayes was addressed above, and it was argued that the conditionally dependent form is the best choice. The Bayes formula involving conditionally dependent probabilities is repeated here:

$$P(H_j/E_1\&E_2\&\cdots\&E_n) = \frac{P(H_j)\cdot P(E_1/H_j)\cdot P(E_2/H_j\&E_1)\cdots P(E_n/H_j\&E_1\&E_2\&\cdots\&E_{n-1})}{\begin{array}{l} P(H_1)\cdot P(E_1/H_1)\cdot P(E_2/H_1\&E_1)\cdots P(E_n/H_1\&E_1\&E_2\&\cdots\&E_{n-1}) \\ +P(H_2)\cdot P(E_1/H_2)\cdot P(E_2/H_2\&E_1)\cdots P(E_n/H_2\&E_1\&E_2\&\cdots\&E_{n-1}) \\ \vdots \\ +P(H_m)\cdot P(E_1/H_m)\cdot P(E_2/H_m\&E_1)\cdots P(E_n/H_m\&E_1\&E_2\&\cdots\&E_{n-1}) \end{array}}$$

The alternative conditionally dependent odds formula (a generalization of the "odds-likelihood" formula) is this:

$$\begin{array}{c}\dfrac{P(H_1/E_1\&E_2\&\cdots\&E_n)}{P(H_2/E_1\&E_2\&\cdots\&E_n)}\\ \hline \vdots \\ \hline P(H_m/E_1\&E_2\&\cdots\&E_n)\end{array} = \begin{array}{c}\dfrac{P(H_1)}{P(H_2)}\\ \hline \vdots \\ \hline P(H_m)\end{array} \cdot \begin{array}{c}\dfrac{P(E_1/H_1)}{P(E_1/H_2)}\\ \hline \vdots \\ \hline P(E_1/H_m)\end{array} \cdot \begin{array}{c}\dfrac{P(E_2/H_1\&E_1)}{P(E_2/H_2\&E_1)}\\ \hline \vdots \\ \hline P(E_2/H_m\&E_1)\end{array} \cdots \begin{array}{c}\dfrac{P(E_n/H_1\&E_1\&E_2\&\cdots\&E_{n-1})}{P(E_n/H_2\&E_1\&E_2\&\cdots\&E_{n-1})}\\ \hline \vdots \\ \hline (PE_n/H_m\&E_1\&E_2\&\cdots\&E_{n-1})\end{array}$$

The alternative, conditionally dependent tabular array format is, of course, this:

j	$P(H_j)$	$P(E_1/H_j)$	$P(E_2/H_j\&E_1)\cdots P(E_n/H_j\&E_1\&E_2\&\cdots\&E_{n-1})$	Row Product	Revised Prob of H_j
1					
2					
⋮					
m					
				sum	1.000

Regardless of which of the three formats we use, i.e., the conditionally dependent Bayes formula, odds-formula, or tabular array, we can choose to think in terms of probabilities and make all our entries probabilities, or we can think directly in terms of odds and make our inputs in these terms. We can do this because it is the ratios of the probabilities that we assign that matter, not the absolute entries.

Suppose, for example, we have decided to use what is the easiest of the three formats, the tabular array, and have made the entries shown in Table 9.14.

TABLE 9.14
Tabular Array (Thinking in Terms of Probabilities)

j	$P(H_j)$	$P(E_1/H_j)$	$P(E_2/H_j\&E_1)$	Row Product	Revised Probability of H_j
1	0.3	0.2	0.4	0.024	0.075
2	0.1	0.4	0.2	0.008	0.025
3	0.6	0.8	0.6	0.288	0.900
	1.0			0.320	1.000

The argument here is that we could have thought directly in terms of odds and made our inputs this way (Table 9.15).

TABLE 9.15
Tabular Array (Thinking in Terms of Odds)

j	$P(H_j)$	$P(E_1/H_j)$	$P(E_2/H_j\&E_1)$	Row Product	Revised Probability of H_j
1	3	1	2	6	$\frac{6}{80} = 0.075$
2	1	2	1	2	$\frac{2}{80} = 0.025$
2	6	4	3	72	$\frac{72}{80} = 0.900$
				80	1.000

As we see, if our assessments are consistent, we will get the same answer either way; and the same would be true if we relied on the (harder-to-use) Bayes formula or the odds-formula.

We therefore have a choice of whether to assign probabilities or odds. It probably is faster (and possibly much faster) to think in terms of odds, but this may do violence to the subtleties involved in dealing with conditionally dependent judgments. That is, if we have a given E_i, then, when we assign one-by-one,

$$P(E_i/H_1\&E_1\&E_2\&\cdots\&E_{i-1}),$$
$$P(E_i/H_2\&E_1\&E_2\&\cdots\&E_{i-1}),$$
$$\vdots$$
$$P(E_i/H_m\&E_1\&E_2\&\cdots\&E_{i-1}),$$

hopefully, we will for each judgment consider carefully the logical consistency of E_i with that particular hypothesis and the previous items of evidence most salient with respect to E_i and that particular hypothesis. For different hypotheses, the salient items of evidence may be different. The problem with thinking directly in terms of odds is that these subtleties may be submerged; we will tend to think only of one group of salient items of previous evidence, and will directly assess the odds with just these items of previous evidence in mind.

Thus, we probably should assign probabilities rather than odds if possible. Occasionally, we will need to convert our probabilities to odds. As we noted in the last problem, when one or more of the assigned probabilities is small (i.e., <0.2), we should normalize the probabilities to see if the resultant set of ratios (odds, that is) is really what we intended -- and if not we should adjust the odds until they do seem to be what we intended.

Further, even if the probabilities we assign are not low, thus not requiring conversion to odds, we simply might prefer to work always with odds rather than bother with probabilities. In spite of the advantages of using probabilities, we might be more comfortable using odds, or might be driven to doing so. This latter situation might arise if there was a steady stream of data and more than three hypotheses. The justifications required for the four or more separate conditional probability judgments required for each piece of data (since there is a judgment for each hypothesis) might become very onerous indeed. Assigning a set of odds might be much easier (and quicker) to do and to justify. (See below, however, for further arguments against using odds.)

An additional point relates to normalizing odds. If we assign odds, we should feel free to assign what looks reasonable without initially worrying about normalizing. (Thinking of whole numbers between "1" and "100" also is useful.) We might, for example, assign values of 85 to 40 to 70. If we like, we can normalize this to 2.125 to 1 to 1.75 or to 1.21 to 0.57 to 1 (or just leave it as 85 to 40 to 70).

Several further arguments against using odds are that (1) the analyst may not have an intuitive feeling for odds and (2) the use of odds rather than probabilities may permit more easily the conscious or unconscious recoupling of the analyst's ego because it may be hard for others to question the odds and hard for the analyst to convert to probabilities that could be questioned.

If the first argument is true for you - if you are an analyst and you don't have much of an intuitive feeling for odds - don't use them (except for the case of checking on small probability assignments for events.)

The second argument needs more development. If a user of Bayes is consciously trying to manipulate Bayes (or is unconsciously doing so), then the use of the odds formulation may make it easier to get away with (or easier to happen) because other analysts who are reviewing the work may have little intuitive feeling for the odds and thus may not be able to argue well against the odds (and it may be hard for the analyst to convert to odds). Reviewing analysts can't calculate the equivalent probabilities for events because there are an infinite number of equivalent probability sets for any odds set for an event. (E.g., if the odds are 2:4:8 for an event, the equivalent probabilities could be 0.2, 0.4, 0.8, or 0.1, 0.2, 0.4, etc.)

The best way for reviewing analysts to fight back may be to normalize, realizing that this won't yield the "actual" probabilities (because the analyst thought in terms of odds) and then proceed to ask the analyst to adjust the normalized probabilities to what (s)he thinks they should be. This could take a lot of time if many judgments had been made by the analyst. Using the above example, if the analyst's odds for E_1 are 1:2:4, normalizing yields (one) equivalent probability set of $\frac{1}{1+2+4}$, $\frac{2}{1+2+4}$, $\frac{4}{1+2+4}$, i.e., 0.14, 0.29, 0.57 (which could be rounded off roughly to 0.15, 0.30, 0.55). The analyst then could be asked to say what (s)he felt the "real" probabilities were. Of course, this could be most unfair to the analyst because (s)he really might have a hard time thinking in terms of probabilities.

I have no better solution to offer, if you do not have a good intuitive feeling for odds but want to question the odds some analyst has used, than to go through this normalization and dialogue procedure.

I frankly don't think analysts are out to recouple their egos. I do think that the difficulties of reviewers in questioning odds may permit more unconscious recoupling of egos than would occur if the analyst used probabilities. If an analyst feels comfortable in using probabilities, this line of argument implies (s)he should do so in order to help reviewers of the analysis.

PROBLEM 41. Group assessments of probabilities: the problem of how to deal with a group assessment of probabilities.

The problem: Occasionally, we may be involved in a Bayesian analysis project where there are a number of analysts making judgments but only one analysis is wanted. The question arises then of how to reconcile the individual probability assignments and come up with a group assessment.

The solution: Student Joseph May suggested several possible solutions:

a. Have a meeting to assign probabilities so that the group produces a consensus for each number.

b. Have each person assign probabilities and use the mean value if the standard deviation is less than some small value (e.g., 0.1). If the standard deviation is larger, then either have a meeting to try to narrow the differences and use the results or use the recommendation of the analyst considered most competent in the area. (If there are several such analysts, their mean value could be used, or they could be asked to meet and narrow their judgments.) Dr. Hopple notes a third alternative:

c. Use an explicit Delphi procedure.

Discussion: Joseph May presented Table 9.16 in support of his suggestions. Suppose there has been no group meeting but that each analyst has made his (her) own assessments, presented here.

Since for $P(E_1/H_j)$ the standard deviation is small (i.e., 0.1), the mean can be used without further question.

Since for $P(E_2/H_j\&E_1)$ the standard deviation is greater than 0.1, there is a need either to call a meeting to try to narrow the judgments or to identify the analyst most competent in the area and use her (his) judgment. (If there are several such analysts, their mean value could be used or only they could be asked to meet and narrow their judgments.) The same applies to $P(E_3/H_j\&E_1\&E_2)$.

TABLE 9.16
Different Analysts' Assessments

Analyst Number	$P(E_1/H_j)$	$P(E_2/H_j\&E_1)$	$P(E_3/H_j\&E_1\&E_2)$	$P(E_4/H_j\&E_1\&E_2\&E_3)$
1	.75	.60	.50	.80
2	.70	.30	.50	.80
3	.80	.80	.50	.80
4	.65	.50	.90	.50
5	.70	.40	.50	.85
6	.75	.55	.50	.75
7	.70	.75	.80	.80
8	.70	.35	.85	.90
9	.75	.40	.60	.80
10	.80	.80	.75	.80
Std Dev	.0458	.178	.158	.100
Mean	.73	.545	.64	.78
Range	.65-.80	.30-.80	.50-.90	.50-.90

The standard deviation for $P(E_4/H_j\&E_1\&E_2\&E_3)$ is at the acceptable limit of 0.1, so by our criterion we will use the mean value. Joe May comments, however, that we might question this if analyst 4 or analyst 8 is <u>the</u> expert on the subject, since these judgments are the high and low ones. If analyst 4 or 8 is <u>the</u> expert, then we might want to go with the appropriate opinion.[6] (If they are the <u>two</u> leading experts, a conference is in order.)[7]

NOTES

<u>Problem 37</u>

1. Jeffrey D. Wise and Gus Schoone. They audited one of my seminars.

<u>Problem 38</u>

2. Dr. Hopple asks if this "doesn't raise the problem of giving the consumer information that will turn him or her 'off' to the use of methods? This is a very serious problem. Decision makers are going to want <u>one</u> result (the revised probability = 0.7), not any 'qualifications'. This problem comes up with almost any use of methods. It is like the beginning student looking only at R^2 in regression analysis. The R^2 may be misspecified, etc., etc."

3. Dr. Hopple notes that this is a "strong argument against the common practice of setting [the] priors = 0.5 (in a 2-hypothesis case). For example, in the early warning discriminant analysis for crises, we set the priors = α at various levels to see what the impact would be."

4. The importance of significant digits when calculating with many items of evidence is a problem here. That is, at the end of Table 9.13 we were down to two significant digits in our latest row product for j = 1. We need to be careful in calculating to not lose more and more significant digits as we continue to multiply previous row products by decimal numbers. The way to not lose significant digits is not too easy to explain. The first thing to do is to note how many digits (i.e., individual entries or places) your hand calculator can hold. Mine has 8, and I well explain what to do for this case. As soon as I reached a row product with 8 digits (which happened in Table 9.11 when I reached E_6 = 0.4) thereafter I would multiply either by the number given or by 10 times the number (e.g., by 7 instead of 0.7 for E_7) and keep track of the powers of 10 (10^{-1} in this case). This procedure would lead me to end up in Table 9.11 with row product entries of

$$3.919 \cdot 10^{-4}$$

and

$$3.265 \cdot 10^{-4}.$$

My goal in multiplying each time either by the number (e.g., 0.7) or by 10 times the number (e.g., 7) was to keep the number showing on the calculator as large as possible but below 9.9999. In this way I would avoid exceeding the capacity of the calculator and also avoid losing significant digits. The entries for all 20 items are shown here for the H_1 row (i.e., j = 1).

		Cumulative Row Product after multiplying either by the number or 10 times the number	Number of digits
$P(H_1) = 0.1$			
$P(E_1/H_1) = 0.8$		0.08	3 digits
$P(E_2/H_1\&E_1) = 0.6$		0.048	4 digits
$P(E_3/H_1\&E_1\&E_2) = 0.4$		0.0192	5 digits
$P(E_4/H_1\&E_1\&E_2\&E_3) = 0.9$		0.01728	6 digits
$P(E_5/H_1\&E_1\&\cdots E_4) = 0.6$	What I	0.010368	7 digits
$P(E_6/H_1\&E_1\&\cdots E_5) = 0.4$	multiplied by:	0.0041472	8 digits
$P(E_7/\quad) = 0.7$	" " 7	$0.0290304 \cdot 10^{-1}$	"
$P(E_8/\quad) = 0.9$	" " 9	$0.2612736 \cdot 10^{-2}$	"
$P(E_9/\quad) = 0.5$	" " 5	$1.306368 \cdot 10^{-3}$	7 digits
$P(E_{10}/\quad) = 0.3$	" " 3	$3.919104 \cdot 10^{-4}$	7 digits

$P(E_{11}/\quad) = 0.6$	" " 0.6	$2.3514624 \cdot 10^{-4}$	8 digits
$P(E_{12}/\quad) = 0.8$	" " 0.8	$1.8811699 \cdot 10^{-4}$	"
$P(E_{13}/\quad) = 0.4$	" " 4	$7.5246796 \cdot 10^{-5}$	"
$P(E_{14}/\quad) = 0.9$	" " 0.9	$6.7722116 \cdot 10^{-5}$	"
$P(E_{15}/\quad) = 0.7$	" " 0.7	$4.7405481 \cdot 10^{-5}$	"
$P(E_{16}/\quad) = 0.4$	" " 0.4	$1.8962192 \cdot 10^{-5}$	"
$P(E_{17}/\quad) = 0.8$	" " 0.8	$1.5169753 \cdot 10^{-5}$	"
$P(E_{18}/\quad) = 0.9$	" " 0.9	$1.3652777 \cdot 10^{-5}$	"
$P(E_{19}/\quad) = 0.7$	" " 0.7	$0.9556943 \cdot 10^{-5}$	"
$P(E_{20}/\quad) = 0.5$	" " 5	$4.7784715 \cdot 10^{-6}$	"

Similarly, for the H_2 row

	What I multiplied by:	Cumulative Row Product	Number of digits
$P(H_2) = 0.9$		0.54	3 digits
$P(E_1/H_2) = 0.6$		0.27	3 digits
$P(E_2/H_2\&E_1) = 0.5$		0.081	4 digits
$P(E_3/H_2\&E_1\&E_2) = 0.3$		0.0567	5 digits
$P(E_4/\quad) = 0.7$		0.02835	6 digits
$P(E_5/\quad) = 0.5$		0.008505	7 digits
$P(E_6/\quad) = 0.3$	" " 6	$0.05103 \cdot 10^{-1}$	6 digits
$P(E_7/\quad) = 0.6$	" " 8	$0.40824 \cdot 10^{-2}$	6 digits
$P(E_8/\quad) = 0.8$	" " 4	$1.63296 \cdot 10^{-3}$	6 digits
$P(E_9/\quad) = 0.4$	" " 2	$3.26592 \cdot 10^{-4}$	6 digits
$P(E_{10}/\quad) = 0.2$	" " 0.5	$1.63296 \cdot 10^{-4}$	6 digits
$P(E_{11}/\quad) = 0.5$	" " 0.7	$1.143072 \cdot 10^{-4}$	7 digits
$P(E_{12}/\quad) = 0.7$	" " 3	$3.429216 \cdot 10^{-5}$	7 digits
$P(E_{13}/\quad) = 0.3$	" " 0.8	$2.7433728 \cdot 10^{-5}$	8 digits
$P(E_{14}/\quad) = 0.8$	" " 0.6	$1.6460236 \cdot 10^{-5}$	"
$P(E_{15}/\quad) = 0.6$	" " 3	$4.9380708 \cdot 10^{-6}$	"
$P(E_{16}/\quad) = 0.3$	" " 0.7	$3.4566495 \cdot 10^{-6}$	"
$P(E_{17}/\quad) = 0.7$	" " 0.8	$2.7653196 \cdot 10^{-6}$	"
$P(E_{18}/\quad) = 0.8$	" " 0.6	$1.6591917 \cdot 10^{-6}$	"
$P(E_{19}/\quad) = 0.6$	" " 0.4	$0.6636766 \cdot 10^{-6}$	"
$P(E_{20}/\quad) = 0.4$			

So, I ended up with

j	Latest (after 20 items) Row Product	Rev Prob of H_j
1	$4.7784715 \cdot 10^{-6}$	0.88
2	$0.6636766 \cdot 10^{-6}$	0.12
	$5.4421481 \cdot 10^{-6}$	1.00

As is very obvious, this procedure yielded 8 significant digits for the j = 1 row product after 20 items of evidence (namely, $4.7784715 \cdot 10^{-6}$) compared to only 2 significant digits for the j = 1 row product after 20 items of evidence (namely 0.0000047, from Table 9.13) when we just kept multiplying by the decimal entry for the probability of each item of evidence rather than worrying about significant digits. By occasionally (and appropriately) multiplying by 10 times the decimal probability of events, we can maintain the number of digits at the capacity of our hand calculator, while at the same time we avoid exceeding the capacity of the calculator.

Problem 39

5. When the conditional probabilities are not very small to begin with, a minor difference will have much less impact. For example it makes only a little difference if we assign

$$\frac{P(E_i/H_1\&E_1\&E_2\&\cdots\&E_{i-1})}{P(E_i/H_2\&E_1\&E_2\&\cdots\&E_{i-1})} = \frac{0.9}{0.8}$$

or

$$\frac{P(E_i/H_1\&E_1\&E_2\&\cdots\&E_{i-1})}{P(E_i/H_2\&E_1\&E_2\&\cdots\&E_{i-1})} = \frac{0.8}{0.8}.$$

Still, as we have seen before (and will discuss again below), the *cumulative* impact even here can be substantial if we have a fair number of items of evidence where similar judgments can be made.

Problem 41

6. Dr. Hopple notes that there is a real question as to whether "expertise necessarily translates into better judgment (cf. literature on models of judges vs. intuitive judgments - clinicians and graduate school committees making predictions)." See NISBETT, RICHARD and ROSS, LEE (1980) Human Inference: Strategies and Shortcomings of Social Judgment. Englewood Cliffs, New Jersey: Prentice-Hall, Inc.

7. There are, of course, a number of articles dealing with rules for arriving at a "group" estimate. See, for example (1974) ROWSE, GLENWOOD L., *et.al.*, "Comparison of Rules for Aggregating Subjective Likelihood Ratios." Organizational Behavior and Human Performance 12: 274-285.

10
Problems Involving the Difficulties Individual Analysts Face in Trying to Assign Probabilities

In this chapter we will discuss problems involving the various difficulties individual analysts face when they try to assign probabilities.

PROBLEM 42. Difficulties in thinking in terms of probabilities.

The problem: According to former CIA analyst Nicholas Schweitzer: "It has been our experience that some people think easily in probabilities, others have to work at it every time, and a few need constant attention and retraining to overcome a distorted or unrealistic feeling for probabilities."[1]

The solution: He went on to say: "The only solution for this problem, aside from a careful initial choice and subsequent replacement of analysts is constant attention to the analysts' assessments and frequent retraining using illustrative items of evidence."[2]

Discussion: One distinguished analyst commented in one of my classes that "It has been said that analysts aren't very good at assigning numerical probabilities. In fact, we haven't had much problem in getting analysts to translate their verbal concepts into numerical form."[3] Schweitzer's statement above, however, implies it isn't always that easy. Some people may find it difficult to represent their judgments about evidence with mathematical precision.

Another class member, Beverly Lohre, who was quite experienced with analysts working with Bayes, noted that it is a problem to get some analysts to understand how to assign conditional probabilities to the hypotheses. In addition, she stated that "Some analysts have found it difficult to reverse the question when they reviewed items of evidence. For example, their intuition strongly believes there will not be a major attack from either the Soviets or Chinese against the other; therefore, they may assign a low probability for an item of evidence being seen. However, they should be assigning a high probability because the item would be seen no matter what the Soviets or Chinese were planning."

She also commented, relevant to our previously discussed

problem of the importance of ratios, that there is a problem in getting some analysts to understand the importance of the ratios between the probabilities of events. Some analysts arbitrarily assign conditional probabilities to events without considering the ratios between them. This may produce revised probabilities which do not reflect the analysts' real judgment.

Her solution (consistent with Schweitzer's advice) to these problems is for someone involved with running the Bayesian project to take the necessary extra time with the analysts until he or she is convinced that the analysts understand the procedure clearly. She said that "Sitting down and discussing with the analysts why they assigned the probabilities they assigned is a particularly useful way of checking to see if they really know what they're doing. Perhaps showing them the formula used would also help them to understand why the probability ratios between the hypotheses and between the events are so important. Of course, in a crisis situation, there may not be enough time to teach Bayesian analysis, but in an ongoing project which will monitor certain events over a specified period of time, this may assist the analysts."

PROBLEM 43. Lack of analyst expertise: the problem of the lack of necessary expertise to assess the probability of events for some hypotheses.

The problem: Given a spectrum of hypotheses, we may feel quite comfortable in assessing $P(E_i/H_j\&E_1\&E_2\&\cdots\&E_{i-1})$ for those H_j dealing with our area of expertise (e.g., military affairs), but find it very difficult to make the assessments for those H_j dealing with areas outside our area of expertise (e.g., H_j dealing with the economy, the political situation, etc.).

The solutions: In decreasing order of desirability, the solutions are these:

(1) Ask other analyst(s) to help in making the assessments $P(E_i/H_j\&E_1\&E_2\&\cdots\&E_{i-1})$ for those H_j where there is difficulty;

(2) Do not consider those events E_i for which there is particular difficulty for some H_j in assigning $P(E_i/H_j\&E_1\&E_2\&\cdots\&E_{i-1})$;

(3) For those events E_i for which there is particular difficulty for some H_j in assigning $P(E_i/H_j\&E_1\&E_2\&\cdots\&E_{i-1})$ make $P(E_i/H_j\&E_1\&E_2\&\cdots\&E_{i-1})$ the same number for all H_j.

Discussion: This problem deals with the possible difficulty in assessing the probability of events for certain hypotheses. For example, a military analyst may feel quite comfortable in assessing $P(E_i/H_j\&E_1\&E_2\&\cdots\&E_{i-1})$ where H_j deals with the possible outbreak of one level of war, but the same analyst may find it much harder to assess the probability of an event given a hypothesis (perhaps the residual hypothesis) dealing with, say, "no war." The "no war" residual hypothesis is so inclusive of possibilities that it can be argued that problem 2 applies - that the analyst is dealing with a recognizably too inclusive, meaningless residual hypothesis.

This argument seems valid, and "no war" can be broken down into categories, e.g.,

> "no conflict but a high level of threatening military maneuvers and words," etc.

(Again, a tree diagram could be used to develop a list.)

The original problem may well remain, however. That is, even with a well-formulated expansion of the "no war" hypothesis, the analyst may be hard-pressed to assign probabilities for events because of his (her) lack of knowledge about the economy, the political situation, etc. There is no single solution to such problems. Judgments normally still must be made, but it may be possible for the analyst to simply say that (s)he needs someone to help out and make the assignment of probabilities of events for those hypotheses outside of the analyst's area of expertise.

If an analyst cannot ask for such help or is refused such help, the next best alternative probably is to explicitly not consider those items of evidence for which there is particular difficulty in assigning probabilities for some hypotheses.

As a final resort, if the analyst is told that (s)he must consider all the evidence, rather than assign undefendable probabilities the best choice may well be to make

$$P(E_i/H_j\&E_1\&E_2\&\cdots\&E_{i-1})$$

the same for all H_j. This effectively gives the evidence no diagnostic value. I frankly don't think that an analyst is at all likely (probability < 0.1) to be driven to this extreme.

PROBLEM 44. Logically "distant" events: the problem of an event E_i logically "distant" from one or more hypotheses, which makes it difficult or impossible to assess meaningfully $P(E_i/H_j\&E_1\&E_2\&\cdots\&E_{i-1})$ for those hypotheses.

The problem: An event (i.e., item of evidence) may be so remotely connected to one or more hypotheses, that it may be very difficult or even impossible to assess the conditional probabilities of its occurrence.

The solution: Use of hierarchical Bayesian analysis.

Discussion: The problem here is that an event E_i may be so "distant," in a logical, causal sense, from one or more hypotheses, that it is difficult or impossible to assess, for those hypotheses, $P(E_i/H_j\&E_1\&E_2\&\cdots\&E_{i-1})$. (Note that the previously discussed problem of a meaningless residual hypothesis is quite different.)

An example is where

H_j = Country A has a major development effort underway for weapon system X, and

E_i = An esoteric technical paper, on a technology slightly related to those needed if there is a weapons development program, written by an obscure scientist in country A.

The solution to this problem is the more advanced form of Bayes mentioned previously, and called hierarchical Bayesian analysis.[4] Developed in the last quarter century, this form of Bayes permits the event to be "linked" to an "indicator" (not necessarily, however, the type of indicator on indicator lists we have referred to before). The "indicator" in turn can be linked to an "activity," which in turn can be linked to the hypotheses. The conditional probability judgments are made only between linked elements, so we first ask not the difficult or impossible-to-assign probability $P(E_i/H_j\&E_1\&E_2\&\cdots\&E_{i-1})$, but instead the easier-to-assign probability of $P(E_i$/Indicator state j) for each indicator state (or setting) j. The indicator will be much closer logically to E_i than the hypotheses are, so the probability assignments will be much easier to make.

The question remains, of course, if we want to use hierarchical Bayes just to take care of the difficulty with one item of evidence. Hierarchical Bayes is a highly structured form of Bayes with very strong assumptions of independence, among other problems. A decision to use hierarchical Bayes should not be made lightly. It is, however, the only solution I know of to this problem. Another solution using normal Bayesian analysis would be quite desirable, and I would be most pleased to hear of one.

PROBLEM 45. Systematic bias.

The problem: The last problem really dealt with very basic problems involved in getting people to understand how they are to go about assigning probabilities. The problem remains, however, of systematic biases - even among the "sophisticated" and "knowledgeable" - in the probabilities that are assigned. Psychological research shows that there are indeed such systematic biases.

The solution: None is offered here. The literature has emphasized the difficulties in assigning unconditional probabilities rather than conditional probabilities. Training improves performance in assigning regular probabilities only to a limited extent.

Discussion: There is a rather large body of literature dealing with subjective estimates of both point probabilities and probability distributions, but emphasis has not been placed on the subjective assignment of conditional probabilities. The literature has been summarized (better than I could ever hope to do) in Sarah Lichtenstein, et. al., Calibration of Probabilities: The State of the Art to 1980.[5] A good normative guide to assigning probabilities (but again not stressing conditional probabilities) is A Manual for Encoding Probability Distributions by Carl-Axel D. Staël von Holstein and James E. Matheson.[6]

PROBLEM 46. Idiosyncratic or situational biases.

The problem: The previous problem applied to situations of systematic biases in assigning probabilities. These biases are the sort everyone will tend to have. The additional problem being referred to here is that there are biases that only some people will have because of personal, idiosyncratic factors, or because of the special situation. There may, for example, be particular biases or distortions involved when persons with dogmatic personalities engage in assigning probabilities.[7] There may be particular biases in assigning probabilities in situations (such as crises) where there is a lot of stress.[8]

The solution: None is offered here. One specialist has pointed out to me that being aware of our biases does not mean that we easily can overcome them. The best general treatments of the psychological problems may well be the superb works of Dr. Frank J. Stech and of Nisbett & Ross.[9]

PROBLEM 47. Conscious manipulation of probabilities.

The problem: On rare occasions, an analyst may manipulate consciously his (her) assigned probabilities to support a favored hypothesis.[10]

The solution: The fact that Bayes does provide an audit trail of the analyst's thinking means that biases can be confronted and argued against. If an analyst has to "go public" with an estimate, a possible way of countering this problem (suggested by Schweitzer) is to present the graphs (of the revised probabilities of hypotheses) of several analysts, but not associate a particular graph with a particular person. It may be helpful in such cases also to present supplementary "intuitive" graphs where the individual is identified so that s(he) can "express strongly-held personal opinions in a forum designed for that purpose..."[11]

Discussion: This probably is a rather rare problem, as Schweitzer reported. He found it in some early studies where specific Bayesian graphs (of revised probabilities of hypotheses) were labeled with the individual's names. The process of making the Bayesian graphs anonymous, while letting the analysts present supplementary intuitive graphs, provided an effective solution.[12]

PROBLEM 48. Trend line biases: the problem of analyst biases emerging from paying attention to the trend lines.

The problem: There may be an unfortunate, subconscious tendency on the part of analysts to assign probabilities in a way that supports the direction of the trend lines. This may be true particularly if the analyst has projected the lines and indirectly reintroduced unconscious ego involvement into the outcome of the analysis.

The solution: Having to justify the probabilities assigned may help to alleviate this problem. In addition, analysts should not project trend lines and should remember that up and down fluctuations are common, and changes in the state of nature are not uncommon (see problem 50 and 51 in chapter 11). The analyst should remember that Bayesian analysis, properly conducted, is a deductive intellectual exercise, where each piece of evidence is to be examined for its logical consistency with each and every hypotheses in turn (along with previous evidence.). Nonetheless, it must be admitted that knowing we have biases and trying to overcome them are no guarantees that we will succeed.

Discussion: It was pointed out earlier that the best way of presenting the results of a Bayesian analysis to a decision maker is to present a graph of the trend lines of each of the hypotheses. One unfortunate side effect of doing this, however, is that the analyst may have a subconscious tendency to interpret evidence to favor one of the trend lines. This may be true particularly if a hurried decision maker asks for the analyst's opinion as to which line(s) can be projected. Decision makers want definitive answers, and analysts do their best to please. So, analysts may subconsciously reintroduce ego-involvement and come to assign higher probabilities to evidence given the favored hypothesis than given other hypotheses. The trend line of the favored hypotheses may "prosper," the decision maker may be happier, but all may be ill-served.

If the analyst is forced to justify the probabilities that have been assigned, prior to presenting results to the consumer, this may help to alleviate this problem somewhat. Such review is not always possible, however. The analyst should be aware of this possible bias, and should not engage in the (mathematically unjustified) practice of projecting trend lines. This practice can only strengthen the bias.

Further, the analyst should remember that up and down fluctuations in trend lines are common, so that any "trend" can be transitory (see problem 50). The analyst also should remember that any "trend" can be interrupted by a not-uncommon change in the state of nature (see problem 51). Finally, the more the analyst tries to make his (her) reasoning clear and justified to him(her)self in engaging in the Bayes process, the better off the analyst will be. The analyst always must remember that Bayesian analysis is deductive in nature, assumes each hypothesis in turn to be true, and then asks for an evaluation of the logical consistency of the evidence at hand with respect to the assumed hypothesis (and the previous evidence when we use the conditionally dependent form of Bayes).

PROBLEM 49. Time-pressure difficulties: the problem that the time required to do a Bayesian analysis may make it difficult to use it in a crisis situation.

The problem: As stated.

The solution: Use of a computerized form of Bayes or use of the tabular array can speed up the process of using Bayes. Thinking in terms of odds, rather than probabilities, also can speed up the process, as we mentioned in problem 40. However, even if the process is more time-consuming than is the case in "normal" analyses, the increased systematization and other advantages may make the effort to use Bayes worthwhile.

Discussion: Richards J. Heuer, Jr., pointed out in class that "any systematic procedure breaks down in a crisis.[13] To help overcome this, we [CIA] have devised a simple interactive program that enables an individual analyst to do calculations literally in seconds or a few minutes." The tabular array approach presented previously in this book also is designed to greatly speed up the process. Also, making probability judgments in terms of odds rather than probabilities can speed up the analysis, as we noted in problem 40 above.

Even if a situation is not a crisis, where there well may be very little time for analysis, the comment often is made that analysts just don't have time to do a Bayesian analysis, given all their duties. There may be an associated feeling in non-crisis situations of "why bother?" Dr. John Kringen, commenting in class on the use of Bayes in both crisis and non-crisis situations, said that "...standard Bayesian procedures (and perhaps any more formal procedures) tend to be squeezed out by either action or inaction. If the situation is 'hot,' the analysts are too caught up in their conventional reporting activities; if the situation is 'cool,' analysts figure what is the point of going through all this statistical mumbo jumbo when 'anyone' knows nothing is going to happen."[14]

An answer to the objections to the use of Bayes in crisis situations is that even if the process is more time-consuming that "normal" analyses, the increased systematization and other advantages may make the effort worthwhile.

For "cool" situations as described by Dr. Kringen, one answer is that the "conventional" wisdom that "anyone" knows nothing is going to happen has been proved wrong often enough to warrant trying to use such alternative techniques as Bayes. Moreover, the more Bayes is used, the greater will be the analysts' proficiency in using it and the easier it will be to use in crisis situations.[15] Practice may not make one perfect, but it can make one faster. (In point of fact, Bayesian analysis has been used for problems of an on-going nature, such as in the Middle East.)

Dr. Kringen made the interesting point, however, that a practical problem is that "the people called upon in a crisis to work on a problem may not be the ones who have been doing Bayes."[16] This, of course, effectively rules out its use - it can't be taught that quickly.

NOTES

Problem 42

1. SCHWEITZER (1978), p. 30.

2. Ibid.

3. He might wish to remain anonymous, so I won't cite him.

4. The best description of this technique is to be found in BARCLAY, SCOTT et. al. (1977) Handbook for Decision Analysis Decisions and Designs, Inc. Technical Report TR-77-6-30, September. McLean, Virginia.

Problem 45

5. LICHTENSTEIN, SARAH, et. al. (1981) "Calibration of Probabilities: The State of the Art to 1980." Decision Research Technical Report PTR-1092-81-6, June. Eugene, Oregon.

6. STAËL von HOLSTEIN, CARL-AXEL S. and MATHESON, JAMES E. (1978) "A Manual for Encoding Probability Distributions." SRI International Project 7078, September. Menlo Park, California.

Another document, which came to my attention recently, and which does deal with how to assign conditional probabilities better, is the following:

MOSKOWITZ, HERBERT and SARIN, RAKESH K. (1980) "Improving Conditional Probability Assessments for Forecasting and Decision Making." Krannert Graduate School of Management, Purdue University, Paper No. 734, July. West Lafayette, Indiana.

Problem 46

7. BRIGHTMAN, HARVEY J. and URBAN, THOMAS F. (1974) "The Influence of the Dogmatic Personality Upon Information Processing: A Comparison with a Bayesian Information Processor." Organizational Behavior and Human Performance 11(2)(April): 266-276.

8. MORGAN, PATRICK (1977) Deterrence: A Conceptual Analysis. Beverly Hills: Sage, p. 186 (quoting SCHNEIDER, BARRY (1974) Danger and Opportunity: Decision-Making, Bargaining and

Management in Three United States and Six Simulated Crises." Columbia University Ph.D. Desertation, New York, pp. 67-68).

9. See, in particular, STECH, FRANK J (1981) "Estimates of Peacetime Soviet Naval Intentions: An Assessment of Methods." MATHTECH, Inc. Final Report, March. Bethesda, Md. AD A098153.

STECH, FRANK J. (1979) "Political and Military Intention Estimation: A Taxonometric Analysis." MATHTECH Final Report, November. Bethesda, Md. AD A081486.

NISBETT, RICHARD and ROSS, LEE (1980) Human Inference: Strategres and Shortcomings of Social Judgment. Englewood Cliffs, New Jersey: Prentice-Hall, Inc.

<u>Problem 47</u>

10. SCHWEITZER (1978), p. 30.

11. <u>Ibid</u>.

12. <u>Ibid</u>.

<u>Problem 49</u>

13. Mr. Heuer, a senior CIA employee at the time, sat in on my seminar the second time it was offered.

14. Dr. Kringen, also a CIA employee, was sitting in on my class at the time. He was not necessarily agreeing with the attitudes described, by the way.

15. Dr. Hopple notes that usage of Bayes in crises is partly a problem of analyst familiarity with the technique and also its institutionalization into the repetoire of techniques.

16. Again, a point made in a class discussion.

11 Problems Involving the Results of a Bayesian Analysis

In this chapter we examine some problems which involve the results of a Bayesian analysis.

PROBLEM 50. Fluctuating revised probabilities: the problem of revised probabilities of hypotheses fluctuating up and down, as each piece of conflicting data, reflective of inconsistent or "hedging" government policies, is processed.

The problem: A difficulty in using Bayesian analysis is that, because of both policy disagreements within governments and "hedging" actions that governments take to prepare for different options, the actions that governments take will appear to be inconsistent, that is, will appear to reflect inconsistent government policies. So, when the data about those actions are processed, the resulting Bayes lines for revised probabilities of hypotheses will go up and down.

The solution: Grin and bear it. Trying to incorporate hypotheses about the policies of different factions, or the hedging options a government may be considering, probably is too complicated to do.

Discussion: There are several reasons for inconsistent data, i.e., data about inconsistent actions. The first, and primary reason, is that governments inherently are made up of conflicting individuals, groups, and organizations. Policy disagreements are natural, and actions (e.g., press leaks, public statement) will reflect differing views. A Bayesian analysis, then, seldom will find all the evidence supporting one particular hypothesis about which "policy" the "government" is pursuing.

The second reason for data reflecting inconsistent actions is that any government's actions are not based solely on one policy option on a subject but rather reflect a consideration of options. That is, there usually are hedging actions -- actions that will help the government if it ends up pursuing a different policy option than the one it intends, at a particular moment, to follow. The government, in short, hedges its bets. If a Bayesian analysis is being used and collection efforts are good, data reflective not only of the policy at the time but also reflective of alternative options will be picked up.

FIGURE 11.1
Tree Diagram for Logical Possibilities of Support and Non-support of a Policy by Four Factions

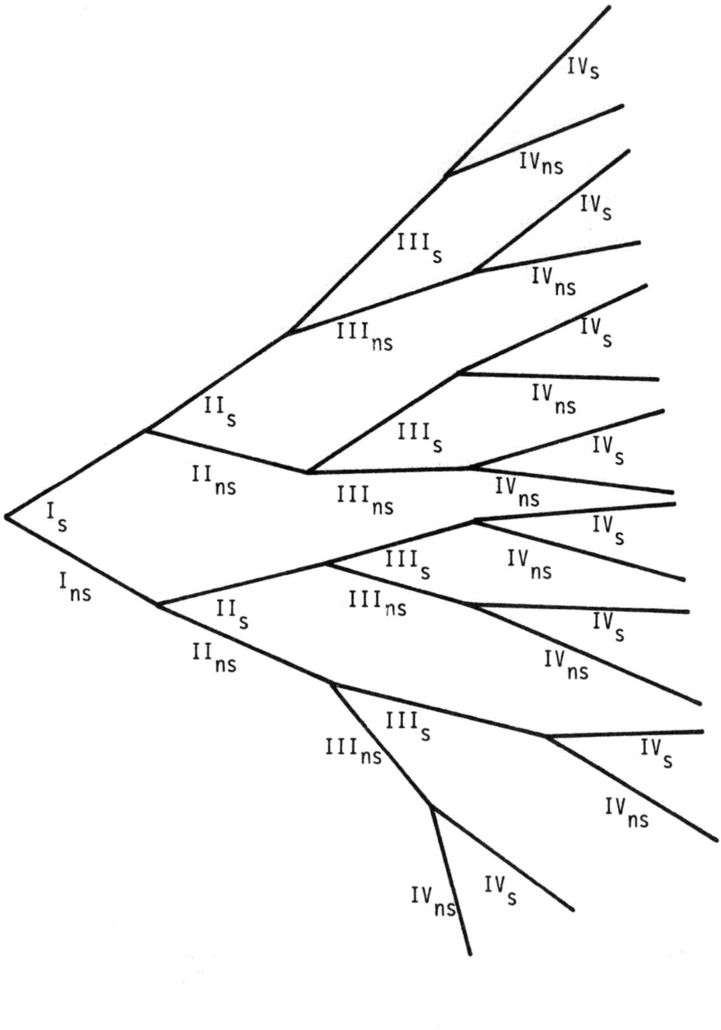

For the above stated reasons, use of a Bayesian analysis will result in conflicting probability judgments based on conflicting evidence, and up and down fluctuations for revised probabilities of hypotheses will result.

A solution which has been suggested is to try to develop hypotheses about the policies of different factions or about what hedging options are being considered. If we try to follow this suggestion, however, we probably will find it is too complicated. There are likely to be too many hypotheses to have to consider. For example, suppose we have just four factions, and for each faction we hypothesize that it is or is not supporting a particular policy (a separate policy for each faction). Then we would end up with a 16 path tree, where I_s means faction I supports a particular policy, I_{ns} means faction I does not support a particular policy, II_s means faction II supports a particular (possibly different) policy, II_{ns} means faction II does not support a particular (possibly different) policy, etc. Figure 11.1 displays the resulting tree.

Each path on this tree represents a hypothesis about a factional pattern of support for up to four different policies. So, when we get E_i, we would have to ask $P(E_i/H_j\&E_1\&E_2\&\cdots\&E_{i-1})$ for j = 1 through 16. This could be too much to do efficiently!

It might be easier to deal with possible hedging options, but if the number of possible hedging options is, say, just three, we still would get a substantial tree when we add these onto our starting hypotheses about what we think the options are.

The only solution may be to "grin and bear it," i.e., to understand that since (1) governments really aren't monolithic and (2) they hedge their bets, our trend line will go up and down when we try to interpret evidence with respect to "the" government policy -- because the evidence actually either (1) is reflective of the policies of different government factions or (2) is reflective of hedging actions.

A closely related problem is the following "non-stationarity of hypotheses" problem.

PROBLEM 51. Non-stationarity of hypotheses: the problem of a change in the state of nature.

The problem: Because actors (e.g., governments) can and do change their minds, the true state of nature with respect to which policy option an actor is pursuing can change. Thus, a hypothesis which reflected the previous, true state of nature will no longer be "true." The reason that this is a problem is that it may be a while, before enough data reflective of the new state of nature are collected to show a shift away from the old hypothesis to one reflective of the new reality.

The solutions: There are at least four ways of trying to deal with this problem. They are as follows:

a. Don't worry about the problem, but instead count on the sensitivity of Bayes to show a change as soon as diagnostic data are received.

b. Drop out all evidence older than a specified period, say six months, so that newer evidence will have greater weight and thus cause a quicker shift. If this is done, I strongly recommend running two or three parallel Bayesian analyses.

c. Project trend lines so that when a shift occurs we will anticipate further movement in the same direction.

d. Try somehow to put in the probabilities of a non-stationarity situation.

None of these solutions is terribly satisfactory. My preference order is a, b, c, d,.

Discussion: In the usual applications of Bayes, one hypothesis about the state of reality (or state of "nature," to use the more common phrase) really is true (e.g., in an urn problem we are drawing from urn I) and this remains true -- the state of nature is fixed (i.e., we continue to draw from urn I) and, while we don't know which hypothesis is true (i.e., whether we are drawing from urn I or, say, from urn II), we make an initial assessment of the probabilities of the hypotheses about the state of nature (i.e., which urn we are dealing with) and then modify the probabilities in light of evidence from the state of nature.

But suppose that, instead of dealing with a physical state of nature (where, for example, there either is oil underground or not) or an artificial problem (where, for example, we are drawing from urn I or not), we are dealing with an actor (e.g., a government.) Actors such as governments can and do reconsider decisions and can and do change their minds about policy options, so the "true" hypothesis, which is reflective of the true "state of nature," can and may actually change through time. (It is the state of nature that has changed (that has been non-stationary) rather than the list of hypotheses about the state of nature.) This means, of course, that the hypothesis which best reflected the old state of nature no longer is correct, and that another hypothesis, reflective of the new state of nature, now is the "true" hypothesis.

The reason for saying there is a "problem" associated with this non-stationary situation is that it may take a while for evidence to accumulate to indicate this change in the state of nature, and further that this indication of a change will be inhibited by the "weight" of the past evidence which favors the previous, no longer true hypothesis about the state of nature. Thus, it may be a while before enough new, diagnostic evidence occurs to lead to revised probabilities favoring the hypothesis which is reflective of the new "reality," i.e., the new state of nature. The result is a sort of lagged fever thermometer.

There are several "solutions" to this problem. None is, I think, very satisfactory.

a. The first "solution" simply is to not worry about the problem, but instead count on the inherent advantage of Bayes (cited earlier) of quickly changing probabilities - revised probabilities can change greatly on the basis of only a few items of diagnostic data. As soon as diagnostic data arrive, we can expect our revised probabilities to shoot up or down. The real question is how soon we will get diagnostic data. If it could be a while, then there might be a problem (possibly a severe one). In an indications and warning situation, a last minute change of mind by a government might be reflected slowly, and we might fail to predict an attack.

Because of this, there are several additional proposed "solutions" that can be (and have been) used. Of the following, my preference is the first.

b. This "solution" is to decrease the weight of old evidence by dropping out the probability judgments of evidence after it is over a certain period of age -- perhaps over six months -- thus letting new evidence have greater weight. (This so called "sliding window" procedure,[1] of course, really lets all evidence subsequent to that dropped have greater weight.) Several points should be made about this procedure. The procedure as used really does two things: it drops out the probability judgments of evidence over a certain period of age <u>and</u> it no longer considers this old evidence in making subsequent probability judgments.

This second aspect of dropping evidence almost never is mentioned because conditional independence has been used so extensively. But if we adopt the suggestion of this book and choose to use the conditionally dependent form of Bayes, we have several choices. First, we can drop out the probability judgments of old evidence and also not consider this old evidence when making conditionally dependent judgments for new evidence. Second, we can drop out the probability judgments of old evidence but still consider this old evidence when making conditionally dependent judgments for new evidence. This second choice may not give as much "weight" to new evidence and may cause us to interpret new evidence in a way more consistent with the no-longer valid hypothesis rather than the now valid hypothesis.

However, before rejecting this second choice, we should note the point raised by Dr. John Kringen in a class discussion. He pointed out that the procedure of dropping out old evidence (where implicit in his argument was the dropping of consideration of this old evidence in making new probability judgments) contains an implicit assumption. The assumption is that evidence loses its value ("decays") at a uniform rate -- that is, that <u>all</u> the evidence over a certain period of age (say six months) is equally non-diagnostic and therefore equally deserving of being dropped out. This almost certainly will not be true. A somewhat time-consuming solution for dealing with this faulty assumption is to run parallel Bayesian analyses and compare the results.

There actually are five logically possible Bayesian analyses that we can carry out for purposes of comparison.

(1) Conditionally dependent Bayes, where the probability judgments for old evidence are deleted and this old evidence is <u>not</u> considered in the probability judgments for new evidence.

(2) Conditionally dependent Bayes, where the probability judgments for old evidence are deleted, <u>but</u> this old evidence <u>is</u> considered in the probability judgments for new evidence.

(3) Conditionally dependent Bayes, where probability judgments for old evidence are <u>not</u> dropped and where we continue to consider

this old evidence when making probability judgments for new evidence.

(4) Conditionally independent Bayes, where probability judgments for old evidence are not deleted, but where (automatically) this old evidence is not considered when making probability judgments for new evidence because of the use of the conditionally independent form of Bayes.

(5) Conditionally independent Bayes, where the probability judgments for old evidence are deleted (and the old evidence automatically is not considered when making probability judgments for new evidence because of the use of the conditionally independent form of Bayes).

We derive the five possibilities from the tree shown in Figure 11.2. Clearly, (5) and (3) are at opposite extremes, and (5) will

FIGURE 11.2
Tree Diagram of Logically Possible Bayesian Analyses that can be Used for Purposes of Comparison

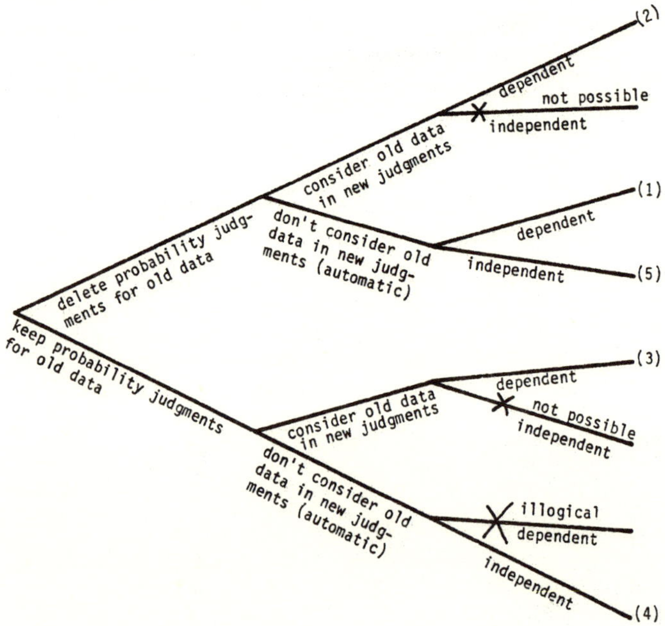

give the greatest weight to recent evidence, while (3) will give the least weight to recent evidence (and the greatest weight to old evidence).

Since I personally favor the assumption of conditional dependence, my suggestion is to run and compare at least analyses

(1) and (3). When doing this, of course, two conditionally dependent judgments will have to be made for each new item of evidence; the judgment for (1) will not consider old evidence, but the judgment for (3) will consider old evidence. There is a real possibility here of one judgment influencing the other, and it seems best to make the judgment for (1) and then the judgment for (3) to minimize this possibility.

By the way, it should be noted that it is almost as easy to compare (1) and (2) and (3) as to compare (1) and (3). This is because, as the tree shows, the conditionally dependent judgments for new evidence made for (3) are the same judgments we make in using (2). The best solution actually may be to compare (1) to (2) to (3). Analysis (3) will put the least "weight" on new evidence, analysis (2) will delete probability judgments for old evidence (giving greater weight to more recent evidence), and analysis (1) will give the greatest weight of the three analyses to more recent evidence. This sort of comparison, however, still may be very time consuming. Computer automation of trend line graphing will be of great help in comparing the results.

In defense of alternative (5), student Norman Allred has pointed out that, if there has been a shift in the true state of nature, then use of the conditionally independent form of Bayes means that new evidence will not be interpreted in light of the old evidence and the analysis may reflect the change more quickly than if alternatives (3) or (2) are used. That is, use of the conditionally dependent form of Bayes in alternatives (3) or (2) means that, for a new E_i, when we ask $P(E_i/H_j \& E_1 \& E_2 \& \cdots \& E_{i-1})$ we will be trying to interpret E_i in light of evidence that tends to support a hypothesis reflective of an old state of nature. Our probability estimate for E_i may then be "distorted" somewhat. If we drop older data, and utilize alternative (1), conditional dependence will be less of a problem but still could be somewhat of a problem if the shift in the state of nature was more recent than the dropped out evidence. That is, if we drop out evidence over six months old and ask, say

$$P(E_i/H_j \& E_{20} \& E_{21} \& \cdots \& E_{i-1})$$

where E_1 through E_{19} are over six months old, but E_{20}, etc. are less than six months old, then, if the change in the state of nature occurred three months ago, we still would be considering data reflective of an old state of nature and this could bias our probability judgments. Still, because of its realistic nature, I continue to support use of the conditionally dependent form of Bayes and suggest we compare analyses (1), (2), and (3). (Furthermore, as we will discuss below, use of the conditionally dependent form of Bayes may inhibit somewhat the effectiveness of deception efforts.)

c. Another proposed solution to the non-stationarity of hypotheses problem is to project the Bayes trend lines for hypotheses. If we have been plotting the Bayes lines and they begin to turn up or down, we might want to extend them on past the last point plotted to suggest the future course of events, arguing that we may have had a change in the state of nature and hence that new evidence is reflecting and will continue to reflect this change. We should be somewhat careful here, because, mathematically speaking, projection of a line is not justified. The plotted revisions of hypotheses reflect all the "information" that we can extract mathematically from the data (revisions, it should be remembered, that already are in advance of our intuition). If the trend lines have had a tendency to shift frequently (perhaps reflecting the competing views of different factions as the last problem mentioned), then we definitely should not project the trend lines.

Frankly, I have grave doubts about ever projecting trend lines, partly because we can turn the non-stationarity argument around against the practice by saying that the "trend" could be interrupted by a "non-stationarity" situation, so it is best not to try to project. Student Thomas McIntyre also pointed out that we have shown that one of the great merits of Bayesian analysis is the faster-than-intuition revision of probabilities. "So," he said, "to project a historical trend in light of having a tool which can drastically change the direction of the curves (trend lines) is to ignore the primary usefulness of the tool."

d. Another "solution" that has been suggested is to try somehow to include the probability of a "non-stationarity of hypotheses" situation in the analysis. It is not clear why, however, if the data at hand are enough to indicate that a non-stationarity of hypotheses situation may exist, the data are not enough to affect the trend lines.

There is one subtlety that has been glossed over in the entire discussion up to this point. The statement has been made that a non-stationarity of hypotheses situation means that the true state of nature has changed. It was then stated that it may be a while before enough evidence reflective of the new state of nature is collected to show a shift away from a hypothesis consistent with the old state of nature to a hypothesis reflective of the new reality.

But is there necessarily a hypothesis in the original list of hypotheses that is <u>clearly</u> reflective of the new state of nature? If our original hypotheses were properly worded so that the alternatives were mutually exclusive and exhaustive, then at least one hypothesis will be reflective of the new state of reality . But student Bob Wright pointed out in class that what might be involved here is a shift in the buried subhypotheses of the residual hypothesis. For example, suppose the residual hypothesis is a "no attack" option. The true state of nature might be one form of "no attack" so that this hypothesis is indeed supported by the data. Later on, however, the government policy might shift to a different form of no attack; the state of nature has indeed shifted, but this would not show up.

This may, however, not be a real problem; our decision maker simply may not care if there are changes in the state of nature with respect to the residual hypothesis. It may be the "desirable" hypothesis, but only hypotheses concerning its absence may be of enough interest to require an analysis of a spectrum of possibilities. In short, subhypotheses about the residual hypotheses may be of no interest to a decision maker. If they are, then it is fairly easy to develop subhypotheses when they are desired. (See problem 2.)

Wright also pointed out, however, that this situation of "no shift" in the favored hypothesis even after a change in the state of nature could apply to a non-residual hypothesis. If we build our hypotheses without seeing that a particular one has certain subhypotheses buried in it, then we almost certainly wouldn't assign our probabilities in a way consistent with a shift in reality from one buried subhypothesis to another.

That is, we actually would be assigning probabilities on the basis of one implicit subhypothesis and might not detect the change in the state of nature to the reality expressed in another subhypothesis simply because we never explicitly conceptualized either of the buried subhypotheses. This seems to be an almost inevitable problem, however. We simply can't consider dozens of hypotheses; even if we could, there still would be buried subhypotheses - they are inevitable. The only "solution" to all this is to introduce subhypotheses if it seems useful to do so, which is a judgment which will be based partly on our perception of the needs of the consumer of our analysis.

PROBLEM 52. "Life-span of evidence" (or "impact of past information"): the problem of when and why and to what extent the probability judgments for "old" items of evidence should be deleted from the analysis (and whether the "old" evidence should be considered in making probability judgments for "new" items of evidence).

The problem: We already have discussed two situations in which it sometimes is suggested that probability judgments for "old" items of evidence should be deleted from the analysis (and where there is also a question of whether the "old" evidence should be considered in making probability judgments for new items of evidence).

The first situation is when we use a sliding window in phrasing a hypothesis (problem 7). The main question is the appropriate age for dropping evidence. The second situation is where we are concerned about the "non-stationarity of hypotheses" problem. Again, there is the question of the appropriate age for dropping evidence.

There may be a third situation when it may be desirable to delete the probability judgments for "old" evidence (and not to consider "old" evidence in making probability judgments for new evidence) even if we neither are using a sliding window nor worrying about a non-stationarity of hypotheses situation. This situation is where some of the "old" evidence would be reinterpreted (i.e., given different probability assessments, if we were to go back and reexamine it). This "old" evidence then can be viewed as "distorting" our revised probabilities of hypotheses, so we might want to adopt a policy of deleting the probability judgments for "old" data (and possibly not considering the "old" data when making probability assessments for new evidence). The question again arises of how old it should be.

Because of the almost exclusive use of the conditionally independent form of Bayes, implicit in the literature's discussion of each of these three situations is the idea that we also will not consider the old data when making probability judgments for new evidence. However, as we saw in our discussion of the non-stationarity of hypotheses problem, we have a choice if we use conditional dependence of considering or not considering the old data when we make our conditionally dependent probability judgments for new evidence.

The solution: In all three situations, what is "old" is a matter of judgment, and can not be answered definitively here. Furthermore, while dropping out the probability judgments for old data may be somewhat helpful, it may be neither necessary nor desirable. If we do want to consider the effect of dropping out the probability judgments for "old" data, then I strongly suggest running the three parallel Bayesian analyses (all using conditional dependence) discussed in the "non-stationarity" of hypotheses" problem above. The first will delete the conditionally dependent probability judgments for old data and also will not consider this old data when making the conditionally dependent probability judgments for new evidence. The second will delete the conditionally dependent probability judgments for old data, but still will consider this old data when making the conditionally dependent probability judgments for new evidence. The third will include the conditionally dependent probability judgments for "old" data and will continue to consider this old data when making conditionally dependent probability judgments for new evidence. Only by comparing the results of these three analyses can we determine the impact of "old" data.

Discussion:

The First Situation

In problems 6 and 7 we dealt with the sliding window concept, where we would phrase a hypothesis like "state A has decided to attack state B within the next 30 days." As time passes, the time frame of the "next 30 days" moves (slides) forward into the future. Some analysts, e.g., Schweitzer, raise the issue of dropping out the probability judgments for older data "evaluated months earlier with regard to an earlier frame of possibilities."[2]

The question, of course, is how "old" the data should be. If the hypotheses talk about something happening in the next 30 days, should we drop the probability judgments for data older than 30 days, for example? There is no obvious logical reason why, if we choose to drop out the probability judgments for "old" data, the age of the data has to match our "sliding window" time period. We could have, for example, a hypothesis saying that "state A has decided to attack state B within 30 days" and decide to drop out probability judgments for data older than 60 days.

Actually, there are two important questions. First, what is "old" data? Second, is it really necessary or desirable to drop out the probability judgments for old data? The answer to the first question is that there seems to be no obvious answer. It is a matter of judgment, and the "judgment" may be more intuitive than analytical. Suggestions that "old" data are data over 3 months or 6 months old seem based on little else but "gut" feelings. As was noted in our discussion of the non-stationarity of hypotheses problem, there is perhaps an unwarranted assumption that "old" data lose their value at the same rate, so that the probability judgments for _all_ evidence older than a particular period (say 6 months) deserves to be dropped.

Since there almost certainly will be doubt about this, and since there is no obvious answer as to what is "old" data, it seems that we really should run the three parallel Bayesian analyses (all involving conditional dependence) discussed in the "non-stationarity of hypotheses" problem above. The first will delete the conditionally dependent probability judgments for "old" data and also will _not_ consider this old data when making the conditionally dependent probability judgments for new evidence. The second analysis responds to the desire to delete "old" data by deleting the conditionally dependent probability judgments for old data, but still will consider this old data when making the conditionally dependent probability judgments for new evidence. The third analysis, which is just the normal Bayesian analysis, will include the conditionally dependent probability judgments for "old" data and also will continue to consider this old data when making the conditionally dependent probability judgments for new evidence. (These judgments will be the same as in the second analysis.) Only by comparing the results of these three analyses can we determine the impact of "old" data. (The first analysis here actually should be done first in order to avoid the "contamination" that could occur from previously considering the old data as is required in the second and third analyses listed.)

The second question, however, is whether it really is necessary or desirable to drop the probability judgments for "old" data, whatever we might (arbitararily) decide "old" means. We have noted many times the advantage that Bayes quickly changes probabilities -- the revised probabilities of hypotheses can change greatly on the basis of only a few items of diagnostic data. As soon as diagnostic data appear, the revised probabilities of hypotheses can be expected to shoot up or down quickly. The non-conservative nature of Bayes means then that new evidence which is diagnostic will take us quickly in the direction it should, and the new evidence thus will "swamp" not only the prior probabilities but also will "swamp" the probability judgments for the previous "old" data. So, the problem of "old" data may not be particularly severe, and deleting the probability judgments for "old" data may not be necessary.

Moreover, we have suggested that an intuitive view of Bayes, consistent with including old data, is that of a lagged fever thermomenter. That is, Bayesian analysis yields an up and down trend line (graph) of current probabilities of hypotheses and is like a fever thermometer, although the analogy is not perfect since the current "temperature" is affected by past "readings," i.e., previous data. Although there may be a "lag" because of old data, the "swamping" aspect implies that the general trend (direction of change) normally will be unaffected. Therefore, again, it can be argued that it may not be necessary to drop the probability judgments for old data. Furthermore, it may be desirable to consider an "old" item of data because data probably do not "decay" at a uniform rate, and a particular item of old evidence might be very important in our assessment of a recent item of evidence.

All things considered, the best thing to do probably is to run, as described above, the three parallel Bayesian analyses.

The first parallel analysis lets us examine the effect of deleting the probability assessments of the old data while also not considering the old data when making assessments for new items of evidence. The second parallel analysis suggested above is a compromise of sorts because it deletes the probability assessments involving the old items of evidence, but continues to consider these old items of evidence when assessing the probabilities of new items of evidence.[3]

The Second Situation

We have discussed earlier in detail the idea of dropping "old" data as a way of dealing with the non-stationarity of hypotheses problem. (It should be noted again that there is no obvious rule for deciding what is "old," and judgments tend to be based on gut feelings.) The comments in the discussion of the first situation apply equally here.

The Third Situation

Finally, turning to the third situation where we may want to drop the probability judgments for "old" data, we have the argument that we should delete "old" evidence, some of which would be reinterpreted (i.e., given different probability assessments if we were to go back and reexamine it). Such old evidence can be viewed as having inertia and distorting our revised probabilities of hypotheses, so we might want to adopt a policy of dropping out the probability judgments for this "old" data. Moreover, a related line of argument points out that including the probability judgments for all data means that the probabilities associated with an event which occurred three or even six months ago, for example, have just as much weight in the final calculation of current revised probabilities of hypotheses as do the probabilities assigned to the most recent event. Shouldn't we be able to count later events more heavily, by dropping the probability judgments for "old" data, the argument goes, especially since the "old" events were relevant to situations in the past, but may not be as relevant to present conditions (even if a non-stationarity of hypotheses situation has not happened)?

The answer and discussion in essence are the same as for the first situation. We can only emphasize once more that parallel Bayesian analyses seem appropriate.

An additional issue that needs to be raised with respect to old evidence is that of deception. Students have pointed out that if we have been desensitized by the actions of an actor through time, then (assuming that we use the conditionally dependent form of Bayes - and consider old data when making the probability judgments for new evidence), old data may be an advantage -- it may help to reveal a pattern in the evidence and limit the amount of change in perceptions that the desensitization procedure was intended to accomplish. For example, if an actor has engaged in certain military actions for a long time, these actions might have had a significant diagnostic impact indicative of hostility long ago, but not have much impact now. The actor may have upped its military preparations to the point where it has to do very little to attack - so a continuation of the actions has little, if any, diagnostic impact. If we then do not consider the old evidence (which might have been diagnostic of hostilities), when we make our probability judgments for new evidence, the impact will be to make conflict look even less likely. For this reason, again the three parallel Bayesian analyses we have described above seem to be called for.

(Also, if we suspect the possibility of such a desensitization program by the enemy, we should focus very hard indeed on finding diagnostic indicators and collecting data on them.)

An argument in favor of <u>not</u> considering old data is that, if deception is being used against us (and, again, we are using the conditionally dependent form of Bayes), then failure to consider old data may break the pattern slightly.[4] A counter-argument is that a complete view of the record may better reveal the deception effort. The conclusion still is that we need to use the three parallel Bayesian analyses and very carefully compare the results.

<u>PROBLEM 53</u>. Revising previously assigned probabilities: <u>the problem of</u> deciding under what circumstances previously assigned probabilities of events should be revised.

<u>The problem</u>: As we saw above in our discussions of <u>problems 10, 26, and 30</u>, we may find out that a previous item of evidence either didn't happen at all or happened differently than was reported previously. If we are using the conditionally dependent form of Bayes, it then is necessary to reassess all the probability judgments for subsequent events. What, then, are the other situations where reassessment may be necessary?

<u>The solution</u>: Unfortunately, a complete listing cannot be given. <u>One additional</u> case can be discussed, however.

<u>Discussion</u>: A complete discussion has been provided above of the <u>need for r</u>eassessment when the conditionally dependent form of Bayes is used and a previous item of evidence is found later

either to have not happened at all or to have been different than described. The question here, then, is what are the other circumstances where a reassessment of previously assigned probabilities may be necessary or desirable. A complete list of such circumstances is beyond our ability to provide. We will discuss one additional case of which we are aware.

This situation, suggested by a student (unfortunately, I did not document which student) is where, in doing an analysis, two items of evidence E_i and E_k have been considered, but no particular interrelationship has been noted. That is, when we assessed, for each j, $P(E_k/H_j \& E_1 \& E_2 \& \cdots \& E_i \& E_{i+1} \& \cdots \& E_{k-1})$, we found that E_i was not particularly critical in assessing the probability of E_k. Suppose later, however, that an item of evidence, E_x, occurs that in a sense "completes the jigsaw puzzle" linking E_i, E_k, and E_x. In other words, in retrospect, E_i and E_k now clearly are interrelated, and the probability of E_k is profoundly influenced by the fact that E_i happened. But, it took the occurrence of E_x for the interrelationship to become clear. Should we now go back and reassess, for each j, $P(E_k/H_j \& E_1 \& E_2 \& \cdots \& E_i \& \cdots \& E_{k-1})$? Furthermore, should we not reassess for all subsequent events E_{k+1}, E_{k+2}, etc. because the completion of the jigsaw puzzle might affect these sequentially later probability judgments?

The answer is clearly one of judgment, and a definitive answer at this point would be a dogmatic one. My personal feeling is that I would <u>not</u> go back and reassess. Bayes, as noted many times, is highly sensitive. I suspect that, if E_x completes the jigsaw puzzle linking E_i and E_k, then the assignment of the probabilities for each j of $P(E_x/H_j \& E_1 \& E_2 \& \cdots \& E_i \& \cdots \& E_k \& \cdots \& E_{x-1})$ would be very diagnostic, so that the completion of the puzzle will have a major impact without the necessity of going back and doing the great amount of work necessary for reassessments. Furthermore, I think that this phenomenon of completing the jigsaw puzzle is one that always is underway. We could be reassessing constantly if we looked back after each item of evidence and asked what sorts of interrelationships among previous events it revealed that we had not discerned previously. In practice, such constant reassessments would require too much time to be practical.

Finally, it can be argued that doing such reassessing does some violence to the deductive spirit of Bayes. It can be argued that we always are asking the probabilities of events given what we know at the time about previous events. Bayesian analysis is, then, to use our previous analogy, like a lagged fever thermomenter, and reassessing in a sense is a rewriting of history. I frankly find this final argument to be weak. If we somehow later find something (E_x) that shows the interrelationship between E_i and E_k, then the probability of E_k is indeed affected by the occurrence of E_i, and we might want to go back and recognize this. Still, in all, the inordinate time which would be required to be engaged constantly in such reassessments would seem to rule out going back.

PROBLEM 54. New prior probabilities: the problem of when and why to abandon the original prior probabilities and start a new analysis with new priors.

The problem: A number of arguments can be given for abandoning the original prior probabilities and starting a new analysis. Perhaps the most prominent of these arguments is that after a long period there will be a lot of "old" evidence in the analysis, some of which would be reinterpreted (i.e., given different probability assessments) if we were to go back and reexamine it. Thus, the argument goes, this old evidence is "distorting" our revised probabilities for hypotheses.

The solution: Those making the above argument suggest that, instead of using a sliding window, we simply "start over" with new prior probabilities.

Discussion: This issue of the purported need to "start over" with new prior probabilities has been raised occasionally, but it is not completely clear that there really is any problem at all. As noted many times, Bayes is a very sensitive instrument, and the prior probabilities will be swamped by diagnostic data. Even if there is a growing body of "old" data which would be reinterpreted if we were to go back and make reassessments, newer, diagnostic data will tend to swamp this old data, too. A sliding window can drop out really old data. On the other hand, if we are under pressure from above to start over or want to for various reasons (e.g., a lot of recent data has been found to be bad, and we don't feel like doing a lot of reassessing), then we should not be overly upset about starting a new analysis. That is, we know the new priors also will be swamped eventually by new data, so it generally is not terribly important whether we start over or not. A definitive list of all the circumstances under which we would want to abandon the original prior probabilities and start over cannot be given here. It is a matter of judgment; the basic point is that this probably is not a critical problem at all.

PROBLEM 55. The "reliability" of Bayes: the problem of the "reliability" of Bayesian analysis as a function of proximity to the actual occurrence of one of the hypothesized events and whether or not we can improve the reliability.

The problem: What we are asking here is simply how good a job Bayes does in "predicting" a hypothesized event as we get closer and closer to its actual occurrence and if we can improve the performance of the technique.

The "solution": We simply don't have enough empirical data to tell how good a job Bayes has done and can do. As for solutions, from our perspective, discussions of the problems of Bayes and suggested solutions can be seen as ways of improving the performance of the technique. No cure-all to the difficult problems of prediction can be expected, however.

Discussion: In the analysis of the problem of unrecognized "buried" hypotheses, we closed with an analysis of what we thought would be the trends in one possible set of hypotheses as time passed. In general, it can be argued that, as we get closer to the actual occurrence of the event specified in one of the hypotheses, Bayes will become more "reliable" or "accurate" because we will be getting a greater volume of evidence and more evidence with high diagnostic value. As we have noted many times, the major advantage of Bayes cited in the literature is that it will yield the "truth" well before the end, and certainly well before intuitive calculations will. Hopefully, then, the sort of picture in Figure 11.3 (suggested by Tom Murray)[5] will apply.

FIGURE 11.3
Diagram Showing how the Use of Bayesian Analysis will more Quickly Reveal the "Truth"

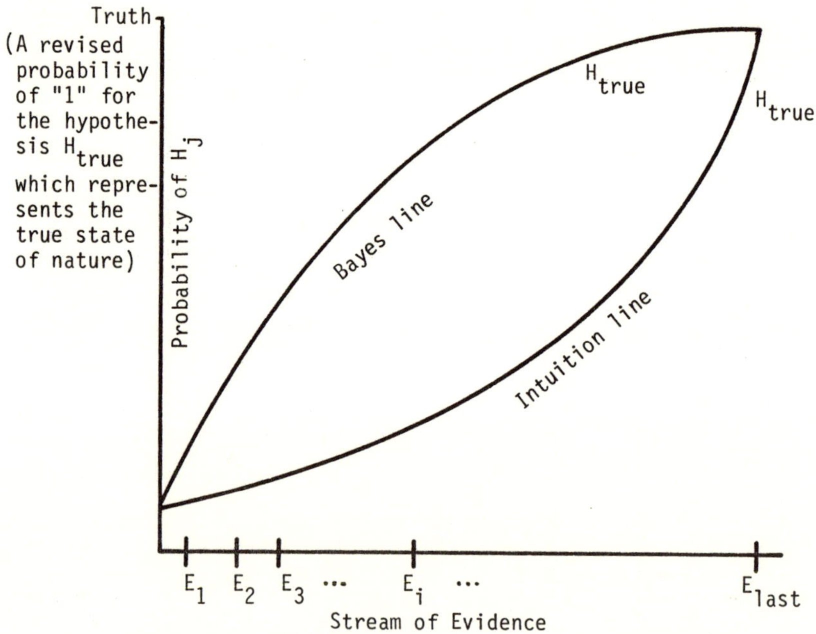

For a number of reasons, however, we might not find this ideal situation occurring. First, because of "noise" or the problem of a quiet period between the conclusion of preparations to attack and the actual attack (during which period it may appear there will be no attack), the Bayes line actually might be like Figure 11.4 for an attack hypothesis.

FIGURE 11.4
Possible Bayesian Graph of the Attack Hypothesis - A "Lull Before the Storm" Situation

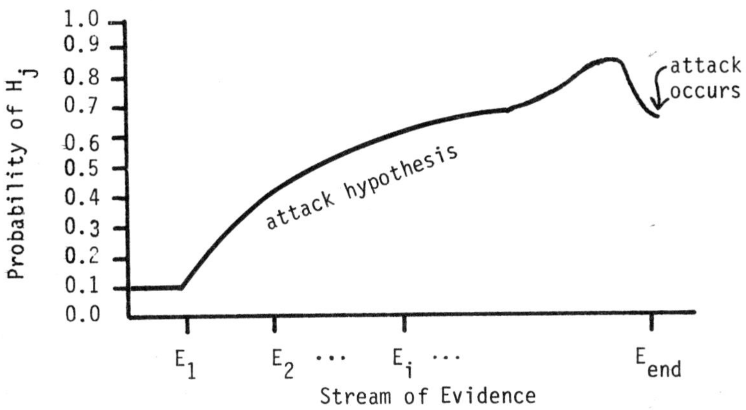

Dr. Hopple points out that this is a "lull before the storm" situation and suggests an investigation of how often it occurs. He notes there is a problem of pre-crisis peak and false alarm - common for extended crises, e.g., Czechoslovakia in 1968, Poland in 1980-1981. In the former we had a peak and attack, while in the latter several peaks and no attack. See Figure 11.5.

FIGURE 11.5
Bayesian Graph of the Pre-1968 Czechoslovakian Invasion Situation

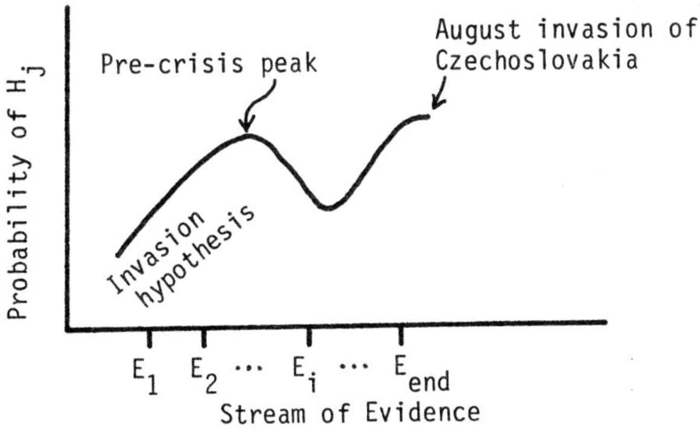

Second, another problem could be a non-stationarity of hypotheses situation. For example, the graph of an attack hypothesis might be like Figure 11.6.

FIGURE 11.6
Bayesian Graph of a Non-stationarity of Hypotheses Situation (e.g., a Last Minute Attack Decision)

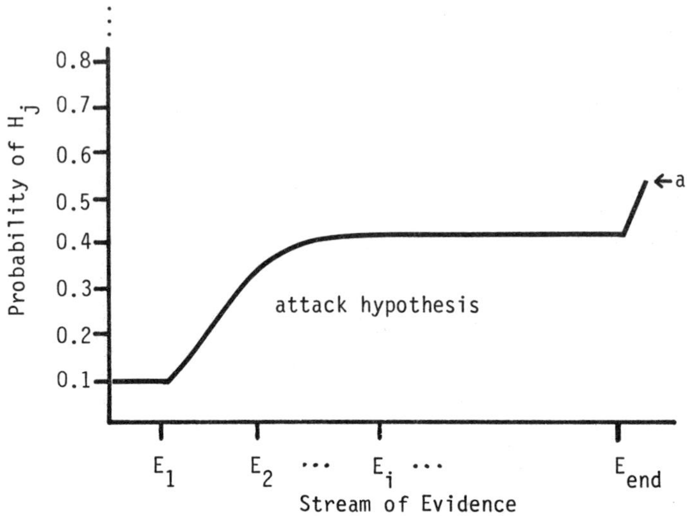

This situation could exist if a sudden change of mind to attack occurred, and there was not much time between the decision and the actual execution of the attack. Then there might not be enough time for evidence to be received to cause the trend line to change sufficiently to show that an attack now was likely. The discussion of the non-stationarity of hypotheses problem revealed no really good solution to this problem.

Third, as was pointed out in class by student Bob Wright, there could be a so called "paradox of warning situation." That is, the Bayes trend line for the hypothesis about the event that eventually occurred may appear to be greatly in error because our own decision makers, receiving this trend analysis of what seemed likely to happen, may have warned the actor (that the hypotheses were about) to refrain from doing what we though the actor was going to do. If the actor then responded by not doing what we feared, our previous Bayesian analysis would appear to have been "wrong". The picture could be as in Figure 11.7. (Note that this is a type of non-stationarity of hypotheses situation; it was mentioned at the end of problem 15.)

Dr. Hopple notes that "this issue could be discussed at more length. It is a recurring problem in intelligence analysis and explains the difficulty of coming up with (the many) intelligence 'successes' to juxtapose against (the relatively few) 'failures.' Good warning prevents the event and thereby prevents us from confirming the warning."

FIGURE 11.7
Bayesian Graph of a "Paradox of Warning" Situation (One Type of a Non-stationarity of Hypotheses Situation)

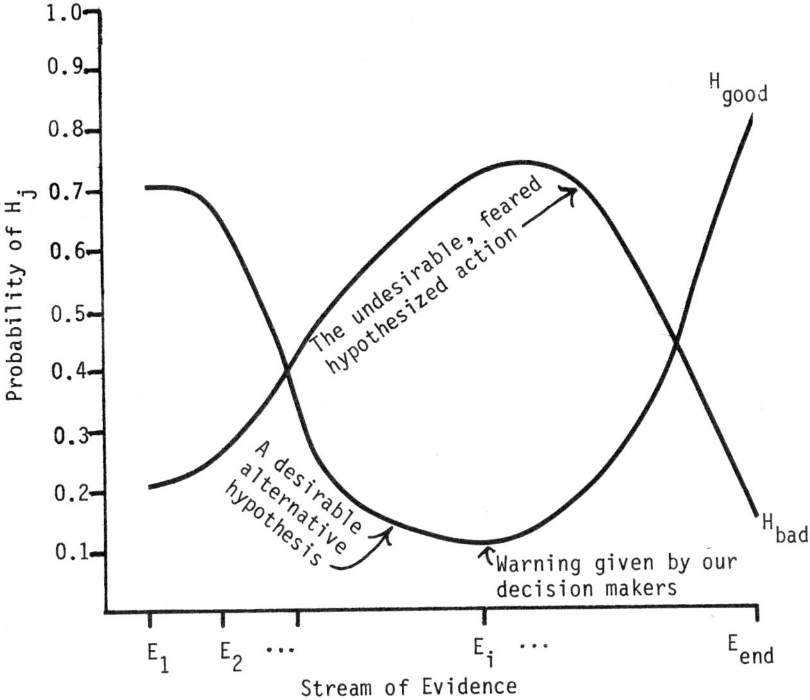

It should be noted here that Richard Betts has written brilliantly on the problems of predicting attacks,[6] and many of his arguments can be interpreted to suggest strongly that Bayesian analysis will be no cure-all, i.e., that the Bayes line will not and cannot unambiguously predict what will happen.

The real truth of the matter is that we simply do not have enough empirical data to tell how good a job Bayes has done or can do with respect to predictions in political/military analyses. As for ways to improve our ability to use Bayes to "predict" events, discussions of problems and possible solutions can be interpreted as ways to try to do better. Only empirical testing can tell if the technique really is working and if it is worth the effort.[7]

<u>PROBLEM 56</u>. Differences between intuitive and Bayesian <u>predictions</u>: the problem of how to deal with differences (particularly drastic differences between Bayesian predictions and predictions based on personal experience), when you are "sure" your analysis is better than the one based on Bayes.

<u>The problem</u>: As stated.

The solution: Student Martin Joyce commented as follows:

"Don't throw Bayes out the window in these cases, because it has provided you with an alternate view -- one either more, or less, radical than your own. If it varies drastically from your own viewpoint, it should force you to re-examine your own subjective analysis to make sure that you have sufficiently examined all previous evidence."

PROBLEM 57. Consumer bias: the problem of consumer bias, particularly either gullibility or cynicism.

The problem: While consumers may have a wide variety of biases about the product presented, it is quite common to run into biases of either gullibility or cynicism.

The solution: We should try to educate the decision maker as to the major problems involved with Bayes, and particularly to the sensitivity of Bayes and the importance of the trend lines of the revised probabilities rather than the exact revised probabilities.

Discussion: The teaching of mathematics in the United States is, in my opinion, a disaster. Instructional materials, teaching methods, and teachers have combined to produce a nation of mathematical illiterates. This is reflected in the decision makers in power, people who generally are quite intelligent, but also generally lack knowledge of mathematics and therefore have little confidence in their ability to understand presentations involving mathematics. All this often is reflected in a particular consumer's bias of either over-confidence in the accuracy of the incomprehensible mathematics (gullibility), or total disbelief in the "artifically precise" nature of the incomprehensible mathematics (cynicism). Both biases are dangerous, but the decision maker may be better off being cynical. I feel this way because of my belief that there is grossly inadequate analysis of the limitations of various mathematical techniques that are highly touted in Washington (D.C.) and that a strong dose of hard questioning of proselytizers of various techniques about the underlying problems of and assumptions of those techniques is long overdue. This book, really, is intended to offer the practitioner (and perhaps consumer) of Bayes some stern warnings about many of the basic problems and pitfalls that must be considered in order to do a good Bayesian analysis.

When presenting the results of a Bayesian analysis to a consumer, then, we should try to educate him (her) about some of the major problems involved with Bayes, and particularly about the sensitivity of Bayes and the importance of the trend lines of the revised probabilities rather than the exact numbers. The analyst who proudly states that, after 23 items of evidence, the revised probability of H_2 is exactly 0.87 deserves to be charged with the sin of false precision (and perhaps shot). The analyst should provide a statement of the high sensitivity of Bayes to minor variations in the assessed probabilities, and present the trend line along with the "best estimate" of 0.87, stressing the approximate nature of 0.87. Also, notes Dr. Hopple, "the analyst should emphasize the particulars of the case."

NOTES

Problem 51

1. SCHWEITZER (1978), p. 29. The term "sliding window" is unfortunate, since the procedure of dropping out old data can be used whether the hypotheses involve (sliding window) time frames or not.

Problem 52

2. SCHWEITZER (1978), p. 29.

3. Dr. Hopple notes that the second paralled analysis also is good because it supports the "multimethod ⟹ validation" idea.

4. Student Bob Wright commented that if we keep evidence too long then "recurring patterns could camoflage real intentions, e.g., Egyptian preparations for the 1973 Arab-Israeli War. (But) discard it too soon and potential diagnostic value could be lost." This "dilemma" can be alleviated by running the three parallel analyses.

Problem 55

5. Suggested in live and taped lectures he has given for the government on Bayesian analysis.

6. There are an increasing number of articles by him dealing incisively with the problems of predicting attacks. See, for example BETTS, RICHARD K. (1978) "Analyses, War and Decision: Why Intelligence Failures are Inevitable." World Politics 31 (Oct.): 61-89.

BETTS, RICHARD K. (1980) "Surprise Despite Warning: Why Sudden Attacks Succeed." Political Science Quarterly 95(4): 551-572. Also see his book BETTS, RICHARD K. (1983) Surprise Attack: Lessons for Defense Planning. Washington, D.C.: Brookings.

7. Dr. Hopple notes that this "relates to the much larger and very basic issue of intelligence systems' effectiveness and performance."

Appendix A: The Use of a Probability Tree in Prediction

Using a probability tree diagram for prediction is quite easy. In Bayesian analysis, as we have seen, the hypotheses come at the beginning of the tree and we revise their initial probabilities in light of the evidence which comes later in the tree. In using a probability tree diagram for prediction, however, we put the hypotheses at the end of the tree, and the factors which influence the hypotheses make up the rest of the tree. We put the factors down in the temporal order in which we expect them to occur or in the causal order in which we believe they impact, one on another, in sequence.

To make the above clearer and to explain what to do subsequently, an example will be helpful.[1] Suppose we are trying to predict whether or not there will be a coup in a particular third world country. So our hypotheses are

C = there will be a coup in the country, and

not C = there will not be a coup in the country.

The hypotheses must be mutually exclusive and exhaustive. (It is customary not to use the H_1, H_2, ··· notation of Bayes, however.)

Suppose that after some thought it is concluded that the factors[2] that are most important in determining whether or not there will or will not be a coup are

(a) the strength of the economy,
(b) the degree of support in the country for a particular general, and
(c) the degree of threat of aggression from an adjacent country.

Further, we feel that factor (a) impacts on (b) which impacts on (c).

We next have to decide how many mutually exclusive and exhaustive parts to break each factor into. Here, we simply dichotomize each factor. So, we will have these possibilities: the economy can become strong or weak, E_s or E_w; the support for the particular general can become strong or weak, G_s or G_w; and

FIGURE A.1
Probability Tree Diagram for the Possibilities of having a Coup or not having a Coup

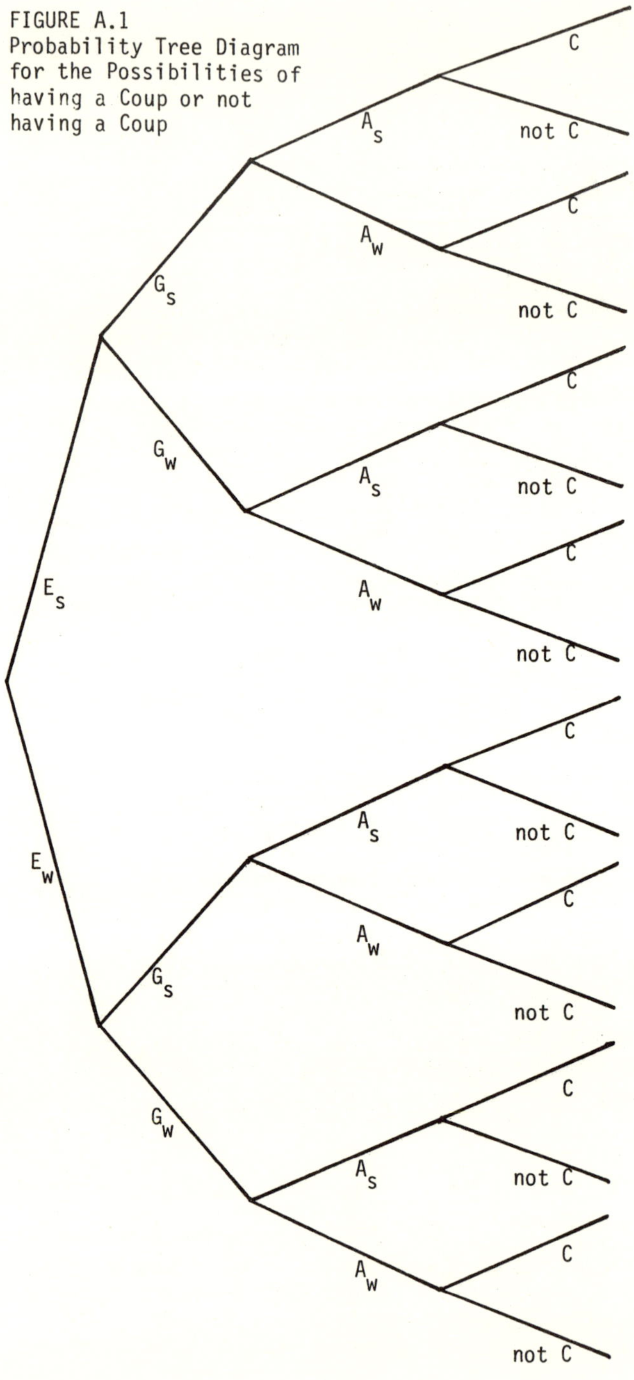

the degree of threat of aggression from the adjacent country can become strong or weak, A_S or A_W.

We build our probability diagram as shown in figure A.1.

We then assign the conditional probabilities to all the branches as shown in Figure A.2 and calculate the probability of a coup (i.e., C) by adding up the probabilities of all the paths containing C (i.e., in this case, ending in C), getting 0.67. The probability of no coup, which we can calculate the same way, is 0.33.

To recapitulate, our steps are as follows:

a) Decide what subject we wish to make a prediction about and decide on a set of mutually exclusive and exhaustive hypotheses specifying the possible outcomes with respect to that subject.

b) Decide which factors will have the greatest influence on the subject the hypotheses concern.

c) Decide on the temporal or causal order in which the factors will occur or impact, one on another, in sequence.

d) Decide on how many mutually exclusive and exhaustive parts to break each factor into.

e) Construct the tree, left to right, starting with the first factor, and having one branch for each of the mutually exclusive and exhaustive parts each factor is broken into. (We remember that in a properly constructed probability tree, from each node the alternatives are mutually exclusive and exhaustive.) Note carefully that <u>all</u> of the possibilities on the tree refer to events that <u>will</u> happen in the future; the tree is completely forward looking.

f) Assign the <u>conditional</u> probabilities to each branch on the tree. (Note that since from each node the probabilities add to "1," the probability of the last branch from any given node is determined once the probabilities for all the other branches are assigned.)

g) Find the (overall) probability of each hypothesis; to do so use Rule 15 and add up the probabilities of all the paths containing that particular hypothesis.

NOTES

1. The example is drawn from Anonymous (1973) "Probability Diagrams." Central Intelligence Agency, Office of Current Intelligence, OCI Notes on Methodology No. 8, Oct 1, Langley, Va. For further discussion of probability diagrams, see Decision and Design's Handbook for Decision Analysis (Barclay, <u>et</u>. <u>al</u>. (1977)).

FIGURE A.2
Assignment of all Conditional Probabilities and Path Probabilities

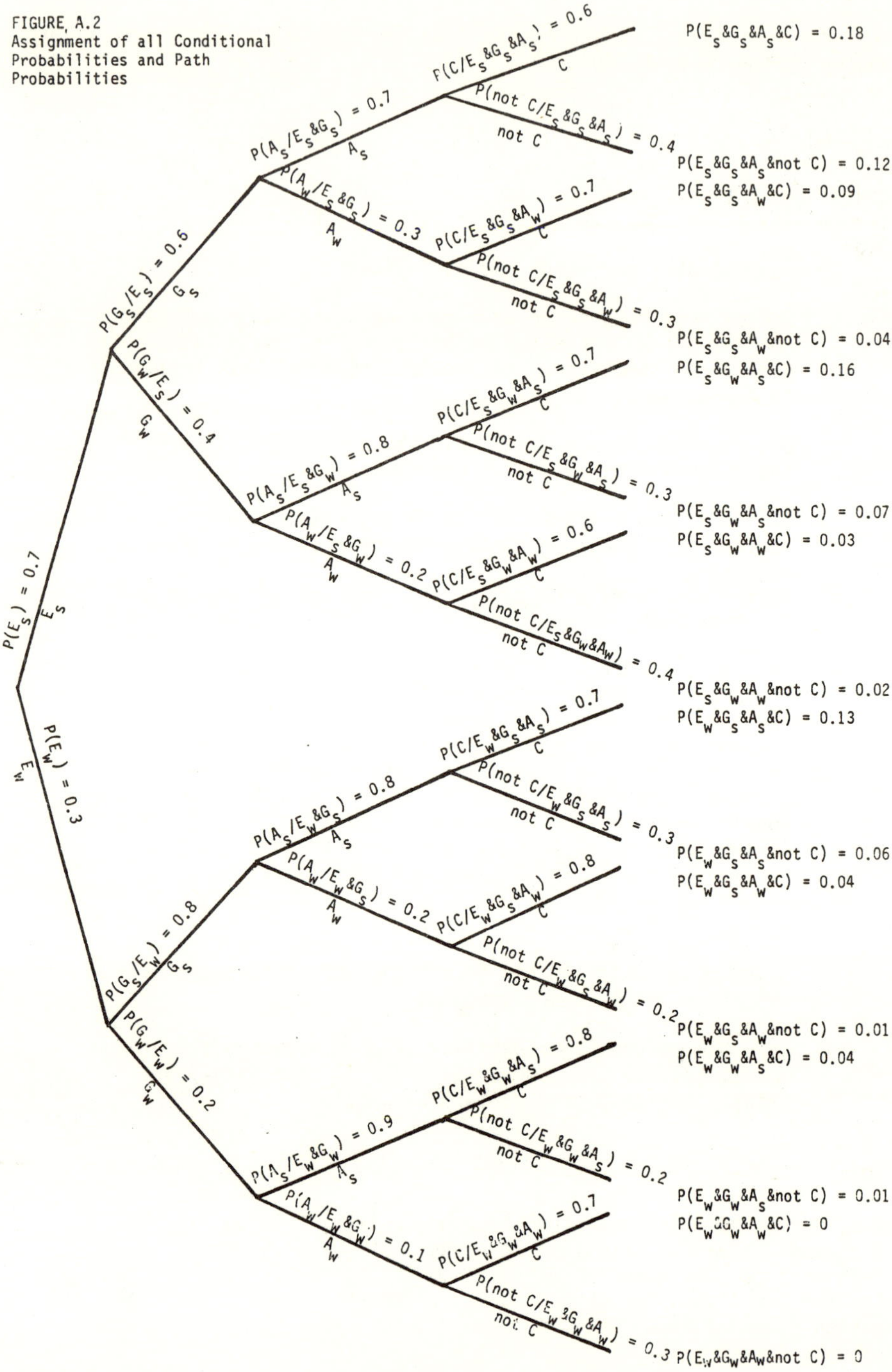

2. I use the word factors here instead of "independent variables" simply to avoid any confusion of the word "independent" in "independent variables" with the very different usage of "independent" in "statistical independence" and "conditional independence." Also, in order to avoid confusion, I have spoken of putting the hypotheses - instead of the "dependent" variables - at the end of the tree.